VOYAGES OF DISCOVERY

Also by the same author

DEAREST FRIEND: A Life of Abigail Adams

Lynne Withey

VOYAGES OF DISCOVERY

Captain Cook and the Exploration of the Pacific

University of California Press
Berkeley and Los Angeles

University of California Press
Berkeley and Los Angeles
First Paperback Printing 1989

Reprinted in arrangement with William Morrow & Company, Inc.

Library of Congress Cataloging-in-Publication Data

Withey, Lynne.
Voyages of discovery.

Bibliography: p.
Includes index.
1. Oceania—Discovery and exploration.
2. Cook, James, 1728–1779—Journeys—Oceania. I. Title.
DU20.W55 1988 919'.04 88-29543
ISBN 0-520-06564-6

Printed in the United States of America

1 2 3 4 5 6 7 8 9 10

BOOK DESIGN BY ELLEN FOOS

for my mother and father

Preface

To speak of European explorers as the "discoverers" of the Pacific is to ignore centuries of exploration and migration by the Pacific peoples themselves. And yet, as the Australian geographer Oskar Spate observes, discovery without recording and preserving the results for others to use and build upon is "discovery" in its most limited sense; and recorded discovery of the Pacific was largely a European achievement.

Creating a written and visual record was an essential part of European exploration from its beginnings, but until the mid-eighteenth century, the record was haphazard at best. Limited by primitive methods of navigation and government policies that discouraged the dissemination of information for political reasons, fifteenth- and sixteenth-century explorers discovered lands that couldn't be found again for lack of accurate charting and often saw their journals and maps suppressed by government officials. By the second half of the eighteenth century, however, a changing political climate, growing interest in scientific investigation, and improved navigational techniques encouraged the governments of England and France to mount voyages of exploration joining scientific and political goals. Although political concerns remained significant as a motivating force, the voyages' scientific purposes were strong enough to overcome the sort of national rivalry that had kept sixteenth- and some seventeenth-century voyages closely guarded secrets. Official accounts of the major eighteenth-century voyages were published and translated into several European languages; the French explorer Louis Antoine de Bougainville visited London to report on his findings to the Royal Society; and, in a gesture that would have been unthinkable a century earlier, the belligerents in the war for American independence issued proclamations ordering their warships not to molest James Cook on his third voyage because of its scientific importance for all nations.

With Cook's voyages, the goals of exploration expanded to include the study of plant and animal life and the customs of the native inhabitants as well as the physical features of land and sea. He was the first to carry a full complement of scientists, both astronomers to fix locations of landfalls and naturalists to study their physical characteristics, and the first to include artists to supplement the written record of the voyage. Equally important for the continuity of scientific exploration, he was the only eighteenth-century explorer who headed more than one voyage to the Pacific. As captain of three expeditions over a ten-year period, Cook built upon the experiences of each voyage in planning the next and employed the same officers and crew members whenever he could, thus training the next generation of explorers. His voyages, full-fledged scientific expeditions dedicated to "discovery" in its broadest sense, established a model that would be followed by explorers for decades to come.

Because Cook was the greatest explorer of the eighteenth century, perhaps of any century, this book takes his career as its centerpiece. But his life and accomplishments are well known, and it is not my intention merely to review them again. Rather, I consider "discovery" as a mutual process of exploration between Europeans and Pacific peoples. As Spate put it, "the drama of exploration and exploitation was played out upon an already peopled stage," and a complete story of Pacific discovery must recognize that the "discovered" peoples were engaged in their own form of exploration among the foreigners who visited their shores, with their own methods of recording their observations, by preserving European objects and adding stories of the explorers to their long and complex oral traditions. A few of the more curious even persuaded explorers to take them along; a handful made it to Europe, or New England, or China, and returned to regale their countrymen with accounts of places as exotic to them as their islands were to the Europeans.

This mutual discovery had a profound effect on both Pacific islanders and Europeans. The consequences were more obvious for the Pacific peoples, who saw their lands depopulated by new diseases, their crafts changed by the introduction of iron tools and woven cloth, their political systems upset by the favoritism Europeans showed to some chiefs over others, and their religious customs challenged by uncomprehending foreigners who violated sacred practices. For Europeans the influences were more subtle and less wrenching, but far-reaching nonetheless; by broadening Europeans' knowledge of the world, the discovery of the Pacific forced them to come to terms with people vastly different from themselves, to recognize the diversity of cultures beyond Europe and the inescapable fact that thousands, if not millions, of men and women lived contented, productive lives without benefit of Christianity or the material advantages of western civilization. For some Europeans, knowledge of Pacific peoples inspired a new burst of missionary zeal, as they felt morally bound to carry the benefits of European religion and culture to the Pacific; others, struck by the simplicity and fulfillment of the islanders' lives, were moved to question the values of their own culture.

It has become commonplace to deplore Europeans' exploration of both the Americas and the Pacific for the destructive changes it brought to native residents; and while the force of what Alan Moorhead called "the fatal impact" is undeniable, the indictment of Europeans' impulse to explore the world ignores both the certainty that no group could remain undiscovered forever and the extent to which at least some of the Pacific islanders welcomed their involvement with Europeans. Discovery was a two-way business, although certainly the advantages in the exchange were heaped on the side of Europeans with their greater technological sophistication.

The more perceptive of the explorers, including Cook, George Forster, and William Bligh, recognized this fact and foresaw some of the negative consequences of their presence in the Pacific; but although troubled by the changes they envisioned, they did not waver in the commitment to continue their work. Theirs was a dilemma not confined to explorers or to the eighteenth century: does the value of knowledge for its own sake justify the uses to which it is put, regardless of the consequences? The answer for

Cook and his contemporaries was never in doubt. For them, as scientists, discovery itself was the highest goal; it would take later generations to use their discoveries for less noble purposes.

I could not have written this book, at least within any reasonable period of time, without drawing upon the work of many other scholars. Like all students of Pacific exploration, I am especially indebted to the work of the late J. C. Beaglehole, who produced authoritative, richly annotated editions of both Cook's and Joseph Banks's journals, a definitive biography of Cook, and a general survey of Pacific exploration that retains its freshness fifty years after publication. My debts to a host of other historians, anthropologists, art historians, literary critics, and others are indicated in the references.

Librarians on three continents helped with manuscript and graphic sources. I am grateful to the staffs of the British Library, the Public Record Office, and the Archives of the Royal Society in London; the Mitchell and Dixson Libraries in Sydney; the National Library of Australia in Canberra; the Sutro Library in San Francisco; and especially Michael Hoare at the Alexander Turnbull Library in Wellington, both for his assistance with that library's collections and for his work on Johann Forster, which is essential to a full understanding of Cook's second voyage.

To my friends and colleagues who have lived with this book and with me during its long gestation, I offer my heartfelt thanks: to Claude Kolm for his help in interpreting the art related to the voyages, to Sandria Freitag for talking out the treatment of issues too numerous to mention, to Henry Horwitz for helping me understand the political background to exploration, and to Linda Doyle, Harvey Ginsberg, Michael Hindus, and Timothy Seldes for their critical reading of many more pages than ever came to light.

L.W.
San Francisco
November 1986

Contents

III Circumnavigating the Antarctic

IV The Search for a Northwest Passage

V Exploration and Empire: The Legacy of Cook's Voyages

Prologue: The Royal Society Plans a Voyage

On an early spring day in 1768, the members of the Royal Society, England's prestigious association of scientists, met to discuss a voyage to the South Pacific. The Society's interests ranged across all sorts of scientific phenomena—its members included such diverse men as the Unitarian minister and chemist Joseph Priestley, Royal Astronomer Nevil Maskelyne, and the American inventor, politician, and writer Benjamin Franklin—but voyages around the world were a bit out of its line. Far from being a simple diversion, however, this voyage would be part of what the Society considered one of its most important projects in recent years.

Sponsored by the English government at the Royal Society's request, the voyage would take astronomers to a little-known part of the South Pacific to observe a rare phenomenon called the transit of Venus: the passage of Venus across the sun's surface, which, if properly timed, could provide the basis for a calculation of the distance between the earth and sun. After completing the observations, members of the expedition would continue across the Pacific in search of lands as yet undiscovered by Europeans.

The Society welcomed at its meeting the two men it had selected as official observers on the voyage. One, a thirty-two-year-old astronomer named Charles Green, was familiar to most as Nevil Maskelyne's assistant in his position as Royal Astronomer. The other was a man few of them had ever met, although some recalled his report on an eclipse of the sun two years earlier. A sailor rather than a scientist, he was James Cook, a noncommissioned naval officer who had spent much of the past five years charting the coast of Newfoundland. Tall and spare, with piercing blue eyes, sharply defined features, and a serious, almost stern demeanor, the forty-year-old Cook gave the impression of a man confident in his abilities but without arrogance. Cook was the government's choice to command the voyage; he was also a self-taught astronomer, and the Society had therefore appointed him, along with Green, as one of its observers.

At first glance, Cook seemed an odd choice to command so important a voyage. The son of a Yorkshire farm laborer, he had been apprenticed to a grocer and then to a merchant and shipowner, entering the Navy as a common seaman at the relatively advanced age of twenty-six. He served in several ships during the Seven Years War of the late 1750s and early 1760s, working his way gradually from seaman to master, the man responsible for navigating the ship and the highest noncommissioned rank in the British Navy. A reserved and studious young man, Cook had spent much of his spare time teaching himself mathematics, astronomy, and the techniques of marine surveying. His ability was well known to his superiors, but in the classbound military system of the eighteenth century it was unlikely that a man of Cook's humble origins would ever break into the ranks of commissioned officers and even more unlikely that a major naval expedition would be commanded by a man of lesser position.

Cook's special skill at navigation and mapmaking had not gone unnoticed in the higher reaches of the Admiralty, however, and it was just this sort of skill that was most needed on a voyage of exploration. Impressed by his Newfoundland charts in particular, the Admiralty selected him for the South Pacific voyage and then promoted him to lieutenant, a rank more consistent with his new responsibilities.

Observing the transit of Venus was a critical event for the scientists of the Royal Society, since the opportunity occurred only twice in a lifetime. Because of its elliptical orbit, Venus makes the transit across the sun's surface twice in the space of a decade, but then not again for about a century. The first of a pair had occurred in 1761, and the second would be visible in 1769; the next transits would not take place until 1874 and 1882.

The Royal Society and other scientific groups throughout Europe had organized observations of the 1761 transit, stationing 120 observers at various points on the globe, but the results had been disappointing, primarily because the observers were not widely enough dispersed. The members of the Society were determined to avoid a repetition of the 1761 failure, in part because this would be the last chance to observe the transit within their lifetimes, but also because they had a scientific tradition to uphold. England was a pioneer in the development of astronomy—the government had established the famous observatory at Greenwich and the position of Royal Astronomer in 1675—and Edmond Halley, one of the Royal Society's members, had been the first to suggest a method of determining the distance between earth and sun by observing the transit of Venus. Halley, better known for predicting the path of the comet that bears his name, published his theory in the Society's *Philosophical Transactions* in 1716, urging the next generation of scientists to seize the opportunity that would be offered by the transits of Venus in the 1760s.

To increase the chances of success in 1769, the Society planned to dispatch more observers and distribute them more widely around the earth. It was especially critical to locate one team as far south as possible, ideally at some point in the middle of the South Pacific Ocean. The difficulty was knowing exactly where. The few known bits of land in the South Pacific—islands discovered by Spanish and Dutch explorers in the sixteenth and seventeenth centuries—had not been charted with any reliability, and no one could be certain of finding them again easily. The best that the Royal Society could do was recommend that their observers sail to the Marquesas, discovered by the Spanish explorer Alvaro de Mendaña in 1595, or to Amsterdam or Rotterdam Islands, discovered by Abel Tasman in 1643. None of these islands had been seen by Europeans since their discovery, and their locations on contemporary maps were approximate at best.

The members also had to reckon with the expense and risk of a voyage to the Pacific. The Society was an old and distinguished organization; founded in 1662, it counted among its former presidents such notable minds as Isaac Newton and admitted new members by election only, in recognition of their scientific achievements. But prestige did not translate into money, and although one of the Society's major purposes was encouraging scientific inquiry, it did not, as an organization, sponsor research. Instead it relied on the individual efforts of its members and others, who reported to the group at its meetings and published their findings in its journal. Topics requiring collective attention, like the transit of Venus observations, were delegated to informal committees, whose members then devised strategies to raise money and recruit the personnel needed for the task at hand. Such methods were generally adequate, for scientific inquiry in the eighteenth century was not the highly specialized and complex business it would become in later times; but mounting a voyage to an essentially unknown part of the world thousands of miles from England was beyond the scope of the Royal Society members.

Their solution was an appeal to the Crown. In February 1768 the Society presented a petition to the King asking that the government provide a ship and crew to transport two astronomers to the South Pacific. The members stated their case in both practical and political terms, pointing out first that accurate observations of the transit would advance knowledge of astronomy, which would in turn improve the science of navigation. Then they appealed to national pride, noting that all other major European nations would be dispatching men to observe the transit and that England had made a reputation in the scholarly world for its achievements in astronomy. The fact that in 1761 France had sent nearly twice as many observers as England was not lost on the petitioners. It would "cast dishonour" upon the nation, they argued, if England did not make a more serious effort to observe the transit of Venus this time. What they did not say in the petition, but was surely obvious to the King and his ministers as well as to the members of the Royal Society, was that such a voyage would also provide an opportunity to search for new land in the Pacific. If successful on that score, the mission would do even more to improve England's political standing among the nations of Europe.

A few years earlier, the Royal Society's petition might well have languished among the King's advisers, but in 1768 the Pacific was a topic of major interest in government circles. England's victory over France in the Seven Years War had established the nation's supremacy at sea and secured its control over most of North America; the next step to consolidate England's position as a world power and prevent a resurgence of French influence was to press into the Pacific, both to improve trade routes to Asia and, possibly, to add more colonies to the empire.

In an attempt to establish a foothold in the Pacific, the English government had sent John Byron on a voyage around the world in 1764. Upon his return two years later—the voyage having produced little in the way of new discoveries—government officials dispatched another expedition under the command of Samuel Wallis. Wallis had not yet returned to England when the Royal Society petitioned for a ship to the South Pacific, but the proposal for another voyage coincided with official plans to continue exploring the Pacific. In fact, the transit of Venus offered a useful excuse for a government-sponsored expedition, because England's earlier voyages had excited the suspicions of the French and Spanish, who feared, not without reason, that these voyages of exploration were merely a cover for attempts to expand the nation's influence in the Pacific. But the significance of the transit was understood throughout Europe, and an expedition sent to observe the phenomenon could be accepted as a purely scientific mission. Then, once the observations were accomplished, the crew could pursue its political objectives.

Not that the scientific mission was unimportant to the King and his ministers. The Royal Society's arguments about the importance of astronomy to the improvement of navigation were well understood in a nation where economic and political power was based on the sea, and the 1760s were a time of growing interest in science generally. King George III himself had an amateur interest in science, but more important, men throughout the higher levels of government, business, and the professions were coming to believe in both the importance of scientific inquiry and the need for government to lend its support to such inquiry. So it was with relatively little discussion that King George granted the Royal Society's petition for a ship and directed the Admiralty to select a suitable vessel, a captain, and a crew.

It was up to the members of the Society, however, to hire the astronomical observers. Their first choice was Alexander Dalrymple, a brash, brilliant, argumentative young Scotsman who had recently returned to England after thirteen years in Asia as an employee of the East India Company. Dalrymple's experiences had fired his interest in exploration, and his study of earlier explorers and geographers had convinced him of the existence of a large continent somewhere in the South Pacific, an idea that had enjoyed widespread popularity for centuries and was beginning to be revived, in part under Dalrymple's assiduous sponsorship.

Dalrymple was at work on a collection of Pacific explorers' journals, but wanted to do more than simply compile the work of others. He aspired to lead a voyage of exploration himself and saw his opportunity in the preparations for the transit of Venus observations. Dalrymple would not be satisfied with less than full command of the expedition, however, as he made clear to the Royal Society. Such an arrangement was perfectly acceptable to the Society, but the Navy would not have a civilian commanding one of its ships. Dalrymple stuck to his conditions, and the Society had to look elsewhere for its astronomers.

Inquiries at Oxford and Cambridge proved fruitless. Nevil Maskelyne had made a voyage to Bermuda five years earlier to test one of the first chronometers, a timepiece designed to calculate longitude at sea, but he was not interested in another. As an alternative, he suggested William Wales, who had volunteered his services as an observer in any part of the world with a warm climate. Instead Wales was dispatched to Hudson Bay, and the Society settled on Charles Green as its representative to the South Pacific.

While looking for a second observer, the members learned that the Admiralty had appointed Cook as commander of the voyage. Recalling his detailed observations of an eclipse of the sun in Newfoundland, they invited him to serve as one of their official observers. Cook accepted the appointment and prepared to sail to the Pacific in his dual role, as sailor and as scientist.

I

FOUNDATIONS
Pacific Exploration Before Cook

1. *The World Beyond Europe*

An expedition to the South Pacific, sponsored by the Crown in the name of scientific discovery, was an unprecedented step and a notable example of the interest in science and geography growing throughout Europe in the second half of the eighteenth century. But the desire for political and economic gain and the lure of the exotic, the unspoken motives of Cook's voyage, were centuries old.

European explorers had been venturing forth in search of new lands for nearly three hundred years, driven primarily by the desire for riches, either in the form of new trade routes or new lands that might prove to be sources of wealth. Their voyages revolutionized Europeans' understanding of world geography, discovering North and South America, charting the coasts of Asia and Africa, and dispelling myths about boiling temperatures near the equator and ferocious sea monsters in distant parts of the ocean.

England, Spain, Portugal, and France had all established colonies in the New World, and voyages across the Atlantic and around the Cape of Good Hope to Asia and the East Indies had become almost routine. But the Pacific remained largely unknown apart from its perimeter (a few Spanish outposts struggled along on the west coast of South America, and trading ships occasionally visited ports on the mainland of Asia), and ships crossing the Pacific were mostly Spanish trading vessels plying a well-established route from the Philippines to Acapulco. Spanish and Dutch explorers in the sixteenth and seventeenth centuries had stumbled across a few islands in the South Pacific, among them the ones the Royal Society considered as sites for the transit of Venus observations, and touched on the northern and western coasts of Australia (then known as New Holland) and the South Island of New Zealand. But even these discoveries were vaguely documented, and it remained unclear whether New Holland and New Zealand were islands or part of a continent.

Lack of knowledge, however, did not mean lack of theories about what ought to exist on the fringes of the known world. Most Europeans could not believe that the vast South Pacific might be an uninterrupted expanse of water, but thought a continent must exist somewhere in the yet-unexplored regions; the more imaginative pictured land stretching across the entire breadth of the Pacific from the South Pole well north into temperate latitudes. Mapmakers and geographers liked to call it "terra australis incognita," the unknown southern land, although some more optimistically labeled it "terra australis nondum cognita," or the southern land not yet known.

The notion that a great continent existed at the bottom of the world dated back to ancient times, when the Greeks argued that a land mass around the South Pole must exist to balance the continents in the northern hemisphere. By the twelfth century, the existence of "terra australis incognita" was widely accepted. Published accounts of Marco Polo's travels lent credence to theories about the existence of a southern continent, for he told of little-known lands southeast of Asia rich with gold and spices. And some Europeans believed that the unimaginably wealthy Land of Ophir, visited by servants of the Biblical King Solomon, must be a continent in the South Pacific. Bad translations and popular imagination blended Marco Polo's stories and accounts of the Land of Ophir into a belief that these lands were terra australis.

Other writings added conviction as well, particularly the *Travels and Voyages* of Sir John Mandeville, written in the 1350s and widely read for hundreds of years afterward. Mandeville, whose book was something of a medieval bestseller—it was translated into eleven languages and existed in at least three hundred manuscript forms before the advent of printing allowed it to be even more widely distributed—claimed to have visited inhabited lands near the South Pole. His "travels" were entirely fanciful, but Europeans of the fourteenth, fifteenth, and even sixteenth centuries found him at least as convincing as Marco Polo, and his book consequently helped strengthen the mythology of the southern continent.

The sea voyages of the sixteenth century, far from dispelling these myths, only added the weight of observation to them. Survivors of Ferdinand Magellan's voyage in the 1520s told of land, which they called Tierra del Fuego, lying south of the Strait of Magellan. Most thought the land was merely a group of islands, but mapmakers for decades afterward confidently drew the outline of a large continent just south of the Strait, culminating in the famous and influential world map by the Dutch cartographer Abraham Ortelius. Published in 1570 and frequently reprinted, Ortelius's map showed a continent extending from the tip of South America across the Pacific to the area south of New Guinea.

The Spanish explorers Alvaro de Mendaña and Pedro Fernandez de Quiros, sailing in the 1580s and 1590s, added new force to these beliefs with their discovery of a group of islands in the southwestern Pacific. Thinking he had found the Biblical Ophir, Mendaña called his discovery King Solomon's Islands, and both he and Quiros thought they saw the outlines of a continent in the distance. For the next two hundred years, Europeans made sporadic and unsuccessful efforts to find the Solomon Islands again. Like other Pacific discoveries in the sixteenth and seventeenth centuries, the Solomons remained lost to future explorers because of inadequate methods of mapping their location.

Belief in global symmetry of another sort spurred the search for a second, equally elusive geographical phenomenon: a navigable passage across North America linking the Atlantic and Pacific. Such a "northwest passage," widely believed to exist since Columbus's time, would provide a northerly alternative to the routes around Cape Horn and the Cape of Good Hope. (Some geographers also argued for a "northeast passage" leading through the Baltic, across Asia, and into the Pacific.) A northwest passage was an especially attractive possibility for England, since it would provide a much shorter route to the Pacific and one free from Spanish or Portuguese interference.

Robert Thorne, an English merchant living in Seville in the 1520s and 1530s, was among the first to see the potential advantages to England of a northern route to the Pacific; in 1527 he wrote Henry VIII suggesting that England initiate a search for a strait across the North American continent, and in 1540 he and Roger Barlow presented their *Declaration of the Indies,* a more fully developed plan that called for a voyage through the northwest passage and south across the Pacific to rich, yet-to-be-discovered tropical lands. Contemporary maps supported Thorne's conviction that a northwest passage must exist, just as they supported the theory of a southern continent. Gerhard Mercator's maps showed open seas up to the North Pole, and in the 1570s Humphrey Gilbert produced a chart showing a strait from the St. Lawrence River to the Gulf of California. That Gilbert drew his map with a view toward promoting his Company of Cathay, formed to sponsor voyages of exploration, did not diminish its influence.

No one acted on Thorne's ideas until the 1570s, however, when Martin Frobisher, sailing under the auspices of Gilbert's company, made three attempts to find a northwest passage. He got no farther than the bay in northeastern Canada now known by his name, but a 1578 map shows "Frosbisshers Straights" running all the way across Canada. Subsequent explorers sailing at the end of the sixteenth and beginning of the seventeenth century—John Davis, Henry Hudson, William Baffin—pushed farther west through Canada's Arctic waters, but fared no better. After about 1630, further searches lapsed, but belief in the existence of a navigable northwest passage persisted.

During the sixteenth century, the Pacific became a "Spanish lake," as Spain established settlements up and down the west coast of Central and South America and in the Philippine and Ladrones islands (now called the Marianas) in the western Pacific. Spanish ships traveled regularly between Asia and Latin America, trading gold and silver from Mexico and Peru for the spices and silks of the East Indies, unloading their cargoes in Central America and carrying them overland to ships on the Caribbean side for transport to Spain. In the latter part of the century, English merchants and investors, prohibited by the Spanish government from trading with Latin America and unable to find any similar source of riches farther north, became increasingly resentful of Spain's stranglehold over this lucrative trade. Some tried to urge Queen

Elizabeth to adopt a more aggressive policy to curtail Spanish power and expand England's overseas trade, but she was unwilling to commit Crown resources to speculative voyages or risk war with Spain. Instead, she tacitly supported the exploits of pirates—or buccaneers, as they were more politely known—who preyed on Spanish shipping in the Caribbean. In wartime their activities were legally sanctioned by government commissions that in effect permitted privately owned vessels to become warships. In the 1560s and 1570s, when England and Spain were at peace, their activities continued without benefit of official authorization, justified in the minds of the buccaneers (and the government officials who looked the other way) as a means of circumscribing Spanish power.

At the same time, a few merchants and gentlemen investors began to think about ways for England to curtail Spanish power and gain a share of Pacific trade without recourse to piracy or deliberate incursions into Spanish territory. In 1574 a group organized by Richard Grenville and William Hawkins proposed a voyage that would explore the southern reaches of South America, an area not yet settled by either the Spanish or the Portuguese, and search for new lands in the Pacific south of the equator. While South America was their principal objective, recent maps showing a huge continent in the South Pacific and accounts of Mendaña's discovery of the Solomon Islands, which had leaked out despite Spanish attempts to keep it secret, raised the possibility of an alternative site for English trade, should the expedition fail to establish a foothold in South America. Grenville and Hawkins proposed a voyage through the Strait of Magellan, across the South Pacific to China, and then back to England by way of the Cape of Good Hope. But Elizabeth, fearing that the scheme would lead to hostilities with Spain, refused to give the promoters a license to sail.

English relations with Spain were deteriorating anyway, however, and three years later Elizabeth approved a similar project, although she rejected a request for financial support from the Crown. Influenced in its conception by the Hawkins/Grenville plan, this voyage was to be commanded by one of the most notorious of the buccaneers, the veteran of several years' successful raiding in the Caribbean, Francis Drake. He was to explore the southern coasts of South America, both Atlantic and Pacific, but beyond that the purposes of his voyage were vague and open to conflicting interpretations. The gentlemen adventurers who sailed with him thought they had signed on for a voyage of exploration, which might include a search for terra australis, a northwest passage, or both, while Drake himself was most interested in plundering Spanish shipping in the Pacific. He kept these plans to himself, however, knowing that Elizabeth would not approve a predatory voyage.

Drake embarked in November 1577 with his flagship, the *Pelican,* later to be renamed the *Golden Hinde,* and four other vessels. They had an unusually easy passage through the Strait of Magellan, but as the ships emerged into the Pacific they were caught in a severe storm and blown south, past rocky islands and finally out of sight of land. Without intending it, they had discovered the southern tip of South America, proving that Magellan's Tierra del Fuego was not part of a continent but merely a group of islands. This was the most significant single discovery of Drake's voyage, although it did nothing to change Europeans' conviction of the existence of terra australis.

Heading north, Drake abandoned all pretense of exploration and sailed boldly into Spanish territory in search of booty, alienating those of his crew who thought of themselves as explorers rather than pirates. He raided Valparaiso, captured a silver-laden Spanish ship off the coast of Peru, stopped briefly to reprovision at Guatemala, and then pushed beyond the limits of Spanish settlement along the west coast of North America, in search of a northwest passage—not out of any great interest in geographical discovery, but because he knew that Spanish ships would be lying in wait to retaliate against him if he went back the way he had come. The fleet stopped for repairs along the California coast near the site of present-day San Francisco, the point farthest north on the American coast yet reached by Europeans. There Drake abandoned his search for a northwest passage, sailing instead for Manila, the Spanish-controlled port in the Philippines. He and his men threaded their way through the islands and reefs of the Indian Ocean, narrowly averting disaster when the *Golden Hinde* struck a shoal in the treacherous waters off New Guinea, and then returned to England by way of the Cape of Good Hope. They arrived home in September 1580, the first Englishmen to sail around the world. The Spanish denounced Drake's escapades as piracy; the English hailed him as a national hero.

Queen Elizabeth herself was ambivalent about Drake's escapades. Following popular sentiment, she rewarded him with a knighthood; but, still attempting to preserve peace with Spain, she also returned much of the stolen treasure to the Spanish government and suppressed information about the full extent of Drake's travels, with the result that his geographical discoveries had little immediate impact. Her diplomatic efforts could not stem the continuing deterioration of English-Spanish relations, however, and in 1585 the two nations went to war.

England's defeat of the Spanish Armada in 1588 crippled Spanish seapower and opened opportunities for England to expand its maritime enterprise, but lack of any clear purpose or concerted support from the Crown prevented effective action. Drake's voyage had illustrated the conflict between those who favored a long-term strategy to discover new lands and sea routes that would eventually permit significant expansion of trade and those who wanted the short-term profits gained from preying on Spanish shipping. The Queen openly supported the former but tacitly sanctioned the latter as well, while refusing to put government money behind any voyage, whatever its goals. This conflict of purpose and lack of support dogged the few English efforts to extend English seapower into the Pacific in the last years of Elizabeth's reign.

When the Queen died in 1603, James I, the first of the Stuart dynasty, came to the throne and quickly moved to conclude peace with Spain. More concerned with domestic matters than with foreign affairs, he offered little encouragement for voyages devoted either to exploration or plunder, and the changing political climate, coming on the heels of the failure of recent attempts to reach the Pacific, led expansionists to focus their attention on North America. Further discoveries in the Pacific were left to the Dutch, who by the early seventeenth century had pushed the Portuguese out of their former sphere of influence in the East Indies, and to the buccaneers, who continued to roam the Caribbean and the Pacific in search of fortune.

Toward the end of the seventeenth century, a series of major political changes in Europe helped revive strategic interest in Pacific exploration. The troubled years of Stuart rule ended with the abdication of James II in 1688, and the English Crown passed to the Dutch prince, William of Orange. William quickly pulled England into the war he was fighting with France, which escalated into a general European conflict over the next decade. After a brief respite at the end of the century, dynastic changes on the continent ignited war again, as France, the most populous and powerful nation in Europe, attempted to gain control over Spain and, in the process, emerged as England's chief rival. As England maneuvered to increase its political power within Europe, issues of expansion in the New World and further exploration of unknown territories, primarily in the Pacific, took on renewed importance. A weakened Spain could no longer control access to the Pacific, and both England and France saw the value of increasing trade and territory in more distant parts of the world. The demands of war prevented any concerted policy of encouraging exploration over the next sixty years, but occasional voyages and an increasing volume of writing about exploration kept alive the belief that England must eventually establish a presence in the Pacific.

As a beginning, in 1697 the English government took an unprecedented step, authorizing a purely exploratory voyage to New Holland and New Guinea under the command of William Dampier, an ex-buccaneer who had achieved instant fame with the publication, earlier that year, of *A New Voyage Around the World*, an account of his twelve-and-a-half-year odyssey across the globe.

The son of a farmer in southwestern England, Dampier had the urge to see the world from an early age. He sailed to France and Newfoundland while apprenticed to a ship's master, and then joined the crew of a vessel bound for the Dutch East Indies. Next he went to Jamaica, and then to the Caribbean coast of Honduras, where he worked as a logwood cutter before falling in with a group of buccaneers raiding Spanish ports along the Central American coast. After a brief visit to England in 1678, Dampier returned to Honduras and signed on with the first of a series of ships that would take him around the world. As part of a buccaneer fleet of nine ships with almost five hundred men, he took part in the sack of Porto Bello, a wealthy Spanish port on the east coast of Panama. Part of the group crossed the Isthmus in search of more loot; on the Pacific side, quite unexpectedly, they captured five Spanish ships. The buccaneers seized this opportunity to cruise the coast of South America, attacking more Spanish towns along the way. At Juan Fernández Island off the coast of Patagonia, they rescued a Central American native, known only as

Will, who had been accidentally marooned on a buccaneering voyage three years earlier. (Will and Dampier both would eventually provide grist for adventure writers, notably in Daniel Defoe's *Robinson Crusoe.*)

When the buccaneers finally returned to Panama in 1682, Dampier cruised the Caribbean for several months until joining another ship bound for Cape Horn and the Pacific. The crew picked up others along the way, until the fleet numbered ten ships and about a thousand men. Upon reaching the west coast of Central America, the captain of one of the ships announced his plans to sail across the Pacific to the East Indies; Dampier decided to go with him. In the Philippines a mutinous crew marooned the captain, who wanted to turn the voyage into a legitimate trading expedition, and took off on a pirate escapade through the East Indies.

Heading south, they anchored on the northwest coast of New Holland, earning the distinction of being the first Englishmen ever to visit that continent. Finding little sustenance on the barren coast—Dampier thought the native inhabitants were "the miserablest People in the World"—they sailed north, stopping at a small cluster of islands south of Sumatra, where Dampier, disgusted at the cruelty and drunkenness of his companions, decided on a desperate move; with six other men, he traveled 150 miles in a small open boat to Sumatra. From there he continued his journey rather tamely as a passenger on a series of commercial ships, visiting several ports in southeast Asia and India before returning to England in 1691.

Dampier's almost unbelievable adventures and his talent for putting them into words made his the most widely read travel book since the entertaining but false adventures of Sir John Mandeville. More significantly, his perceptive and detailed observations of the lands he had visited caught the attention of English government officials and of the members of the Royal Society.

Not the typical buccaneer, Dampier had joined their ranks more out of curiosity and a sense of adventure than greed for Spanish gold. He was interested in natural phenomena of all sorts—plants and animal life, tides, currents—and kept a detailed journal throughout his voyages, even carrying it across the Isthmus of Panama sealed in a piece of bamboo for protection from wet weather. The members of the Royal Society discussed his observations and published a summary of his book in their *Philosophical*

Transactions; the Lords of the Admiralty, impressed by his knowledge, invited him to submit a proposal for a voyage.

They were influenced too by other books published about the same time, including volumes of buccaneers, some of whom had sailed with Dampier at one time or another, and the account of Abel Tasman, the Dutch explorer who had first discovered New Zealand and parts of New Holland in the 1640s. (Tasman's volume was published in an English translation for the first time in 1694.) These books portrayed the Spanish as weak and unable to control their colonies, which were constantly in danger from Indian insurrection. Dampier thought Spain could no longer manage so large an empire, and another of the buccaneers argued that England should initiate trade with South American ports by force if Spain continued to prohibit foreign commerce.

Dampier proposed a voyage to New Holland and New Guinea, areas he thought were good prospects for further discoveries. The Admiralty approved his plan, conferred upon him the rank of captain, and gave him command of the ship *Roebuck.* It was not quite what Dampier had in mind; he had requested two ships fitted out for a three-year voyage and got instead one small, leaky vessel barely fit to venture beyond the English Channel, much less into the Pacific. But he began collecting his crew and mapping out his plans nevertheless.

Dampier planned to sail through the Strait of Magellan and across the South Pacific to Australia and New Guinea, which he, like most people at the time, thought were part of the same land mass. Having explored this region, particularly the as yet unseen east coast of New Holland, he would proceed through the islands of the East Indies and back to England around the Cape of Good Hope. Neither Dampier, nor the Lords of the Admiralty, nor anyone else concerned with the voyage recognized the similarities between his plan and Drake's, for Drake and his contemporaries' voyages had been all but forgotten as a new generation set out to discover the world beyond the Americas.

Dampier was an expert navigator but had no experience of command, a deficiency that created trouble almost as soon as he left England. He battled constantly with his lieutenant, George Fisher, and before they were across the Atlantic Dampier confined Fisher in irons, a move that resulted in a near-mutiny among his crew. When the expedition reached the coast of Brazil, Dampier had Fisher thrown into jail without any provision for his subsequent release or transport back to England, a breach of naval procedure that reflected his lack of experience and would come back to haunt him later.

From Brazil, Dampier followed the usual route through the Strait of Magellan and across the South Pacific to the East Indies. Many others had traveled this route before him, and he made no new discoveries of significance until he was almost to New Guinea, where he sighted an island much larger than any yet charted in the Pacific. Dampier named his discovery New Britain; in fact what he had found, although he did not realize it, was two islands divided by a narrow channel, known today as New Britain and New Ireland.

From this point Dampier intended to sail south in search of the east coast of New Holland, but the condition of his leaky ship, now desperately in need of repair, convinced him to head for the Dutch port of Batavia in the East Indies, where the *Roebuck* could be temporarily patched up. Despite the repairs, the *Roebuck* sank just off Ascension Island in the mid-Atlantic. Dampier and his crew subsisted five weeks on rice and turtles before being rescued by an English ship headed for the West Indies, and they did not get back to England until early in 1702, almost five years after their departure. There Dampier discovered that George Fisher had made his way home from Brazil more than a year earlier, giving him ample time to spread his version of the dispute with Dampier among influential Navy officials. Dampier, court-martialed for the loss of his ship and his treatment of Fisher, received as thanks for his efforts the loss of all compensation for his voyage.

But although his voyage did not live up to either Dampier's or the Admiralty's expectations, it was not a total loss. New Britain was an island large and fertile enough to be valuable to England; even if the island itself did not yield tangible riches, it might make an excellent base for trade with other parts of the Pacific and Indian Oceans. More significant in the long run, Dampier collected quantities of plants from the tropical islands he had visited, which he turned over to the Royal Society, inaugurating a tradition of scientific observation and collection among Pacific explorers. Drawings of some of these finds illustrated the book he published on the voyage in 1703. And his observations on winds, storms, and currents shed light on the systematic relations among these conditions for the first time, helping to make future navigation in the Pacific more predictable.

War broke out between England and Spain in 1702, putting an end to any further government-sponsored expeditions for the time being. Dampier, however, simply shifted back into his buccaneer's role; in 1703 he set out again for the South Pacific, this time commissioned by Queen Anne as a privateer to prey on Spanish shipping. This voyage was scarcely more successful than the last. Several mutinous seamen deserted to captured Spanish ships, and the quarrelsome captain of a ship accompanying Dampier, the *Cinque Ports,* decided to go his own way. By the time the *Cinque Ports* anchored at Juan Fernández Island some weeks later, a seaman named Alexander Selkirk was so disgusted with this captain that he decided to jump ship and take his chances alone on the island with nothing more than some clothes and bedding, a gun, powder, bullets, tobacco, hatchet, knife, kettle, a Bible and a few other books, and some mathematical instruments.

Back in England, in 1708 Dampier joined another privateering voyage, this one sponsored by a group of Bristol merchants. He was not captain this time, but served as pilot under Woodes Rogers, an experienced commander. It was a happy combination, for Rogers had the talent for command that Dampier lacked, and Dampier was at his best when he could concentrate on technical and scientific matters without the necessity of managing men.

Dampier had learned about the incident of Alexander Selkirk, and after rounding Cape Horn, he and Rogers sailed directly for Juan Fernández to take on provisions and try to discover the man's fate. Although it had been four years since Selkirk chose his solitary exile, they found him in good health and spirits, living in a hut of pimiento trees lined with goat skins and subsisting on a diet of goats and crayfish the size of lobsters.

The Englishmen took Selkirk aboard and continued up the South American coast on what turned out to be one of the most successful privateering voyages of the war. They returned to England in 1711 rich men, and Rogers added to his fame and fortune by quickly publishing an account of the voyage. The parallel stories of Selkirk and Will, both marooned on the same island and both rescued by groups including Dampier, as Rogers pointed out, ensured the book's popularity with a curiosity-craving public.

Peace returned to Europe two years later, discouraging further Pacific expeditions; privateering voyages like Dampier's once again became illegal, and Spain, its American empire still intact, renewed its policy forbidding nearly all foreign trade with its American settlements. In 1720 the South Sea Company, a joint-stock venture organized ostensibly to open new trade routes in the Pacific, collapsed amid financial scandal, which further dampened enthusiasm for Pacific ventures. For the time being, English interest in expansion of trade and empire remained focused on North America, where thriving colonies offered certain markets and the possibility of French incursion into territory claimed by England posed a more immediate threat than Spanish hegemony in Latin America.

The waning of official interest in the Pacific did not diminish popular fascination with the region, however, nor did it dampen the enthusiasm of travel writers, mapmakers, and theorists on world geography. In the half-century between Dampier and Cook, they kept alive English curiosity about the Pacific.

Books about travel, especially to the Pacific, proliferated in the first half of the eighteenth century and enjoyed a steadily increasing audience. Dampier himself had started the trend; his first book went through three editions in nine months, encouraging him to publish a second volume two years later. The books continued to do well, and his publisher became a specialist in travel literature. Over the next decade, eight new collections of travel accounts were published in London, and others were reprinted. Between 1660 and 1800, more than one hundred collections of voyages appeared in print, many of them in multiple editions and some translated into several languages. By the 1720s, travel books were second only to theology, the staple of seventeenth- and eighteenth-century publishing, in number of titles published annually.

Writing accounts of their adventures became an accepted and often lucrative task for returning travelers, and printers vied with each other to publish the latest works. The booksellers Awnsham and John Churchill, seeing an opportunity for profit, in 1704 put together a four-volume, sixteen-hundred-page collection of voyages translated from seven languages, including several never previously published. John Harris, a minister who also wrote on scientific subjects and later became secretary of the Royal Society, published a similarly weighty collection in 1705, titled *Navigantium atque Itinerarium Biblioteca.* Sponsored by a syndicate of booksellers, Harris's work was intended to compete with the Churchills' for a share of the market in travel literature.

Both the scope and the lavish style of publication priced these volumes beyond the average reader's budget, but after 1700 the introduction of cheap editions of books, circulating libraries, and book clubs made books increasingly accessible to people with small incomes. Where a clergyman or schoolmaster of the late sixteenth century had to spend the equivalent of one to three weeks' income to buy Hakluyt's *Voyages,* one of the earliest collections of travel accounts, or as much as a day's pay for an inexpensive edition of one of Shakespeare's plays, by the eighteenth century the availability of serial editions had reduced the cost of books substantially. Under this arrangement, begun in the seventeenth century and greatly expanded after 1730, publishers issued books a few sheets per week, stitched in blue paper covers and priced at a few pennies an issue. The "number books" or "subscription books," as they were often called, included works on travel—a collection of travel accounts called *A View of the Universe,* published in 1708, was the first of its type to be issued serially—history, biography, religion, and even an edition of Samuel Johnson's *Dictionary.*

Readers could also keep up with the latest travel accounts by reading popular magazines, like the *Gentleman's Magazine* or the *London Magazine,* both established in the 1730s, which frequently published excerpts from books on travel. And those who could not afford to buy books at all, and did not wish to pay the fee to join a subscription library, could always repair to one of the coffeehouses, which kept the latest popular books, newspapers, and magazines for their patrons to read. One London coffeehouse advertised that a copy of Woodes Rogers's book was available to be "seen and read Gratis" as an inducement to potential customers.

So popular were books about travel, especially travel to exotic places, that accounts of real adventures could hardly fill the demand, and enterprising writers added to the literature by making up tales of exploration. Some of these books were obvious hoaxes, but others were more subtle fabrications, intended to be accepted as truth. At a time when plagiarism was routine and copyright laws nonexistent, writers commonly copied and elaborated the accounts of others until it became nearly impossible to distinguish truth from fiction, even in books ostensibly based on fact.

Daniel Defoe was the quintessential master of this genre. His most famous work, *Robinson Crusoe,* loosely based on the adventures of Alexander Selkirk and Will, was published as a deliberately imaginary account, although he followed the conventions commonly used by real-life voyagers in writing about their travels.

But Defoe published other books that were received as truth, at least by a substantial portion of his readers.

A New Voyage Around the World by a Course Never Sailed Before, published anonymously in 1724, five years after *Robinson Crusoe,* like the earlier book owed a great deal to Dampier. Defoe not only borrowed Dampier's title, but much of his style as well, describing the physical characteristics of islands and people, as Dampier did, in preference to the long discourses on winds and currents so common in most accounts of voyages. His account of his first meeting with the inhabitants of a South Pacific island perfectly parodied real voyagers' descriptions of such encounters, from men swimming around the ship, to the Englishmen's efforts to get them on board, to the exchange of gifts.

Despite the realism of his account, *A New Voyage* fooled its readers for just a few years, but two other books, *The Life, Adventures, and Piracies of the Famous Captain Singleton* and *Robert Drury's Journal,* both set in Madagascar and based in part on the experiences of real people, were more successful hoaxes. *Drury's Journal,* a remarkably realistic account of Madagascar, was not only accepted as true but commonly used as a source on that country for nearly two centuries. Defoe's style was widely copied. At least eight imitations of *Robinson Crusoe* were published by the middle of the eighteenth century, including one in which the hero meets Dampier in Brazil, becomes involved in a sea battle, visits Juan Fernández Island, and seizes a Spanish galleon.

Writers whose subjects had nothing to do with travel sometimes adopted the conventions of travel literature to catch public attention. Jonathan Swift's *Gulliver's Travels,* a satire on English politics first published in 1726, was also loosely based on Dampier's travels. Gulliver, like Dampier, went on four voyages, all to parts of the world that Dampier had visited; Lilliput, the home of six-inch people, was located off the coast of Australia, and Brobdingnag, land of giants, was east of Japan. Although the people Gulliver encountered were clearly ridiculous, Swift's style of writing cleverly captured the manner of the travel writers, to the point of professing the accuracy of his observations in a preface.

Even some religious books were written in the form of travel accounts, with titles like *An Historical Geography of the Old and New Testaments* and *A Treatise on the Situation of Paradise to which is prefix'd a Map of the Adjacent Countries.* The extreme example of the type was a paraphrased version of the Bible titled *The Travels of the Holy*

Patriarchs . . . our Saviour Christ, and his Apostles, as they are related in the Old and New Testaments: with a Description of the Towns and Places to which they Travelled, and how many English miles they stood from Jerusalem. With travel books apparently threatening to outsell religion, one English bishop blasted the popular taste for books of voyages, which he claimed were filled with "monstrous and incredible Stories," because they encouraged a "morbid taste" for "extravagant fiction."

Most travel books were written purely to entertain their readers and profit their authors, but some were deliberately intended to draw attention to the world beyond Europe and develop interest in the possibilities of exploration. The great compilations, in particular, were written partly to revive serious interest in exploration and link it with English expansion; Defoe and Swift had similar interests. Defoe, who as a young man had worked as a merchant in Spain, urged William III to establish settlements in South America, one on the west coast of Patagonia and another on the Atlantic coast just north of the Strait of Magellan. Shortly before England went to war with Spain in 1702, he published a pamphlet advocating a maritime war to break Spanish power in the Pacific, rather than a continental war that would, he believed, yield nothing of great importance to England. Swift published a pamphlet in 1711 that supported many of Defoe's sentiments of ten years earlier.

On the whole, writing about exploration replaced action for the first six decades of the eighteenth century, and no obvious successors to Drake and Dampier appeared on the horizon. But the arguments of Defoe and Swift and others who would have England adopt a more aggressive posture on world expansion enjoyed a flurry of popularity when war once again broke out in Europe in 1739, spurred by conflict between England and Spain over freedom of the seas.

The Spanish practice of stopping legitimate English merchant ships, searching them, and seizing their cargoes if they included anything that might have been obtained from Spanish ports angered English merchants, sea captains, and many politicians, who preferred to ignore the fact that this policy was prompted by the considerable amount of illegal English trade and piracy in the Caribbean. A particularly gruesome case involving Captain Robert Jenkins—who testified before Parliament that a Spanish raiding party had boarded his ship, strung him up on the yardarm, and then chopped off his ear—provided the necessary pretext for those in England who wanted war with Spain. The conflict escalated into a general European war the following year and continued until 1744. The end result was a further weakening of the Spanish empire and strengthening of France as England's chief rival for empire.

As part of the campaign against Spain, England sent a squadron of eight ships under Commodore George Anson to raid Spanish ports on the west coast of South America. The plans for Anson's voyage harked back to the buccaneers' exploits; he was to attack Spanish ports on the west coast of South America and then raid Panama. His goal was not simply to plunder and weaken Spanish control over their colonies, however, but also to locate suitable bases for future English trade in the Pacific. Once he had completed his raids, Anson was to sail across the Pacific and return home by way of the Cape of Good Hope.

Despite careful planning and the knowledge gained from the experiences of his predecessors, Anson failed completely in his ambitions for conquest in South America. A poor crew accumulated from the dregs of the Royal Navy, the less than ideal condition of several of his ships, and a long, tortuous passage around Cape Horn left his squadron in a severely weakened state by the time it entered the Pacific. Two ships didn't even make it around the Horn, but headed back to England instead; and a third—the *Wager,* named for the First Lord of the Admiralty, who had proposed the expedition—was wrecked off the coast of Patagonia. Scurvy killed two-thirds of the remaining crew, making any serious attack on Spanish territory out of the question. Anson launched a few minor raids on Spanish ships and then limped across the Pacific to Macao, on the coast of China. By this time only his flagship survived.

Then Anson salvaged his disastrous voyage in a single spectacular feat: with his one weakened ship and his fragment of a crew, he captured the Manila galleon, which crossed the Pacific from Acapulco to Manila once a year, loaded with treasure, en route to Spain. The haul included more than a million gold coins and 35,000 ounces of silver, with a total value of nearly £400,000, or the equivalent of £4 to £8 million today.

Anson's prize made a substantial contribution to reducing England's war debt, while conjuring up romantic images of pirate escapades. Tales of his adventures filled the newspapers after his return to England, and three different accounts of the voyage were rushed into print within a few months. The official version, written by Richard Walter and Benjamin Robins, was published in 1748, three years after Anson's return. It went through four editions within a few months and eleven more by 1776, was excerpted in several newspapers, and was translated into French, Dutch, German, and Italian. The failure of Anson's original plans and the loss of three ships and several hundred men were forgotten in the popular enthusiasm over this spectacular capture of Spanish treasure.

The excitement over Anson's voyage demonstrated how little English exploration of the Pacific had changed in two hundred years. From Drake's voyage in the 1570s to Anson's in the 1740s, England's ventures into the Pacific had been largely a product of its rivalry with Spain. Most voyages were undertaken as deliberate acts of war, as in the case of Anson and the Rogers/Dampier venture, or as expeditions of piracy thinly disguised as exploration, like Drake's circumnavigation. (The great exception was Dampier's second voyage around the world, sponsored by the government as a legitimate peacetime mission of exploration, although with clear political overtones.) The voyages became more sophisticated in their planning and their political motives more explicit, but their essential purpose remained little changed: to strengthen English power within Europe by increasing its influence in distant corners of the globe.

2. *Old Ideas and New Strategies*

Anson's voyage was an isolated expedition, an anomaly in the midst of a half-decade when England's expansionist ambitions focused on North America. But popular interest in the Pacific, far from diminishing, grew steadily stronger, although seizing a share of Spain's lucrative trade with its Latin American colonies was no longer the primary concern. Instead, over the course of the eighteenth century, English merchants, adventurers, and the public at large rediscovered centuries-old geographical theories, finding in them the possibility of new lands and new trade routes that would strengthen England's trade in the Pacific without treading on Spanish or French territory. The revival of such theories and a growing interest in scientific discovery joined with the changing political climate to encourage a more reasoned approach to establishing an English claim to the Pacific. In the process, Pacific voyages changed from wartime to peacetime missions, their objectives from capturing gold to discovering new land, and their captains from military men or pirates to explorers.

Despite strong popular and political interest in the Pacific, however, for twenty years after Anson's voyage there was more theory than action on the issue of English penetration of the Pacific. Inspired in part by Anson's voyage, John Campbell published a revised and enlarged edition of John Harris's *Navigantium atque*

Itinerarium Biblioteca between 1744 and 1748. Campbell's explicit purpose in reissuing Harris was to spark interest in exploring the Pacific with the ultimate goal of expanding English trade. Walter, the chronicler of Anson's voyage, took a similar view, pointing out in particular the importance of establishing a base on the coast of South America near the Strait of Magellan. In peacetime, such a base would be a convenient resting place for merchant ships, he argued, and in time of war, it "would make us masters of those seas."

During the same period, interest in finding a northwest passage, dormant since the disastrous voyages of the sixteenth and early seventeenth centuries, also revived. A "Letter from Admiral Bartholomew de Fonte," published in a popular London magazine in 1708, set off the revival, describing a voyage ostensibly made in 1640 from the Pacific to the Atlantic through a strait across the North American continent. Although the account was widely considered to be a fake, it was reprinted frequently.

In the 1740s, an Ulster landowner and member of the Irish Parliament named Arthur Dobbs single-handedly pushed along the movement for Arctic exploration, publishing a book on Hudson Bay that included an abstract of de Fonte's account and a series of articles accusing the Hudson's Bay Company, which had a monopoly on the North American fur trade, of shirking its responsibility to search for a shortcut to Asia. Dobbs also resurrected another northwest passage myth, the story of a Greek pilot named Apostolos Valerianos who, under the name Juan de Fuca, sailed for the Spanish on a voyage up the west coast of North America in 1592. De Fuca claimed to have discovered a passage between latitudes 47°N and 48°N (the approximate location of Seattle) leading to a land rich in gold and silver and then on to the Atlantic.

In part on the strength of this tale, Dobbs convinced the Admiralty to sponsor an expedition to Hudson Bay. When that voyage failed to find a northwest passage, he submitted a petition to Parliament, which resulted in an act passed in 1745 offering £20,000 to the discoverer of a passage. In 1746, unable to persuade the government to sponsor another voyage, Dobbs organized an expedition with private backing to undertake further searches in Hudson Bay. This effort was no more successful, and he eventually lost interest.

By this time, however, public interest in the northwest passage was gaining strength. A popular geography by Emanuel Bowen, published in 1747, devoted considerable space to the possibility. Campbell's revision of Harris's *Navigantium atque Itinerarium* included accounts of early Arctic voyages as well as Vitus Bering's 1728 voyage for Russia, during which he discovered the strait between the North American and Asian continents that now bears his name. The French mapmakers Joseph Delisle and Philippe Buache combined actual discoveries with the imaginary accounts of de Fonte and de Fuca to produce completely inaccurate maps, published in the early 1750s, that were received critically in France but favorably in England. And in America, in 1753 and 1754, Benjamin Franklin, who was impressed with Dobbs's ideas, sponsored two voyages from Philadelphia to the Arctic. Their lack of success did not diminish Franklin's interest, and in 1762 he wrote a defense of the de Fonte story.

Before any of these ideas for expanding trade in the Pacific could be thoroughly tested, however, England was once again involved in a major war. In the mid-1750s, the precarious peace of the preceding decade came unglued, and in 1756 war broke out between England and France. Unlike the earlier wars of the eighteenth century, this one, called by Europeans the Seven Years War and by Americans the French and Indian War, was fought almost entirely over issues of empire. And, in another departure from the past, it was a contest between England and France. Spain entered the war in 1762 as a French ally, but by then the conflict was almost over, and the consequence of Spain's involvement was to weaken still further its position as an imperial power.

Much of the fighting took place off the coast of North America as France and England battled over their American colonies. England emerged clearly victorious, although the peace treaty permitted France to retain much of its American territory. But England's power in Europe remained far from unchallenged, and pressure mounted within the government to take steps to expand the country's mercantile influence as a hedge against the threat of future French expansion. The old interest in Pacific trade and in establishing a foothold in South America surfaced once again; to the visionary, the prospect of an entirely new continent yet to be discovered suggested even more ambitious possibilities for English expansion.

A French geographer, Charles de Brosses, helped revive interest in the search for a southern continent with the publication of his *Histoire de Navigations aux Terres Australes* in 1756, one of the most thorough and scholarly compilations of Pacific voyages up to that time. Like John Campbell in the 1740s, he believed commerce, rather than conquest, should be the goal of his country's foreign policy, and he too viewed the Pacific as the most promising arena for future expansion. But de Brosses added a new dimension to the discussion—or, more accurately, revived an old one. Where English writers had focused on the need to establish a base in South America, de Brosses advocated a renewal of efforts to discover the great southern continent as the most promising way of expanding French trade and influence in the Pacific.*

Though de Brosses's book was not immediately translated into English, his work struck a responsive chord among those few who read it in England, and in 1766 the first volume of an English version appeared under the title *Terra Australis Cognita.* It was prepared by John Callender, who changed all references from France to England and issued the book as though it were his own original work. At about the same time, Alexander Dalrymple, the young Scotsman who would later propose himself as leader of the Royal Society's expedition to the South Pacific, returned to England after thirteen years in India.

While working for the East India Company, Dalrymple studied old books and documents on early English trade in the East Indies and on Spanish exploration of the Pacific. In the course of his studies, he discovered a Spanish manuscript describing the 1607 voyage of Luis Vaez de Torres through a strait between New Guinea and New Holland. Torres's discovery had never been published, and eighteenth-century maps usually connected New Guinea and New Holland. Dalrymple's find encouraged him to think beyond Asia and the East Indies to the South Pacific, and by the time he returned to England he was firmly convinced of the existence of a southern continent and its potential as a region for English expansion.

To buttress his argument for the existence of a southern continent, Dalrymple marshaled diverse bits of evidence, ranging

* De Brosses also coined the term "australia" to refer to his presumed continent, and "polynesia" (Greek for "many islands") as a general label for the islands scattered across the South Pacific. Neither term was used until the nineteenth century, however. The continent of Australia was known as New Holland until Cook claimed the eastern portion of it as New South Wales, the name also given to England's first colony on the continent. Subsequent colonies were known as Western and South Australia, and by the middle of the nineteenth century the name came into general use for the English colonies as a group.

from the timeworn arguments of theoretical geography (the need for a land mass to balance the land of the northern hemisphere), to assumptions about race and migration patterns in the South Pacific (the fair-haired people observed by some explorers on South Pacific islands must have come from terra australis), to the actual facts of past discoveries (Juan Fernández Island and New Zealand, he believed, were the eastern and western edges of the continent). In a remarkable display of theorizing from little evidence, he concluded that the continent must extend 5,323 miles from east to west and have a population of fifty million. Dalrymple's ideas, which he would develop at length in two compilations of voyages published in the late 1760s and early 1770s, met a sympathetic audience in England, as a new approach to English expansion in the New World gained favor.

For more than a century, despite its occasional voyages into the Pacific, England had concentrated its expansionist activities on establishing hegemony over North America, much as the Spanish had over South America. The Seven Years War was fought largely for this end, as England attempted to drive France out of America and take sole possession of the continent. The English government promoted the colonization of North America, in theory, to exploit the natural resources of the area, establish bases for the fish and fur trades, and create new markets for the products of English trade. Restrictive trade laws requiring American colonies to ship their goods in English-owned ships and to import goods only through English merchants were attempts to ensure that England would reap the exclusive benefits of its North American colonies. But colonies, as the English government was beginning to learn in the mid-1760s, created headaches as well as benefits; and the American colonies, in particular, were not producing the profits that armchair theorists of overseas expansion had predicted. Increasingly, some voices argued for the building of a "second British empire"—one based on the development of markets, rather than colonies. The markets that would form the basis of such an empire ought to be those of greatest potential profit; the spices and tea and silks of Asia, not the lumber and fish of North America. It should be an empire based in the Pacific rather than the Atlantic.

This concern about expansion of empire and England's new strength as a European power, supported by a new king (George III had come to the throne in 1760) with a personal interest in scientific discovery and by a public fascinated with distant places, opened the way for serious and sustained exploration of the Pacific.

Within a few months after the return of peace to Europe in 1763, English government officials began planning a voyage that would, they hoped, solve some of the mysteries of the Pacific and establish a firm English presence there. Both the unusual haste of the men responsible for the voyage and the secrecy surrounding their actions suggested the importance they attached to their actions.

In the first week of March 1764, the Lords of the Admiralty selected the *Dolphin,* a thirteen-year-old, five-hundred-ton frigate, to be specially fitted out for what they described as an experimental voyage to the West Indies to test the usefulness of copper sheathing as a protection for ships' hulls. Two weeks later, they appointed the veteran naval officer John Byron as captain of the ship, ordering him to prepare to sail as soon as possible. By the end of the month, Byron and a skeleton staff had taken up their quarters on board, and the employees at the Woolwich shipyard, where the *Dolphin* was berthed, worked overtime to get the ship ready for sea. In the meantime, the Admiralty picked a second ship, the *Tamar,* to accompany the *Dolphin,* and authorized the recruitment of larger than normal crews for both vessels—150 men for the *Dolphin* and 100 for the *Tamar.*

The size of the ships' crews was the first indication that the *Dolphin* and *Tamar* were destined for something more than a simple voyage to the West Indies. At about the same time, the Admiralty issued strict orders that no foreigners were to be admitted to the shipyards or docks under any circumstances. A few days later, just before the ships were ready to sail, their destination was changed abruptly from the West Indies to the East Indies, and Byron was named commander in chief of all ships employed in the East Indies, an appointment that allowed him to add extra provisions to his ships without going through the usual Navy channels and without, therefore, arousing curiosity about the unusually large quantity of provisions stowed on board.

But even these new orders were a ruse. Byron's secret orders directed him to sail to the Falkland Islands, then through the Strait of Magellan, and finally north along the coast of the Americas, searching for a northwest passage. Upon finding the passage, he was to sail through it into the Atlantic and home to England.

Only the Lords of the Admiralty, King George, and a few other highly placed advisers to the King knew the ships' true destination. So secret were Byron's orders that even the officers of the East India Company were ignorant of the real purpose of his voyage. Believing Byron's appointment and "official" destination genuine, they asked that he deliver some important documents to India, forcing the Admiralty to come up with an inventive excuse to deny the request.

The secrecy surrounding the voyage was intended to head off suspicion among the French and Spanish, for any indication that England entertained thoughts of extending its influence from the Atlantic to the Pacific might be enough to upset the peace so recently achieved. And indeed the other European powers would have had cause for concern, had they known what the English government had in mind. Although the motives behind Byron's voyage were complex, the desire to establish an English base for further exploration and trade in the Pacific was foremost among them.

The Falkland Islands, discovered in 1592 by John Davis and last visited by Dampier in 1683, seemed ideal for the purpose. Located in the extreme southern Atlantic, not far from the South American coast, the islands would be a convenient stopping place for ships headed either to the Pacific via Cape Horn or around the Cape of Good Hope into the Indian Ocean. The First Lord of the Admiralty, the Earl of Egmont, may have exaggerated when he called the islands *"the key to the whole Pacifick Ocean,"* but a base in the Falklands would at least be a start toward establishing England's influence in the Pacific. The only difficulty was finding them again; neither Davis nor Dampier had charted them accurately.

The Honorable John Byron, second son of a nobleman, had entered the Navy at the age of eight and achieved a distinguished record by the time he assumed command of the *Dolphin.* He had already made one voyage to the Pacific as part of Anson's crew and demonstrated courage and an instinct for survival on that ill-fated expedition. Just seventeen when he joined the voyage, Byron had the misfortune to sail in the *Wager,* the ship that sank off the coast of Patagonia. One of the few who survived the wreck and the subsequent struggle for survival on a barren offshore island, he eventually made his way to a remote Spanish outpost and spent three years moving from one Spanish settlement to another before finally taking passage back to England in a French ship. After his return, he rose quickly in the Navy and was a logical choice for the first major Pacific expedition since Anson's.

Byron's officers included several men who would later distinguish themselves as explorers, including Philip Carteret, a lieutenant in the *Tamar,* who would become captain of his own expedition around the world. Charles Clerke, a handsome, impetuous, twenty-one-year-old midshipman much given to practical jokes, had already served nine years in the Navy, having begun his career as a captain's servant at the age of twelve. Like Byron, Clerke had narrowly escaped death at sea; in 1761, during one of the battles of the Seven Years War, he was stationed at the top of his ship's masthead when enemy fire shot the mast away, sending Clerke plunging into the sea. Stunned and nearly drowned, he managed to crawl up the ship's chains to safety. Clerke would sail on all three of Cook's voyages, rising to command the third after Cook's death. The stolid John Gore, an American, was at thirty-three considerably older than most of the other members of the crew. Round-faced and slightly balding, he was a thoroughly practical seaman, skillful at navigation and chart-making, if not always very imaginative. He would also make four circumnavigations of the globe, succeeding Clerke as captain of Cook's third voyage after Clerke too died at sea.

The *Dolphin* and *Tamar* left England in June 1764. After a brief stop at Rio de Janeiro the expedition continued south, searching for islands but rarely straying far from the coast of South America. Approaching the barren, rocky shore just north of the Strait of Magellan a few days before Christmas, the crew saw smoke along shore, the first sign of habitation since Brazil. Byron recalled that it was just this stretch of coastline where he and his compatriots in the *Wager* had seen men on horseback waving at them with white handkerchiefs. Then he had been unable to investigate; now, with wind and sea calm enough for safe anchorage, he decided to satisfy his curiosity.

True to Byron's recollection, as the ships approached shore a crowd of horsemen galloped down to the beach, shouting and waving as if to encourage the Englishmen to land. Byron ordered the *Dolphin*'s boats hoisted out, gathered a small group of officers, and rowed to meet the horsemen, who numbered close to five hundred. They were a formidable sight—inches taller than most Europeans, broad-shouldered and stocky of build, with long, unkempt hair and faces streaked with red, white, and black paint. Their clothes were long, heavy cloaks of animal skins; and most astonishing, their voices made a "prodigious noise," even at some distance from shore.

Although the horsemen did not appear to be armed and continued, with their gestures, to urge the Englishmen closer, Byron approached cautiously. His companions were intimidated by the men's size, but in fact these people turned out to be quite docile, their loud voices expressing enthusiastic curiosity rather than aggression. As Byron prepared to disembark, they made "the most amicable signs which we were capable of understanding, or they of giving," Charles Clerke recalled. Leaving his crew on the beach with orders not to move until called, Byron approached the horsemen.

Face to face, they were even more astonishing, "the nearest to Giants I believe of any People in the World," Byron thought. One of his lieutenants, John Cumming, stood six feet two—an exceptional height for a European of his day—but even he appeared "a mere shrimp" next to these men and women. The native men, for their part, seemed equally surprised at the diminutive stature of their guests. After persuading them to sit down, Byron placed strings of beads around their necks and gave them lengths of colored ribbon, which pleased them so much they could hardly be dissuaded from expressing their thanks with exuberant, bone-crushing hugs.

Remarkable though this spectacle of several hundred "giants" was, it did not come entirely as a surprise to Byron or his crew, for the legend of a race of giants in this part of South America originated in stories brought back by Magellan's crew in the 1520s and had been a staple of travel stories ever since. Even the name of the region—Patagonia—was derived from the tales of the giants, who were dubbed *patagones,* "big feet," by Magellan's crew. One of the group, Antonio Pigafetta, an Italian endowed with a lively imagination and a talent for writing, published a popular account of the voyage that fixed the notion of a race of giants firmly in the public imagination. He recounted meetings with two groups of giant men: the first, he claimed, were cannibals, with voices like bulls and steps so long that the sailors could not overtake them no matter how fast they ran. Farther south, according to Pigafetta, they encountered a man so tall that Europeans reached only to his waist. He and his companions painted their faces with bright colors and clothed themselves in animal skins; they, too, ran "swifter than horses," as Magellan's men discovered when they were so rash as to try to capture one of them.

Drake's voyage had helped keep the legend alive. His chaplain, Francis Fletcher, described the Patagonians as "giants," but did not specify their height; Drake's nephew, writing fifty years later, elaborated Fletcher's story and considerably magnified his "giants." Although subsequent explorers disputed the stories of Patagonian giants—as recently as 1741, two Englishmen had described them as nothing more than average-sized men between five and six feet tall—influential writers of the eighteenth century often repeated the early stories. De Brosses took a scientific approach, reviewing the evidence on the Patagonians' size and concluding that a study of these people would be useful to help determine the effects of climate on human size. Callender's version of de Brosses's, Harris's *Navigantium,* and Dalrymple's writings all repeated the legends. Byron's version, embroidered by the popular press, only confirmed what people acquainted with travel stories were already prepared to believe.

After concluding their encounter with the Patagonians, Byron and his men turned east toward the Falkland Islands. In mid-January they reached the most westerly of the group, where they discovered a harbor so large and well protected that, according to Byron, "all the Navy of England might ride here together very safely." Byron named it Port Egmont after the First Lord of the Admiralty. They stopped long enough to plant a vegetable garden, in the optimistic hope that future English sailors stopping here would find fresh food, and to replenish their food supplies with the island's abundant fowl. Their bird catching was so successful, Clerke noted, that the ship was crammed with birds and all hands were diverted from their usual tasks to pluck feathers.

After claiming the islands for England, Byron directed his ships to reconnoiter the northern coast; but he soon saw enough of the barren, craggy shoreline to conclude that prospects for further discoveries were not good enough to justify battling the adverse winds and inclement weather any longer. Getting trapped in a narrow bay and nearly wrecking the *Dolphin* was the final blow.

Byron decided to turn back immediately to the Strait of Magellan, where, in a few weeks' time, he was due to meet an English storeship with fresh supplies.

In deciding to turn back, Byron displayed perfectly sensible judgment. He could have spent weeks and courted disaster many times over had he chosen to explore every bay and inlet of the Falklands' torturous coast, and he was quite correct in assuming that there was little else worth discovering in the South Atlantic. But in turning back when he did, Byron missed altogether the struggling little French settlement on East Falkland Island. The colony, if such it could be called, had been settled just a year earlier by an expedition headed by Louis Antoine de Bougainville, a naval officer who, like Byron, had seen service in North America during the Seven Years War and then turned explorer. The French government had sent Bougainville to the Falklands for precisely the same reason that the English government had sent Byron: to establish a base from which to undertake trade in the Pacific.

Byron learned about the infant colony when he made his rendezvous with the storeship, since its captain had with him an account of the French settlement. He hastily penned an account of his own rediscovery and dispatched it back to England, where plans were already underfoot to discredit the French occupation and establish an English colony.

Having done what he could to state the English claim to the Falklands, Byron turned his attention toward the Pacific, and in September the *Dolphin* and *Tamar* entered the Strait of Magellan. A few days later they noticed another ship following them, flying French colors. All three ships anchored in close proximity that evening, and the French sent boats to make contact with the Englishmen. Byron, who suspected that the ship was also bound from the Falkland Islands, told the *Tamar*'s officers to speak with the French crew, but not to invite them on board. Had he exchanged greetings with the French officers, Byron would have learned that the captain of the ship was Bougainville and that he and his men had indeed come from the Falklands. They had sailed to South America only to take on wood and water before returning to France, so within a few days the French and English ships parted company.

The Strait of Magellan was a deceptively difficult passage. Although theoretically it could be traversed in three weeks or less, narrow channels, rocky shorelines, and fierce winds were more likely to keep ships battling their way west for two or three months. The *Dolphin* and *Tamar* spent seven cold, miserable weeks in the Strait, entering the Pacific on April 9. Byron was determined not to sail north in search of a northwest passage, as his orders dictated, but to head directly across the Pacific instead. Later he would justify his decision on the grounds that his ships were too battered and leaky to survive a long voyage north; and indeed they were in serious need of repair, with provisions just barely adequate for a long voyage despite replenishment from the storeship. Within six months Byron would write that their food was barely edible: "all our Beef & Pork stinks abominably & our bread is quite rotten & full of Maggots & Worms."

But Byron had another motive as well. More interested in finding a southern continent than a northwest passage, he hoped to rediscover the Solomons and solve the mystery of terra australis. So little land did he find, however, that the ships swept across the Pacific in only slightly more time than they had taken to get through the Strait of Magellan. Pursuing a course just north of the major land groups in the South Pacific, Byron discovered only a few tiny islands and overshot the Solomons altogether. Compounding his bad luck, most of the islands he found were either ringed by coral and high surf or populated by hostile men and women, making it impossible to land in either case, so the ships got very little in the way of fresh provisions to replenish their food supplies and stave off scurvy, the inevitable consequence of a diet of biscuit and salt meat. By the time the expedition reached Batavia, a Dutch port in the East Indies, not only the ships but many of the men too were in a sadly weakened condition.

Rejuvenated by their stay in Batavia, the Englishmen continued their voyage around the Cape of Good Hope and reached England on May 9, 1766, slightly less than two years after their departure—a record for a voyage around the world. One of Byron's last official acts before entering port was to confiscate the journals kept by his crew members, a matter of routine on Navy voyages that was intended to guard against security leaks and guarantee that only the official, government-sanctioned version of the voyage would reach the public. On this occasion, such a precaution was considered especially important because of the Falkland Islands issue; it would be necessary for the English government to maintain that its ships had arrived in the islands prior to those of the French, an impossible fiction to support if sailors' journals were freely published.

Already Byron's report on the Falkland Islands and the French announcement of their settlement had set off a tempest of claims and counterclaims, as both nations tried to confirm their right to this isolated outpost. In the months before Byron reached England, the Admiralty had dispatched an expedition, headed by John McBride, to establish a settlement in the Falklands. He was instructed to tell any foreigners in the islands that the territory belonged to England, though nothing was said about how he was to oust the several score Frenchmen who had been occupying their island for two years. Undaunted by this detail, in January 1766 McBride established his band of settlers at Port Egmont, where they found the *Dolphin*'s garden flourishing. The two rival settlements, on two different islands widely separated by rugged terrain, were each far too weak in numbers to exert any sort of force, so they simply ignored each other while the controversy over their futures raged in London and Paris.

In the meantime, Spain also joined the fray, claiming possession of the islands by virtue of their proximity to South America. Spain had no real use for the islands, but worried, with obvious good cause, that the French settlement would bring England into the South Atlantic, a development Spain was anxious to avoid. Consequently Spain put pressure on its ally, France, to give up claim to the Falklands, and the two countries negotiated a transfer of the French colony to Spain.

It remained for Spain, however, to try to force the British out. After two minor skirmishes between English and Spanish ships near the islands, in the spring of 1770 a Spanish ship opened fire on the colony at Port Egmont, and the settlers, hardly a match for Spanish warships, quickly surrendered. England and Spain very nearly went to war over the incident, but a diplomatic solution was achieved several months later when Spain issued a declaration repudiating the attack and the English government gave its secret, oral assurances that the colony would be abandoned.

From the English government's point of view, this diplomatic squabble over the Falkland Islands was the most significant consequence of Byron's voyage. For the English public, however, the rediscovery of the Falklands was nothing more than a minor incident; the controversy with France and Spain, shrouded in official secrecy, barely made an impression. Of far more interest was

Byron's discovery, not of islands, but of people; the exotic, frightful, and fascinating Patagonian giants. Hardly as spectacular as Swift's Brobdingnagians, nevertheless the Patagonians created a sensation among a public that devoured travel accounts as fast as they could be published with a gullibility that accepted tales of giants as perfectly plausible.

Although Byron's official account of his voyage, edited and considerably elaborated by John Hawkesworth, did not appear until 1773, newspaper accounts playing up the giants hit the streets almost immediately after Byron landed. Within a few months, an unauthorized account, apparently based on the journal of Charles Clerke, was published, complete with an appendix describing the giants; and the London *Chronicle* printed twelve articles about them in the space of two years. In the hands of journalists, the Patagonians took on even greater stature and more ferocious mien. Hawkesworth transformed Byron's "frightful" men into "hideous" ones, and Clerke's anonymous account claimed that Byron had appeared barely taller than the seated Patagonians as he distributed gifts to them. His book included an engraving of this scene, as well as one of an English sailor standing waist-high next to a Patagonian.

A few people remained skeptical, noting the curious silence of most of the ships' crews on the subject of giants; the French Minister of Foreign Affairs, admittedly a biased observer, commented that Byron must have seen the Patagonians through a microscope. And the essayist Horace Walpole, amazed at the credulity of his countrymen, used the fuss over the giants to satirize the full range of his country's foibles, from its difficulties with the American colonies (Parliament should see that giants wear only "legally stamped" sheep's skins), its involvement in the slave trade (enslaving giants would "give a little respite to Africa"), and its religion (the evangelical preacher Whitefield "intends a visit, for the conversion of these poor blinded savages"), to the controversies of academics (who argue over whether the giants are true aborigines).

But nothing added more to the credibility of the giants than their elevation to a topic of discussion for the Royal Society. Always alert for descriptions of unusual natural phenomena, and particularly interested in exploration of distant places, the members of the Royal Society discussed the popular reports of the giants at length and publicized them with a seriousness that gave them the weight of scientific approval. Seeking more definitive information about the Patagonians, the Society's secretary, Matthew Maty, asked Charles Clerke for a detailed account. Clerke responded with a letter describing the Patagonians as men of "gigantic stature," averaging nine feet tall if not more. Clerke's letter, which carried conviction since it was the first written account from any of the crew (other than Clerke's own anonymous journal), was duly read to the Society at its meeting on November 3, 1766, and published in its *Transactions* the following year.

The willingness of the highly respected Royal Society to believe, and indeed endorse, tales of giants astonished the members of its French counterpart, the Académie des Sciences. The French naturalist La Condamine, familiar with Bougainville's description of the Patagonians as no taller than ordinary men, dismissed the whole affair as a fable promulgated by the English government to draw attention away from the more serious purposes of Byron's voyage and from the plans underway to mount a new expedition to exploit a rich mine discovered by Byron in South America. La Condamine's mine was as fictitious as the height of the giants, but otherwise he was close to the mark in his interpretation.

The English government did indeed wish to divert attention from the Falkland Islands and the political consequences of Byron's voyage; and Byron, while stopping short of any specific statements about the Patagonians' size in his own journal, was perfectly willing to cooperate. The press, recognizing a salable topic, played its part by embellishing the truth with reports that played to popular notions about exotic, faraway lands. The gullibility of the Royal Society is more difficult to understand, although the publicity it gave to the issue may well have been the work of a few members. As for Charles Clerke, his propensity for practical jokes was well known among his shipmates; and to fool the Royal Society was an opportunity too good to be missed.

3. *Paradise Discovered*

So preoccupied were the English with the Falkland Islands and the Patagonian giants that few took notice of Byron's failure to find a northwest passage or, for that matter, anything else of much interest. His discoveries in the South Pacific, however, meager though they were, did succeed in shifting official attention from the North to the South Pacific, from the quest for a northwest passage to the quest for a southern continent.

Of particular interest to the Admiralty was his description of conditions at sea about midway across the Pacific. Here the water was relatively calm, without the great swells characteristic of the open ocean, and enormous flocks of birds filled the skies. Both were typical signs of nearby land. Had it not been for the sickness of much of his crew, Byron claimed, he would have sailed south to find the continent he was certain must be there; but the desperate need to get to a known port as quickly as possible prevailed, even over so promising an opportunity for discovery.

Byron's report, along with the need to get supplies to the recently established colony in the Falklands, encouraged the Admiralty to prepare another voyage immediately. Orders to fit out the *Dolphin* for a return voyage were issued in late May 1766; the *Tamar,* deemed unfit for further service, was replaced by the *Swallow,* although it was hardly in better condition. A third ship, the

Prince Frederick, would carry supplies to the Falkland Islands and then return to England. In June, government officials learned that the French planned to use their settlement in the Falklands as a base from which to search for the southern continent—precisely what the English themselves had in mind. This intelligence made them speed up preparations for the *Dolphin*'s return voyage and redouble their efforts to keep the plans secret.

In mid-June, the Admiralty appointed Samuel Wallis as captain of the expedition, Byron having had enough of Pacific voyaging. At thirty-eight, Wallis already had a solid career as a naval officer behind him. He had commanded three different ships, serving most recently at the siege of Québec. A reliable and well-respected officer, nevertheless he had no particular qualifications for the task of exploration. Philip Carteret was promoted to captain of the *Swallow,* although his effectiveness as Wallis's second in command was diminished by his being kept in the dark about the exact purpose of the voyage. Until the ships reached the Strait of Magellan, Carteret thought it most likely that he and the *Swallow* would be ordered back to England after their stop at the Falkland Islands.

The Admiralty could not maintain complete secrecy about its plans, and word of the impending voyage caused considerable consternation in France and Spain. The French and Spanish ambassadors collaborated in hiring as spies two men scheduled to sail on the voyage, to obtain information about the location of proposed English bases. With their minds riveted on their nations' rival claims to the Falklands, however, the ambassadors failed to realize that the expedition's true purpose was discovering a continent in the South Pacific and that the proposed stop at the Falklands was clearly secondary. Instead, the Spanish ambassador told his superiors in Madrid that the voyage was intended to establish a settlement in the Falklands (which, of course, had already been done) as a base for illegal trade along the coasts of Peru and Brazil.

By July 1766 the three ships were ready to sail, although Carteret complained bitterly that the *Swallow* was inadequate for a long voyage and lacked some essential supplies—notably a forge and iron for repairs to the ship's hardware.

Although a number of the crew from the *Dolphin*'s previous voyage had signed on again, including John Gore, most of the officers were making their first Pacific voyage. The first lieutenant in the *Dolphin* was William Clarke, an ill-tempered and sickly man of about thirty, whose high-handed treatment of the crew earned him the nicknames "Mr. Knowall" and "Old Growl." The second lieutenant, Tobias Furneaux, in contrast, was a man of such affable disposition and calm demeanor that he was sometimes too lenient toward the men in his charge. Like Clarke and most of the other men who sailed on voyages of exploration in the late 1760s and 1770s, Furneaux had seen battle off the coast of North America in the Seven Years War, earning his promotion from midshipman to lieutenant after taking charge of his ship when his commanding officer was killed in an uneven fight with more heavily armed French ships. He would serve on the American coast again, during the American Revolution, after commanding the *Adventure,* companion ship to the *Resolution* on Cook's second voyage.

George Robertson, master of the *Dolphin,* at thirty-five was one of the oldest of the officers, and had earned most of his experience at sea in the merchant marine. An unusually curious and perceptive man, he kept a journal that is one of the best travel accounts written by any of the eighteenth-century explorers. The other mates were Richard Pickersgill and Robert Molyneaux. Both men proved to be talented navigators, with a special skill at making maps, although Molyneaux's effectiveness as a crew member was hampered by his fondness for strong drink. Both would later sail with Cook, and Pickersgill, just seventeen when he joined the *Dolphin,* would eventually command his own voyage of exploration.

One of Wallis's aims was to test the stories about Patagonian giants (although this was not part of his official instructions), and he took along measuring rods for that purpose. It was December by the time he and his crew reached the shores of Patagonia, having parted company with the *Prince Frederick* and its store of provisions for the settlement at Port Egmont. True to Byron's account, men clad in animal skins and mounted on horseback galloped along shore, keeping pace with the ships, waving their cloaks and bellowing greetings. At daybreak the next morning, Wallis went ashore with Carteret and several of the other officers in two armed boats; he ordered Robertson, who was left in charge of the ship, to keep the guns loaded "in order to fire a few Shoat [shots] amongst this Gigantick Race of Men, If they prove troublesome."

Once on shore, however, Wallis discovered the Patagonians to be docile and considerably less gigantic than Byron's reports indicated. Taking out his measuring rods, he found that most stood between five feet ten and six feet, although several were taller; Carteret recorded the tallest as six feet seven. One was later bold enough to come aboard ship. Some of the crew thought this "giant" a woman, because her face was smoother and skin somewhat lighter than most of the others, but the similarity of hair and dress among these people made it difficult for the Englishmen to distinguish their sex. Robertson expressed his admiration for her cloak, and she held out one side for him to examine more closely, but "carefully wrapt upp what would have soon ended our Despute."

The reports of Wallis and his men did not destroy the myth of the Patagonian giants, however. Carteret wrote a careful account of his observations to Matthew Maty, which put a stop to the Royal Society's discussion of the phenomenon, but a letter one of the *Dolphin*'s carpenters sent to London claimed the Patagonians averaged seven and a half feet, with some as tall as eight feet. Byron asserted that Wallis must have seen a different group of people, which helped foster a new fiction about a race of giants living in the mountainous regions of Tierra del Fuego, some distance from the sea. As late as 1819, Byron's grandson, the poet, could write half seriously of going to South America in search of the "gigantic cavalry" of Patagonia.

From Patagonia, the *Dolphin* and *Swallow* headed into the Strait of Magellan, where fierce, frigid winds and rough seas slowed their pace so severely that it took four months to get through the passage. After struggling to stay together all that time, the ships became separated just as they reached the Pacific; the *Dolphin,* a much faster vessel, sailed briskly ahead, all sails set, while an easterly current sucked the *Swallow* back into the Strait. By the time Carteret managed to recover the lost ground, the *Dolphin* was nowhere to be seen.

The two ships did not meet again. Instead, the captains pursued their search for a southern continent separately, Wallis following a track slightly south of Byron's and Carteret, cursing Wallis for what he believed was a deliberate desertion and his rickety ship for all its defects, sailing still farther south.

Six weeks into the Pacific, the *Dolphin* had encountered no signs of land, and the months aboard ship with little in the way of fresh food or exercise began to take their toll on the men's health. Wallis was unusual among ship captains in his attention to diet and cleanliness—his men carried their hammocks on deck to air every day and scrubbed the ship with vinegar—but despite his precautions, the crew looked "very pale & Meagre," and several were suffering from scurvy.

Within another week the first sign of land, and with it the promise of solid ground underfoot and the first fresh food in months, raised everyone's spirits; but closer investigation revealed only a tiny island, surrounded by surf so high that landing was clearly out of the question. A larger and more promising island discovered the next day was so heavily guarded by its inhabitants, who brandished spears and lit fires at every possible landing place, that getting ashore appeared equally impossible. Eventually Furneaux persuaded the islanders to supply coconuts and fresh water in exchange for beads and nails, but as soon as the transaction was concluded, they took the Englishmen by the hand and led them back to the boats.

The next two weeks brought only more of the same—small atolls rising no more than a few feet above the ocean and ringed with impenetrable coral reefs or guarded by hostile inhabitants—until some of the crew glimpsed a mountain in the distance, shrouded in clouds. Beyond it, far to the south, rose what appeared to be a range of peaks. The elated sailors, believing they were at last approaching the fabled southern continent, sailed on through the night with every expectation of sharing in the most significant discovery since Columbus's time.

Hazy skies obscured their view the next morning, but the sight was enough to convince them that their eyes had not deceived them: a massive mountain, perhaps seven thousand feet high and several miles broad, rose from the sea a few miles ahead. Thickly covered with trees, the land had the appearance of a brilliant green carpet, rumpled and creased where streams had carved out valleys from the higher altitudes to the sea. For weary and sickly sailors who had risked their boats in high surf to get a few coconuts just two weeks earlier, it was a sudden, unexpected gift. Their assumption that the mountain was merely the tip of a peninsula, or perhaps an island lying just off the coast of the southern continent, made the discovery appear even more significant.

Wallis, recognizing the importance of this find, would name the island—for the hazy visions of a continent in the distance turned out to be just that—King George's Island in honor of his sovereign. Its native name was Tahiti.

As the *Dolphin* sailed closer, dozens of canoes paddled out from shore. They stopped at some distance—"within pistol shot," Robertson noted, "and lookt at our ship with great astonishment, holding a sort of Counsel of war amongst them." The Englishmen tried to demonstrate their friendly intentions by dangling beads and trinkets over the side of the ship, while the islanders responded by waving plantain branches, but none of the canoes ventured closer. Instead more put out from shore until at least eight hundred men in 150 canoes lined up to study their visitors. Emboldened by numbers, a delegation finally ventured aboard.

The Englishmen's first concern was to get across their need for food and their willingness to barter for it. Some of the sailors imitated animals, grunting and squawking like pigs and chickens, while others went through the motions of eating. The islanders responded with similar noises, and a few returned to their canoes, apparently to get food to trade. Others were more interested in the ship's iron hardware. Yanking on rings and stanchions, they were perturbed to find these curious shapes firmly bolted down. The sailors offered nails as alternative souvenirs, which so pleased the islanders that none would leave without some bit of iron for himself. Some became "surly," as Robertson put it—the word would become a favorite term to describe unruly Pacific islanders—and Wallis ordered a cannon fired over their heads to restore order. The frightened men jumped overboard and swam to their canoes, although not before one of them had managed to steal a sailor's hat.

The following day, Wallis sent John Gore in one of the ship's boats to find a suitable place to anchor. As Gore sailed closer to shore, scores of canoes surrounded him, making signs for the men to land. The Englishmen, those aboard the *Dolphin* as well as those in the boat, feared an attack, and Wallis ordered a cannon fired over their heads while Gore and his crew turned back toward the ship. The Tahitians tried to cut them off, paddling furiously to get between Gore's boat and the *Dolphin,* but the boat was faster and forced most of the canoes to give up the chase. The few who managed to keep up pelted stones at the boat's crew, injuring several men. In retaliation Gore fired his musket and wounded one of the Tahitians, while the others jumped overboard and swam to shore.

The incident did not keep the curious Tahitians from flocking to the coast to get a glimpse of the ship. Thousands of men and women lined the beach, according to English estimates, leading Robertson to label Tahiti "the most populoss country I ever saw." Their formidable numbers made Wallis even more cautious about sending out more boats or attempting a landing, until the prospects of getting ashore became an all-consuming topic among the restless sailors, still confined to their close quarters and seemingly endless dinners of salt meat and moldy biscuit. The optimists among the group, dismissing all questions of potential danger from the islanders as scarcely worth worrying about, remained convinced that they would soon find an anchoring place and get all the provisions they needed. Others, impressed by the vast numbers of people they had seen and swayed by images of South Pacific islanders as fierce savages, believed they would never be able to obtain provisions, much less go ashore, without bloodshed. Any move toward land, they feared, would bring thousands of canoes to surround them.

The optimists, Robertson among them, placed great faith in the power of European weapons to subdue any number of islanders, but the pessimists watched the crowds grow larger on shore and concluded that several hundred canoes could easily overwhelm the *Dolphin,* superior firepower notwithstanding. Much better, they argued, to sail for Tinian in the Ladrones Islands, one of the few definitely known spots in the Pacific, where ships had regularly found refreshment since the days of the Spanish explorers two centuries earlier. Fortunately for the sick, none of the men in command paid heed to the pessimists, for Tinian, as they learned later, was more than four thousand miles away.

After several days of sailing slowly along the coast, searching for a place to anchor, Wallis and his crew discovered a lovely sheltered bay with a river pouring into the ocean at one end and a beach of powdery gray-black sand stretching nearly the full distance around it. Called Matavai by the Tahitians (Wallis would name it Port Royal Bay), it looked like an ideal anchoring spot, and Wallis sent Gore and Robertson to investigate further—in two boats this time, so they could protect each other in case of native interference. And indeed the Tahitians did try to keep the Englishmen away from shore, sending a fleet of canoes to warn them off, but Gore and Robertson ignored them.

Once the *Dolphin* had anchored, Tahitian canoes came out readily enough with provisions for trade, but the atmosphere of mistrust built up over the preceding days made bartering difficult. The islanders refused to part with their goods until they had payment (nails, beads, or mirrors) in hand, behavior that the English found "insolent." Hostility mounted until Tahitians and Englishmen nearly came to blows; only Wallis's strict order that the islanders were not to be harmed prevented bloodshed.

Robertson and Gore, meanwhile, set out again to find a way of getting on shore to obtain fresh water, now in seriously short supply on the *Dolphin.* It was a formidable assignment. The assembly of men and women on the beach had reached several thousand; at least a hundred canoes were beached near the river; and another two hundred or so, many of them enormous double canoes sixty or seventy feet long with high, curved prows and sails, floated between the ship and shore. As the English boats headed toward land, the largest canoes followed them closely, their crews jeering noisily at the Englishmen. Some of the Tahitians even attempted to board the boats. Such behavior, coming on top of the incidents on board ship, made Robertson suspect a plot to attack the English.

He finally gave up all hope of getting closer to the island and ordered the boats back to the ship, setting off a great commotion among the Tahitians, who had assumed that Robertson and his men intended to land. Some of the men in the canoes renewed their efforts to board the boats. Robertson directed his crew to point their muskets at the intruders to keep them off, but with no effect; "they Laughd at us," he recalled, "and one struck his prow right into our Boat stern, and four of the stoutest fellows immediately Jumpt on the prow of the canoe, as if they meant to board us, with their paddles and clubs in their hands." He then ordered one of the marines to fire over their heads, but the burst of fire startled them only momentarily. Angered, Robertson told his men to shoot directly at the two boldest members of the would-be boarding party. This time his order had the desired effect; the canoes departed hastily and stayed well away from the boats, but at the cost of one Tahitian killed and another wounded.

Word of the disaster traveled quickly among the islanders, and yet it did not keep them away from the *Dolphin* for more than a few hours. By the next day, the canoes were back with hogs, fowl, and fruit, which they traded for a pittance. A three-inch nail would purchase a twenty-pound hog; fowl and fruit could be had for beads. Trade proceeded circumspectly for the most part, although some of the Tahitians tried to take nails without giving anything in return. Merely pointing a musket or even a spyglass at them, however, was now sufficient to put a stop to petty thievery. The islanders understood clearly that musket fire had been responsible for the death of their countrymen, as they demonstrated more than once in the presence of the *Dolphin*'s crew. They would cry out "bon-bon" in imitation of the guns, strike their chests and foreheads, and fall backward, lying rigidly with their eyes open in a fixed stare.

This uneasy truce continued another two days, with the Tahitians continuing to harass and insult the English at every opportunity and the English becoming increasingly short-tempered at their inability to take control of the situation. Annoyed at the Tahitians' constant petty thieving, Wallis eventually banned them all from the ship. His order put a stop to the theft, but it also cut off the *Dolphin*'s supply of fresh food.

Getting a sufficient supply of water remained a problem, and Gore hit upon the plan of sending casks ashore in the islanders' canoes to be filled and returned to his boat. But he soon discovered that the Tahitians had no intention of returning all the barrels. Instead, they delivered about forty gallons and tried to entice the Englishmen ashore to get the rest. When water failed to lure the sailors out of their boat, the Tahitian men lined up some of their most attractive young women, who proceeded to undertake "every lewd action they could think of" to get the Englishmen on shore. They also brought food to the waterside and made signs for the sailors to come and eat. But, still uncertain of the Tahitians' intentions and intimidated by their numbers, Gore turned his boat back to the ship. In derision, the women threw apples and bananas and yelled insults at the retreating men.

Tempers wore thin at the frustration of being so close to plenty and yet unable to get it. Some of the sailors were more convinced than ever that it would be impossible to land without bloodshed, now that they had killed one of the Tahitians, and they were quick to blame Robertson for their predicament. Wallis was in poor health, which made it difficult for him to maintain control of his crew, let alone plan a strategy for getting them ashore. Clarke was also too sick to manage his normal duties, although not too sick to hand out advice, according to a disgruntled Robertson. The two men—never on good terms with each other—argued over what their next move should be. Robertson won out, with his plan to send two boats toward a low-lying point of land at one end of the bay, where it appeared that they could land a crew to refill the water casks. The *Dolphin* would follow the boats closely enough to protect them with its cannon.

As the boats drew close to the point, dozens of canoes paddled out to meet them, and the *Dolphin,* attempting to guard the boats, ran too close inshore and struck a coral reef. With the entire crew on deck trying to free the ship, Clarke blasted Robertson for getting them into this predicament, declaring that they would soon have hostile Tahitians swarming across their decks. In reality the accident was not so serious as it first seemed and the *Dolphin* was soon freed; but later, after the ship was safely anchored once again, dozens of canoes carrying at least a thousand men and several young women surrounded it. At a prearranged signal, the women stood up in the prows of their canoes, stripped naked, and made provocative gestures obviously intended to distract the sailors and entice them out of the ship. "As our men is in good Health & Spirits," seaman Francis Wilkinson observed, "it is Not to be wonderd that their Attention should be Drawn to A Sight so uncommon to them Especially as their women are so well Proportiond." Only a few noticed the piles of stones in the canoes.

Moments later a large double canoe approached the ship, carrying what appeared to be several of the island's most influential inhabitants. At a sign from one of them, stones flew through the air, pelting the ship's deck and bruising many of the crew. The friendly greetings, the offers of trade, and especially the provocative acts of the young women had all been part of a clever ruse.

At first the trick succeeded. The Englishmen were so stunned that they did not realize what was happening to them, and it took some time before they could pull themselves together to fight back. By then, Wallis estimated, three hundred canoes with about two thousand men and women surrounded the ship, while several thousand more watched from shore. Surprise and sheer numbers had gained the Tahitians an advantage over superior weaponry.

The advantage was brief, however. Wallis directed his men to fire small shot in an attempt to disperse the canoes, with little effect. He then ordered the cannon loaded with grapeshot and fired directly into the mass of canoes. The cannon fire terrified the Tahitians, and they retreated quickly, reassembling about a mile from the ship once the initial confusion passed. Hoping to forestall future attacks, the *Dolphin*'s officers fired two cannon at the large double canoe that had initiated the attack, ripping it in half and killing several of those aboard. Another volley killed several more.

Robertson, who felt more sympathy for the Tahitians than most of his countrymen did, tried to imagine the reaction of the people who had been watching the battle from shore, Expecting a great booty of "nails and Toys, besides the pleasure of calling our great Canoe their own," they instead came down from their observation posts to the waterfront to see canoes destroyed and friends and relatives dead or wounded—and in a manner particularly bloody and brutal, unlike anything they had witnessed before. Some of the Englishmen, secure in their belief in European superiority, thought the Tahitians would now view them as gods come to punish them for their sins, a point of view that said more about English attitudes than Tahitian. Others could not shake their fear of native hostility, believing that the islanders would attack again to revenge their dead. But for the moment nothing at all happened. The canoes vanished, and the crowds of onlookers disappeared from the beaches.

The next morning, the *Dolphin*'s men were able to land for the first time without interruption. Around noon a handful of canoes with plantain branches fixed in their bows slowly approached the ship, stopping several hundred yards away while the leader of the delegation made a speech. His companions then threw their branches into the water and continued to paddle toward the *Dolphin,* their eyes fixed on the sailors standing on deck. "If any of us lookt surly," Robertson recalled, "they immediately heald up the tope of the plantain tree, and forced a sort of smile, then laid doun the plantain tree top and showd us what they hade got to sell."

Interpreting this ceremony as a form of surrender and feeling secure in their military superiority, the Englishmen adopted the attitude of conquerors. Where once they had despaired of getting fresh provisions, now they dictated the terms under which they would barter for food, allowing only two or three canoes to approach the ship at a time. Minor disagreements aside, trade proceeded without incident, which the often self-righteous Wilkinson attributed to the success of his and others' efforts to teach the Tahitians that "Honesty is the Best Policy." In reality the Tahitians, who now clearly understood the limits of their weapons, behaved in a manner carefully calculated to maintain peace and garner whatever benefits they could from the English presence.

With English supremacy apparently established, Wallis sent Furneaux and a group of armed marines to take official possession of the island, symbolized by planting a Union Jack on the beach. After the delegation returned to the ship, several Tahitians gathered and stared at the flag for some time. They seemed afraid to approach it, but eventually two old men left the rest of the group and walked solemnly toward the flag, carrying the ubiquitous plantain branches. With much ceremony, they knelt before the flag and placed their branches on the ground. Hundreds more followed them, repeating the same routine. Suddenly a gust of wind caught the folds of the flag, whipping it vigorously over the heads of the people kneeling before it; instantly they ran off and would not even turn to look behind them until they were about fifty yards away. They appeared frightened, almost "as if a Great Gun hade been fired at them," Robertson remarked.

Afterward the Tahitians continued to place plantain branches before the flag, and toward the end of the day the two old men who had initiated the ceremony added a pair of pigs to the pile of offerings. Later they placed the pigs in a canoe and paddled out to the ship with them, making signs for the crew to take the animals on board. The sailors offered the usual nails and trinkets in return, but the old men would accept nothing, leading Robertson to conclude that they wished to exchange the pigs for permission to remove the flag.

Neither Robertson nor his fellows comprehended the meaning of the scene they had witnessed. The Tahitians looked upon the flag as a symbol of the Europeans' power—as indeed the Englishmen themselves would have viewed it, had they thought about its meaning in those terms. To the Tahitians, however, the flag was something more than mere symbol. They believed that inanimate objects, as well as people and animals, could be invested with a special kind of power they called *mana*. The flag, they thought, carried the *mana* of the conquering strangers, and they went through their elaborate ceremony both to conciliate the victors in the previous day's battle and to try to gain control over the flag's powers. They interpreted its sudden flapping as a sign that their efforts had succeeded. (The Tahitians treated the flag as a prized object for years afterward, decorating it lavishly with red and yellow feathers; English sailors visiting the island in the 1790s were amused to see it used in a religious ceremony.)

The Englishmen continued to view the Tahitians warily, despite the ritual surrounding the flag and the islanders' eagerness to trade, and they felt confirmed in their suspicions when a party went ashore to collect water early the next morning. Hundreds of Tahitians quickly assembled on the beach; about 150 canoes rounded the point; and a crowd of women and children congregated on top of a hill facing the bay. Wallis immediately ordered the watering party back to the ship. Fearing another attack and convinced that it was "necessary to conquer them in the beginning," he ordered guns fired, first at the canoes and then at the crowds on shore. The islanders retreated up the hill, later called Skirmish Hill by the English, and Wallis had his men fire another round at them. Though the bullets fell short of hitting anyone, they landed close enough to force the crowd to retreat. Then Wallis sent the carpenters ashore with orders to break apart as many canoes as they could find. A few Tahitians tried to stop the destruction, but the marines sent to guard the carpenters drove the islanders away with musket fire.

This show of force had the desired effect, as the Tahitians, now convinced that the Englishmen's guns could reach them anywhere, renewed their conciliatory ceremonies. Once again they appeared on the beach bearing plantain branches and pigs, but this time the attempt at a truce stalled over a misunderstanding about the exchange of gifts. In addition to the pigs, the Tahitians brought yards of a cloth called *tapa,* a fragile, papery fabric made by soaking and pounding bark. Furneaux, again leading the English shore party, accepted the pigs, leaving hatchets and nails on the beach in return, but did not take the cloth, since the *Dolphin*'s crew had no use for it. The Tahitians interpreted their visitors' behavior as a sign that they were still displeased, and so they would not take the hatchets and nails but continued to wave their plantain branches, even after Furneaux and his men returned to the *Dolphin.* Finally some of the officers realized the problem, and sent two boats back to get the cloth. Immediately the Tahitians broke into smiles and gathered up the nails and hatchets.

Observing throughout this exchange that the sailors could not take their eyes off the young women in the crowd, the Tahitian men decided to take advantage of their visitors' interest. Leading the girls out of the crowd, the men lined them up in rows and made signs for the Englishmen to take their pick. In case there should be any misunderstanding, "the old men made signs how we should behave to the Young women," but the sailors made signs in return that they were not so ignorant as the Tahitians might think. The officer in charge of the boat's crew quickly scrubbed the men's hopes, however. With orders to bring back only the cloth, not a party of young women, he declined the Tahitians' offer and herded his protesting crew into their boats.

Back on board the *Dolphin,* they could talk of nothing but the Tahitian women. "All the sailors swore they neaver saw handsomer made women in their lives," Robertson recalled, "and declard they would all to a man, live on two thirds allowance, rather nor lose so fine an opportunity of geting a Girl apiece." Even the sick were now determined to get ashore. After days of uncertainty and hostility, the sight of the women offered renewed incentive to make peace with the islanders. The Tahitians, for their part, quickly realized that they had a commodity potentially more valuable than hogs and vegetables and barrels of water. The officers' immediate concern, however, was establishing a regular, reliable trade in water and provisions and controlling it closely, avoiding any further violence. Traffic in women was not part of their plan.

Wallis and Clarke were still too ill to venture ashore or take charge of relations with the Tahitians, but with the help of the other officers they devised a system intended to obtain the supplies they needed without risking another attack. The ship's gunner, one Mr. Harrison, was appointed chief negotiator for the English; each day, accompanied by several marines, he went to the beach at the spot where the river flowed into the bay. Here he waited for the islanders to approach the opposite bank of the river with the goods they had to trade. Although dozens appeared, Harrison would trade with only two: an old man and his young son, who were permitted to cross the river with their goods. The rest, supplied with food to trade and stools to sit on, watched the transactions between the Englishmen and the two Tahitians; but whenever any of them attempted to cross the river to join in the bargaining, the marines forced them back.

To some extent, the two men acted as agents for other Tahitians who had provisions to trade, but on the whole this carefully controlled system, while promoting a sense of security among the English, restricted the flow of provisions. Many islanders refused to barter if they could not deal with the English directly, and as a consequence, several days after regular trade had been established, the ship's crew still ate salt meat at some meals despite the plentiful supplies of fresh food on the island.

Eventually, about a week after the truce and nearly two weeks after the English had first reached Tahiti, Wallis allowed his men to go ashore, although only under the strictest supervision. The sick went first, in the expectation that fresh air and sunshine would speed their recovery; they were confined to tents pitched on an island in the river, with two officers and four marines standing guard to prevent any of them from wandering off. The more able-bodied were then organized into work parties under orders not to stray out of sight of the officers or venture to trade with the islanders. The restriction on trade extended to the officers as well. Robertson, piqued at being denied the opportunity to gain a greater variety of local goods and explore the island, complained to Harrison about the severe restrictions, only to be told that Clarke, in particular, remained convinced that the Tahitians were a "treacherous people."

Other members of the crew soon adopted a more trusting attitude, spurred by the charms of the Tahitian women. After two days of exchanging provocative glances and presents of nails and trinkets, one of the marines—"a dear Irish boy," Robertson described him—accepted a woman's invitation without troubling to engage in any preliminaries or seek privacy. When the pair had finished, his envious comrades promptly thrashed him, allegedly "for not beginning in a more decent manner, in some house or at the back of some bush or tree." In his defense, the young man proclaimed his desire for the "Honour of having the first." But he enjoyed his distinction only briefly. By the end of the day "the old trade," as Robertson called it, flourished.

The beauty of the island women and the eagerness with which they offered themselves to their visitors made Tahiti a fantasy world for men who had spent months confined to a ship at sea. Though they were "not so fair as our English Ladies," their very un-English delight in casual sex more than made up for any flaws in appearance. The women exacted a price for their favors, to be sure—a 30-penny nail was the going rate—but their obvious pleasure in sex made encounters with them seem more like acts of friendship than of prostitution. Moreover, their behavior had the full support and even encouragement of the Tahitian men. Wilkinson interpreted the women's eagerness and the men's encouragement as a sign that the Tahitians recognized the superiority of Europeans and wanted "A Breed of English Men A Mongst them," but the Tahitians, who valued sexual fidelity among married couples while condoning sexual freedom among the unmarried, viewed the young women's relations with the Englishmen as simply another form of trade. Barred from free trade in provisions, they quickly learned they could get as much iron as they wanted by bartering their women.

The trade in women was so brisk, in fact, that it caused the price of foodstuffs to rise sharply, and Harrison, still responsible for all trade, worried that he would soon be unable to supply the crew with sufficient food. The Tahitian women could easily obtain nails and simple iron tools; the men, as a consequence, began requiring more substantial payment before they would part with pigs and fowl. Robertson and Furneaux suggested keeping the crew on board ship, but the surgeon argued persuasively that time spent on shore was important for the men's health. As an alternative, the officers tried to prevent the men from raiding the ship's supply of nails and other trade goods, but the sailors managed to find their nails elsewhere and the trade with women continued.

Within a few days it became apparent where the men were getting their nails when one of the carpenters reported to Wallis that every cleat in the ship was gone and two-thirds of the men had to sleep lying on deck because their hammock nails had been removed. Wallis called his men on deck and demanded to know who had been responsible for stripping the ship of most of its large nails. No one spoke, and Wallis angrily revoked all shore privileges. Later in the day Robertson overheard an argument that indicated most of the ship's crew had been involved in removing the nails. Several proclaimed they would rather suffer a dozen lashes than lose their shore liberty, while others accused their fellows of escalating the price of pleasure by offering the women larger nails when smaller ones had sufficed earlier. Finally they had a kind of mock trial and condemned six for their excesses in stealing nails. Two, however, cleared themselves "by proving that they got double value" for the larger nails, and finally the blame settled on one man, a seaman named Francis

Pinckney. Pinckney admitted to removing cleats holding the ropes attached to the mainsail, keeping the nails and throwing the cleats overboard. Acting on Robertson's advice, Wallis decided to make an example of the man and ordered him to run a gauntlet of his fellow sailors, who were armed with switches; but the men, well aware that Pinckney was paying for all their crimes, treated him so lightly that the angry Wallis accused them of encouraging rather than punishing thieves. He threatened to confine them all to the ship, but within a day the familiar routine was back in force.

As the days passed without further hostilities, the Englishmen became bold enough to invite some of the Tahitians on board the *Dolphin.* One of the first, a young chief who was invited to stay for dinner, found the English rituals of eating fascinating. The custom of removing both men and food some distance from the floor was mystifying to one for whom mats spread on the ground were the only furniture needed. The use of implements rather than fingers to carry food to mouth was equally peculiar. But, wishing to understand these customs and win favor with his hosts, the chief paid close attention to the use of silverware and imitated his hosts perfectly. Seeing the Englishmen wipe their mouths with their pocket handkerchiefs made him a bit uneasy, however, for he had nothing similar to use. Robertson, observing his concern, offered a corner of the tablecloth for a napkin—much to the disgust of Clarke, who declared that he could no longer sit at the same table with men who used such terrible manners. Robertson thought Clarke was the one guilty of bad manners, and tried to put the chief at ease by appropriating another corner of the tablecloth for a napkin himself; and the Tahitian, wishing to make amends, "made Signs to poor Growel who was still on the fret that he would bring him a fine young Girl to sleep with him," an offer that pleased Clarke enough to make him forget his petulance.

The officers, as amused by the chief as he was by them, named him Jonathan—the Englishmen made no effort to learn the Tahitians' names—and invited him to visit again. The next time they outfitted him with a complete suit of English clothes, and the lessons in how to put on shirt, breeches, jacket, socks, and shoes offered "plenty of devertion" for both "Jonathan" and his hosts. Later, "Jonathan" took great pleasure in showing off his strange new garb to his friends on shore, but he did not return to the ship again. Robertson surmised that the Tahitians, afraid their young chief was becoming too fond of the English, put pressure on him to stay away.

A few days later the *Dolphin*'s officers entertained an even more distinguished visitor: a tall, stately woman whose name was Purea, but whom the Englishmen called simply the "queen" from the prestige she appeared to enjoy among the people of the island.

Of all the parts of the ship, Purea and her entourage were most interested in the galley. The sight of pigs and fowl roasting on spits amused them greatly; accustomed to cooking all their food in pits dug into the ground, heated with stones and well covered with leaves, they found the notion of cooking food in the open air remarkable. One of the men took hold of the spit and turned it around and around, unable to understand quite how it worked. Even more startling were the copper pots, shined until they gleamed. "This," Robertson observed, "Seemd to Surprize them the most of any thing which we showd them."

Purea returned the officers' hospitality by inviting them to a feast at her home. Wallis, still weak from his long illness, nevertheless felt sufficiently recovered to venture on shore for the first time to partake of the "queen's" hospitality. The sickness on board ship had not been lost on the observant Purea, and when the officers arrived at her home she and her companions first insisted on giving all the recuperating men a thorough massage. Though he felt pummeled and kneaded until the cure seemed worse than the disease, to his shock Wallis felt much better afterward.

The therapy completed, Purea presented the men with a gift of yards of Tahitian *tapa* and directed her ladies-in-waiting to take the cloth and dress the men Tahitian-style. The officers, feeling uncomfortable about disrobing and draping themselves in the coarse native fabric, did not react to this exchange of costume with the same good humor that "Jonathan" had displayed toward his English outfit, although Wallis, not wishing to insult Purea, accepted the gift and the lesson in how to wear it. When it came time to leave, Purea insisted—over Wallis's objections—on accompanying the Englishmen back to the ship. Still concerned for his health, she carried the captain over every rough spot in the path back to the shore "with as much ease as I could . . . a child."

Over the next few days, Purea and the officers became much attached to each other. She entertained Robertson at her home too, giving him a suit of the native clothing and offering one of her young female companions as a bedmate for him. Robertson, under orders to return to the ship by nightfall, eased his way out of this proposition by promising to return on another voyage and "Sleep with her in my Arms." When Wallis decided it was time to continue his voyage, Purea tried to persuade the Englishmen to stay; and the night before their departure, she and several companions slept on the beach, to make sure they would not miss their opportunity to bid farewell to the *Dolphin.*

On the morning of July 27, almost six weeks after the Englishmen first discovered Tahiti, the *Dolphin* sailed out of Matavai Bay. Purea and her friends, who had tried so persistently to drive the English away when they first arrived, now wept at the sight of the wind-filled sails carrying the ship from their island. And the Englishmen, who had feared the Tahitians while scorning their inferiority, now also felt sadness at leaving the people who had turned out to be gentle rather than ferocious.

Though most still thought of them as primitive and uncivilized—happy-go-lucky children, in a way, free of the inhibitions of modern civilization, but also lacking its refinements—the more perceptive Robertson was forced to acknowledge that the islanders were clever and intelligent as well. They were "Smart Sensable people," he concluded; and though his fellows might not all have agreed, it was this combination of quickness and curiosity, warmth and gentleness, freedom from wordly cares and constraints, that made the Tahitians so attractive to the crew of the *Dolphin* and to the legions of Europeans who would follow them.

4. *Preparations*

The *Dolphin* returned to England in late May 1768, having discovered nothing more of any consequence on its uneventful voyage from Tahiti across the Pacific by way of the East Indies and the Cape of Good Hope. An impartial observer might have argued that the expedition was a failure, for it had not revealed the fabled southern continent despite some promising signs of its existence in the distance south of Tahiti. Robertson, for one, criticized Wallis for his refusal to explore Tahiti more thoroughly or search for the land they thought they saw to the south.

In official circles and among the English public, however, these failures paled into insignificance compared to the importance of Wallis's discovery of Tahiti, which offered promise as the base England so sorely needed to continue exploring and eventually exploiting the presumed riches of the Pacific. Well located, with docile inhabitants—thanks to Wallis's aggressive measures—and abundant supplies of food, Tahiti was an ideal point for reprovisioning ships traversing the Pacific, and because Wallis had on board an officer trained in the most current scientific methods of navigation, its location was charted correctly. As for the shadowy land seen in the distance south of Tahiti, Wallis's decision not to search further was less important than his sighting of something that seemed almost certain to be the long-sought-after southern continent. It was now only a matter of sending another expedition to finish his work.

And the next expedition, quite fortuitously, was almost ready to sail.

Only the day before the *Dolphin* anchored in the Thames, James Cook had accepted the Royal Society's invitation to serve as one of its official observers on the expedition he was preparing to lead to the South Pacific, and his ship, the *Endeavour,* had just been hauled out of drydock, ready for final provisioning. By his timely arrival Wallis had solved the last remaining major problem Cook faced: where to go to observe the transit of Venus. Tahiti was well within the region where the transit would be fully visible, had a sunny climate, and was accurately enough located by Wallis so Cook should have no trouble finding it again. And Wallis's belief that a continent lay just south of Tahiti encouraged those who thought the most important part of Cook's voyage would not be his contributions to astronomy, but his subsequent search for terra australis.

Cook's instructions, modified in light of Wallis's discoveries, directed him to sail to "King George's Island" to observe the transit and then proceed south to find the presumed southern continent. The official purpose of this second phase of the voyage was both political and scientific: discovering new lands and adding to knowledge of distant places already discovered but little known "would redound greatly to the Honour of this Nation as a Maritime Power, as well as to the Dignity of the Crown of Great Britain," and would also "tend greatly to the advancement of . . . Trade and Navigation."

The president of the Royal Society, James Douglas, the Earl of Morton, added his own "hints" for Cook and his companions, which reflected his confidence that they would indeed find a continent. After explaining in detail how the transit observations should be conducted, Morton suggested ways of determining whether newly discovered land was part of a continent—high mountains some distance from shore, large rivers with sandbars, and a large population were good clues—and provided a long list of points to observe in any new lands Cook might find. He also offered guidance on relations with the native inhabitants of strange lands. Thinking perhaps of Wallis's experiences at Tahiti, Morton deplored violence against native peoples as a "crime" and argued that Europeans had no right to occupy other peoples' lands uninvited. Native inhabitants were justified in being suspicious of any Europeans who appeared on their shores and should be treated gently even if they opposed an English landing. Europeans could demonstrate their technological superiority in many ways short of violence, such as shooting birds; and, in general, Morton concluded, "there can be no doubt that the most savage and brutal Nations are more easily gained by mild, than by rough treatment."

The man entrusted with this ambitious undertaking was not, like Byron and Wallis, an experienced captain of upper-class background and good political connections, but the son of a farmer, who had enlisted in the navy as an ordinary seaman and had never before had command of a ship. Noted for his skill in marine surveying, he had spent most of the previous five years in a small ship charting the bleak coastline of Newfoundland.

James Cook was in 1768 master of the *Grenville,* one of a small fleet assigned to chart the coast of Newfoundland and its offshore islands as part of a larger project to map the territory acquired from France as a result of the Seven Years War. It was a difficult task because of Newfoundland's long, convoluted shoreline and harsh climate, which limited surveying to the summer months. Cook and the other members of the fleet had been working on the Newfoundland charts since 1763, staying three or four months at sea each summer; part of the fleet wintered at one of the tiny settlements along the Newfoundland coast, but Cook returned each fall to London, where he transformed his rough sketches into finished charts and arranged for their publication. In the spring of 1768, he was preparing to return for yet another season in the North Atlantic when the Lords of the Admiralty announced his appointment to lead an expedition to the South Pacific.

Cook was born in 1728, in the north Yorkshire village of Marton-in-Cleveland. He was the second of seven children. At age eight, he moved with his family to Great Ayton, another farming community a short distance away, where his father had obtained a job as a farm manager. The elder Cook's employer paid young James's tuition at the local school, a one-room stone structure in the center of town where the sons of the more prosperous local farmers and shopkeepers learned the rudiments of reading and writing and enough basic arithmetic to keep simple accounts. The senior Cook's employer paid for young James's education, evidence of an enlightened benevolence, or a belief that the boy had more than usual intelligence, or perhaps both.

The life of a farmer held little appeal for Cook, and at the age of seventeen he went to work for a grocer in Staithes, a fishing village on the Yorkshire coast about fifteen miles from Great Ayton. Cook lived with his employer in a house overlooking a harbor crowded with the fishing boats that employed most of the men living in Staithes. Whether it was the influence of living by the sea or restlessness for a life more active than this tiny village could offer, within a year and a half Cook moved a few miles south to Whitby, the major seaport on the Yorkshire coast.

Whitby in 1746 was a city of more than ten thousand people, built along a deep, narrow harbor bisecting the sheer cliffs that run the length of that stretch of Yorkshire coast. The town extended the length of the harbor, its brick houses jammed together on long, narrow streets, hemmed in close to the waterfront by the cliffs rising sharply behind. The ruins of an ancient abbey overlooked the town from the cliffs on the opposite side of the harbor.

Here Cook apprenticed himself to John Walker, a merchant and shipowner who made his living primarily by shipping coal from the mines of northeastern England to London, Norway, the Netherlands, and other Baltic ports. Walker took a liking to Cook, and the young man rose quickly to positions of more responsibility on successive voyages over the next several years. Between voyages Cook lived with his employer, a common enough arrangement with apprentices and other employees; but, rather uncommonly, he devoted much of his spare time to the study of navigation. In a quiet corner of the Walker home arranged for him by a sympathetic housekeeper, he read geometry, trigonometry, astronomy, and the recently published manuals of navigation that used those sciences to instruct seamen in such practical matters as how to determine latitude and longitude at sea.

Eight years and several voyages after Cook began his apprenticeship, Walker appointed him mate on one of his new ships. Three years later he offered Cook the command of the ship, but instead of accepting this promotion Cook enlisted in the Royal Navy. At twenty-six, he was old for a beginner in the Navy; most sailors started out as teenagers, some as children of eight or ten. The very act of enlisting set him apart too, for conditions on Navy ships were so horrible that most sailors were recruited by brute force—snatched or "pressed," in the language of the time, from the bars and alleys of waterfront towns. To go from the security and status of a merchant ship captain to the hardships of a Navy seaman was surely an odd decision—some would have said foolhardy—and yet Cook's life up to this point had already exhibited a pattern of restlessness, of searching for something beyond the limits of his experience.

For an ambitious and adventurous young man, the decision was perhaps not so rash as it first seemed. It was June 1755, and England was on the verge of war. Promotion could be rapid in wartime, and the Navy was not quite so class- and tradition-bound as other branches of government service; navigational skill was undeniably necessary in sailing ships, and ability might hope to be rewarded despite humble background. Cook was a self-confident if modest man, and he might reasonably have expected that he would not remain an enlisted man for long. And the Navy offered the chance to break away from the dull routine of the coal trade—short voyages to the Baltic and down to London, the monotony of familiar routes and ports broken only by the dangers of the shoals and storms so common in the North Sea. Certainly Walker, with whom Cook maintained a warm friendship for the rest of his life, did not try to change his protégé's mind.

Cook's expectations of advancement were well founded, for he was promoted to master's mate a month after his enlistment. If he harbored hopes of adventure, however, he was disappointed. For his first two and a half years in the Navy, he got no farther from the English coast than he had in Walker's coal ships. His first ship, the *Eagle,* was moored at Spithead, near Portsmouth, and preyed on French shipping just off the coast of France. Action was sporadic, and by the time the *Eagle* was sent to North America in the summer of 1757, Cook had been reassigned to the *Solebay.* He was now ship's master, but the *Solebay*'s mission was even less exciting than the *Eagle*'s; based at Leith, on the east coast of Scotland, it was to patrol the Scottish coast to prevent smuggling from France and Holland.

In October 1757 Cook became master of the much larger *Pembroke,* a 1,250-ton, 64-gun warship. The following February, the *Pembroke* sailed for Canada in the fleet of Admiral Edward Boscawen. The fleet's orders were to take first Louisbourg and then Quebec, the major actions that would put an end to the war.

After the English victory at Louisbourg in July, the *Pembroke* was assigned with several other ships to patrol the Bay of Gaspé and the Gulf of St. Lawrence to prevent an attempt to retake the fort. It was dull work, particularly during the long winter when the ships were berthed at Halifax, Nova Scotia, with little to do. Cook occupied himself, as he had during the winter evenings in Whitby, with study.

The day after the fall of Louisbourg, he went ashore and met Samuel Holland, an engineer attached to the English army, at work surveying the newly acquired territory. Struck by the potential usefulness of his techniques to sea charting, Cook asked Holland to teach him. The surveyor's methods were based on the same principles of geometry and trigonometry that were by now familiar to Cook, but he used them in a rather different way and with different instruments than the navigator charting a coast or determining location at sea. Holland found Cook an eager pupil, and the two men spent much of their time together during the next several months.

That fall Cook drew a chart of the Gaspé Bay, which was published the following year—his first printed map. During the winter months, he and Holland worked together on a chart of the St. Lawrence, refining sketches that Cook had drawn earlier. Cook continued his work mapping the St. Lawrence the following spring and summer, as the *Pembroke* sailed from its winter base at Halifax toward Québec, the English forces' next objective now that Louisbourg was in English hands. Victory there would almost certainly mean France's total surrender. Mounting an attack against Québec, however, was a good deal more complicated than the action against Louisbourg; troops had to be transported four hundred miles up the St. Lawrence River, which became progressively narrower and more treacherous. Several of the fleet's ships, including the *Pembroke,* were directed to chart the river, transport troops to staging areas near Québec, and prevent French ships from bringing supplies to the besieged city. On the last point, they were not always completely successful. Among the French ships that slipped through the English blockade was one carrying Louis Antoine de Bougainville.

For the remainder of the war, Cook was occupied primarily in the unglamorous but critical work of charting Canadian territory. Toward the end of 1759, he transferred to another ship, the *Northumberland,* and continued shuttling back and forth between

Halifax and Québec, taking part in the mopping-up operations following the English victory there and continuing to work on his charts. They formed a major part of the detailed map of the St. Lawrence published in London by the noted engraver Thomas Jeffreys early in 1760.

Cook's skill and experience at chart-making did not go unnoticed. The captain of the *Northumberland,* Alexander Colville, praised the young master's ability in a letter to the Admiralty, and within months he was dispatched back to Newfoundland to work on the project that would map its entire coastline.

In the meantime, Cook, now thirty-four and home on English soil for the first time in four and a half years, took a wife: twenty-one-year-old Elizabeth Batts from the village of Barking, in Essex. Little is known of her or of how she and Cook met, beyond the fact that it was a whirlwind courtship; they were married six weeks after he left his ship. Elizabeth's mother lived in the same waterfront district of London where Cook boarded when on shore leave, and so it is likely that the couple became acquainted there. However they met, the brevity of their courtship reflected more the matter-of-factness of the seaman who knows he will not stay in port long than any romantic impetuosity. At thirty-four, Cook was ready to marry, and as a sailor, he could not afford the luxury of a lengthy engagement.

Cook and his wife had barely four months of married life together before he embarked for Newfoundland in April 1763. It was long enough, however, to conceive a son, born the following October, about a month before Cook returned from his first season in Newfoundland. The next years followed a regular pattern: Cook sailed for Newfoundland each spring and returned in the fall, spending the winters with his family and preparing his charts for publication. The day he arrived in London after his second season in Newfoundland, in December 1764, Elizabeth gave birth to a second son, Nathaniel. A daughter, also named Elizabeth, was born while he was at sea in the summer of 1767.

Cook's work in Newfoundland was unusual because he employed the techniques of land surveyors, learned during the long winters in Canada, along with the more traditional methods of marine charting. As often as possible, he anchored his ship and went ashore, setting up his surveyor's instruments and establishing precisely the contours of the coastline. The result was one of the most accurate sea charts produced up to that time.

It was not so surprising, therefore, that the Admiralty thought of Cook when it came time to select a commander for a voyage to the South Pacific. His skills as a navigator and mapmaker, which had come to the attention of influential men at the Admiralty despite his relatively low rank, would be critical for a voyage of exploration. Equally significant for the potential success of the South Seas voyage, his ship had been remarkably free of disease and accident during the five seasons at Newfoundland. A strict but humane man, Cook resorted to punishment much less often than most naval officers, and many of his crew had stayed with him for the duration of the Newfoundland project.

Shortly before naming Cook commander of the voyage, the Admiralty chose the ship he would command. Though Cook had no hand in the selection, he could hardly have objected to the result: a 368-ton, 106-foot-long vessel built in Whitby for the coal trade. Repaired and renamed the *Endeavour,* the ship was exactly the sort Cook had been accustomed to sailing before he entered the Navy. More to the point, the *Endeavour*—though square, slow, and ungainly in appearance compared to the Navy's warships—was well suited for a long voyage, having been built to carry large cargoes relative to the size of its crew and to withstand the storms of the North Sea. One of the problems of sixteenth-century voyages of exploration had been the limited capacity of the ships of that era, which required such large crews that little space was left for provisions. Dutch technology in the seventeenth century created ships that could be sailed with fewer men, but English naval vessels still carried a high ratio of men to cargo space, for obvious reasons; in wartime, manpower was more important than the ability to remain at sea for extended periods. Cook's voyage was the first exploring mission to employ a vessel designed for its cargo capacity instead of a naval warship. Joseph Banks, a young scientist who sailed with Cook, would remark later that the *Endeavour* was a "heavy sailer . . . a necessary consequence of her form; which is much more calculated for stowage, than for sailing." But even Banks, who had little experience at sea, admitted that the *Endeavor* was an excellent ship for this sort of voyage. What it lacked in speed it made up in space and safety, considerations of paramount importance on a voyage of uncertain duration into unknown waters.

On May 27, 1768, Cook assumed command of the *Endeavour*, then anchored at the Navy's shipyard at Deptford, and plunged into the task of outfitting his ship.

The Admiralty had authorized a crew of seventy: twenty-one officers and artisans, including a second lieutenant, master and two mates, boatswain and two mates, gunner, carpenter and mate, surgeon and mate, cook, three midshipmen, a steward, two quartermasters, an armorer, a sailmaker, forty able seamen, and eight servants. Later, they added a third lieutenant, twelve marines, and two other men, bringing the ship's company to eighty-five.

Cook did not have complete control over the selection of the crew on this voyage, as he would later, although the Admiralty consulted him in making their choices and he managed to get berths for five men who had been with him on the *Grenville* in Newfoundland. Another seaman, Isaac Smith, was Elizabeth Cook's cousin. Others came from the *Dolphin:* John Gore as third lieutenant, whose experience with Byron and Wallis made him the *Endeavour*'s ranking expert on the South Pacific; Robert Molyneaux as master; Charles Clerke, who had spent the four years since Byron's voyage in the West Indies, as master's mate; Richard Pickersgill as the second master's mate; and Francis Wilkinson and Francis Haite as seamen. These men were important to Cook, for they constituted a group of experienced, dependable sailors—those from the *Grenville* personally known to him, and those from the *Dolphin* knowledgeable about the seas where they would be sailing.

Cook inherited one other veteran of the *Dolphin* as well: a goat. She had provided milk all the way around the world for Wallis's officers, and would do the same for Cook's.

On the whole, the ship's company was a young group. At the time they sailed, the sailors' average age was about twenty-five. Cook, at forty, was among the oldest; only three other men were his age or more. At forty-nine, sailmaker John Ravenhill was the oldest. Twelve-year-old Isaac Manley was the youngest. He sailed as Molyneux's servant, but was promoted to midshipman by the end of the voyage. Eventually he would become an admiral.

The men came from all parts of the British Isles and from other parts of the world as well, including Scotland, Ireland, Wales, Venice, and New York. Nearly all were experienced seamen, despite their youth. If the general quality of English seamen at this time is any guide, most were illiterate, some of rather doubtful character; but they were more carefully picked than the sailors for a more ordinary expedition, and they would display less of the discontent and insubordination so common on naval vessels. And many would choose to sail with Cook again.

There would be other men too, in addition to the sailors and marines, aboard the *Endeavour*. Among them was Charles Green, the astronomer selected by the Royal Society to observe the transit of Venus. Green, like Cook the son of a Yorkshire farmer, had served as assistant to Nevil Maskelyne for the previous seven years. John Reynolds traveled with Green as his assistant. Together with Cook, and with the assistance of some of the officers—Green instructed several of them in the techniques of astronomical observation—these were the men responsible for the primary scientific mission of the voyage.

Shortly before the ship was scheduled to sail, an intrepid, wealthy young man with a passion for botany also sought permission to go, in order to study the plant and animal life of the South Seas.

Joseph Banks—twenty-five, handsome, Oxford-educated, heir to a large fortune—had already been to Newfoundland, though not with Cook, in pursuit of his scientific interests, and he was determined that his next step would be a voyage around the world. When friends tried to dissuade him from this unorthodox idea, urging him to undertake the more conventional "Grand Tour" of Europe instead, Banks scoffed, "Every blockhead does that; my Grand Tour shall be one round the whole globe." Convinced that money could buy nearly anything he wanted, Banks was not one to give up easily. When, as a student at Oxford, he discovered that there was no professor to lecture on botany, he imported one from Cambridge at his own expense. Now, when the proposed voyage to observe the transit of Venus offered just the opportunity he wanted, he applied to go along, again at his own expense, and had no doubts that his request would be granted.

The Royal Society, which numbered Banks among its members, urged the Admiralty to permit Banks to go as a passenger, for although they had conceived of this voyage primarily as one devoted to astronomy, they could hardly ignore so excellent an opportunity to add to their knowledge of biology as well. Cook had no objection, for he liked Banks and appreciated the value of having a naturalist on board.

Banks would not travel alone, however, or with a single assistant as Green did. He brought with him eight men besides himself; two naturalists, two artists, and four servants, as well as two of his pet dogs and enormous quantities of baggage. The two naturalists were both Swedes, former students of Linnaeus, the distinguished biologist who had developed the first workable system of classifying plants and animals. Daniel Carl Solander, aged thirty-five, had emigrated to London in 1760. A plump, pink-cheeked man, gregarious and cheerful, he was popular with the English and grew to like London so well that he turned down an opportunity to become professor of botany at the St. Petersburg Academy of Sciences. He made his living as an assistant in the British Museum and was, like Banks, a fellow of the Royal Society. The second naturalist, Herman Spöring, served as an assistant to Banks and Solander.

The artists, Sydney Parkinson and Alexander Buchan, were not a rich man's frivolous addition to his entourage but an essential part of a scientific team in the age before photography. Their principal task was to draw the specimens that Banks, Solander, and Spöring collected; although the naturalists intended to preserve some of their specimens and take them home to England, it would not be practical to do so with all of them. Banks also expected to dissect certain animals, and the artists would preserve a record of this work. In addition to their scientific drawing, Banks wanted the artists to sketch scenes of the people and places they visited as a record of the voyage. Parkinson, a twenty-three-year-old Quaker, a quiet man, well liked by his shipmates, had principal responsibility for the scientific drawings. Buchan was to spend most of his time drawing people and landscapes. His untimely death, however, just after the *Endeavour* reached Tahiti, left Parkinson with the full artistic responsibility. A prolific worker, he was capable of the task, producing more than twelve hundred sketches and paintings—the most detailed pictorial record of any eighteenth-century voyage of exploration.

By the time Cook went aboard the *Endeavour* to supervise personally the outfitting of his ship, most of the repairs and modifications necessary for the voyage were nearing completion. Cook's major task was to oversee the provisioning of food and supplies sufficient to last for a voyage of at least two years.

Navy officials had formulas both for the number of men needed to run any given type of ship and the quantities of food required to feed them. The *Endeavour,* however, was more generously supplied than most ships because its voyage would be longer than normal and would cover little-known parts of the world where the availability of additional provisions was uncertain. In addition, the unusual nature of the voyage and its scientific mission induced the Navy to try some experiments of its own, particularly testing the effects of certain kinds of foods on sailors' health.

The Victualling Board, the Navy's agency responsible for outfitting ships, ordered its suppliers to provide the *Endeavour* with 34,666 pounds of bread and 10,400 pounds of flour; 120 bushels of wheat and 10 of oatmeal; 4,000 pieces of beef and 6,000 pieces of pork, all dried and salted to preserve them for the duration of the voyage; 800 pounds of suet, 2,500 pounds of raisins, 120 gallons of oil, 500 gallons of vinegar, 1,500 pounds of sugar, and 20 bushels of salt; 1,200 gallons of beer and 1,600 of spirits, to be supplemented by wine purchased at Madeira or the Canary Islands; 40 bushels of malt; and 7,860 pounds of sauerkraut.

The malt and sauerkraut were among the Navy's experiments. Both, it was thought, might be helpful in preventing and treating scurvy, the most common and one of the most serious of shipboard diseases. Caused by a deficiency of Vitamin C, scurvy produced swelling in the gums and legs, skin sores, and extreme listlessness. It was reversible, but could be fatal in the absence of treatment. Although vitamins were unknown in the eighteenth century, it was recognized that men on long sea voyages were particularly susceptible to scurvy because their diet lacked fresh food for months at a time. Exactly what foods might be used to combat the disease was not understood, however, and theories abounded.

One of the most popular preventives at the time was a concoction made from malt, propounded by an Irish physician and former naval surgeon, David MacBride, in his book *Experimental Essays.* The Victualling Board doubled its initial provision of forty bushels of malt before the *Endeavour* sailed and gave Cook detailed directions for its use. The ship's surgeon was to grind a quart of malt and mix it with three quarts of boiling water, allowing it to stand for three to four hours. This mixture, known as "wort," was then mixed into a paste with sea biscuit or dried fruit.

Sauerkraut was also thought to be effective against scurvy, and the Navy supplied a quantity sufficient to serve two pounds per week to all hands for a year. Cook himself believed in the efficacy of sauerkraut and went to considerable lengths to convince his crew that it was good for them. He also believed in the healthful properties of "portable soup," a dried meat powder that could be combined with water to produce broth. Cook asked the Admiralty for enough to last the duration of the voyage, and received a thousand pounds with directions for its use. He was to boil it with oatmeal and serve it on "banyan" days, the three days a week when sailors ate no meat.

As a third potential preventive, the *Endeavour* carried syrup made from oranges and lemons. The best possible remedy for the disease, the juice of citrus fruits had been proposed as early as 1753 by Dr. James Lind in his *Treatise of the Scurvy;* but unfortunately it was difficult to keep either fruit or juice fresh on long voyages, and the process of making the syrup provided for Cook's voyage destroyed most of its useful properties.

Throughout June and July, the Navy continued to issue orders for provisioning the ship. Cook made frequent requests of the Admiralty Secretary and Navy Board, asking for, among other things, empty casks for wine and water, medical supplies for the surgeon, four swivel guns in addition to the eight the *Endeavour* already carried, and a green floor cloth for the captain's cabin. He devoted his most careful attention, however, to the scientific instruments to be carried on board. Members of the Royal Society itself took charge of selecting the astronomical instruments, supplying two reflecting telescopes; a quadrant with a one-foot radius; an astronomical clock from the Royal Observatory at Greenwich; another clock and a brass sextant, both made especially for the voyage; two thermometers; a watch lent by Nevil Maskelyne; and, the most unusual item on the list, a portable observatory built under the supervision of Maskelyne and Cook. It was a tentlike apparatus of canvas stretched over poles, which could be collapsed for easy storage and then raised quickly, at any site, to provide shelter for the instruments. In addition to these instruments, Cook requested and received mathematical and surveying equipment, including a surveyor's theodolite and plane table, a brass scale, a pair of proportional compasses, glass for tracing, and paper and colors—all for use in determining the ship's position and in charting whatever land they might find.

More than any previous explorer, Cook was well prepared to chart his discoveries and fix their locations accurately. He sailed at a time of rapid advances in methods of navigation; his ship was equipped with every available type of scientific instrument; he had the services of professional astronomers; and he himself had a far more sophisticated understanding of astronomy, mathematics, and surveying techniques than most ship captains.

Since the sixteenth century, sailors had understood how to measure latitude at sea using simple quadrants—instruments designed to measure the angle between the horizon and the sun or moon. Refinements of these instruments in the eighteenth century made it possible to determine latitude with increasing accuracy, but establishing longitude remained an intractable problem. The best sailors could do until well into the eighteenth century was "dead reckoning": beginning from a known point, they estimated the ship's speed, adding the distance covered each day to the original longitude. Even small errors, when compounded over a period of days or weeks, could produce wildly inaccurate results.

The British government considered the problem so critical that in 1714 Parliament had authorized a reward of £20,000 to anyone who invented a reliable method of determining longitude; after more than fifty years, the prize money still had not been awarded. Although the principles of calculating longitude were clearly understood, putting them into practice required complicated astronomical observations and mathematical calculations. In theory, it was simply a matter of comparing local time with the time at any fixed point—Greenwich was the standard used by English navigators—and then translating time into degrees of longitude. (Four minutes equal one degree of longitude.) The difficulty, however, was maintaining correct Greenwich time over long distances, for the pendulum clocks of the eighteenth century were useless on pitching, rolling ships at sea, and spring-operated watches of the sort that Cook carried with him could maintain their accuracy for short periods only. Several men in England, notably a Yorkshire carpenter named John Harrison, were working on nautical clocks, known as chronometers, that would solve the problem. Nevil Maskelyne, accompanied by Charles Green, had tested one of Harrison's chronometers on a voyage to Bermuda in 1763, with good results, but he was nevertheless not convinced of the clock's reliability and did not provide one for Cook's voyage.

Maskelyne and many other scientists were more hopeful about another method, which relied on astronomical observation rather than mechanical inventions. Known as the lunar distances method, it required sailors to calculate the distance between the moon and the sun or another fixed star, note the exact time of the observation, and compare the local time with the time of the same astronomical occurrence at Greenwich.

Such a comparison was theoretically possible because of known principles about the moon's motion, but the practical application of these principles required a set of tables predicting the relative position of sun and moon at Greenwich at regular intervals, which a sailor could use for quick comparison with his own observations. The first accurate tables were compiled by a German scientist, Tobias Mayer, in the 1750s; he sent his results to Maskelyne, who included them in his *British Mariner's Guide* in 1763. This book made it possible for sailors to use the lunar-distances method for the first time, but the complexity of the calculations required severely limited its use. Four men were needed for the job: one to measure the distance between the moon and the sun or another fixed star; two to measure the altitude of the moon and star simultaneously; and a fourth to time the observations with a watch that had been set for the correct time by the sun earlier in the day. All of these observations then had to be corrected for the effects of lunar parallax and refraction, a process that could take as much as four hours and several pages of mathematical calculations. Few seamen—even officers—had the skill and patience to accomplish such procedures with accuracy. Byron had relied on dead reckoning, even though Maskelyne's book was available at the time of his voyage; Wallis was familiar with the book but couldn't understand it. Fortunately his purser, John Harrison, did, and used it to establish the longitude of Tahiti correctly.

Shortly before Cook sailed, Maskelyne published the *Nautical Almanac,* the first in an annual series of publications that would continue until 1908. The *Almanac* included a simplified version of Mayer's tables, which eliminated much of the mathematical calculation involved. Even so, the author of one of the most popular handbooks for sailors still thought "Dr. Masculines Method" was much too difficult for ordinary sailors. Cook and Green, however, had enough training in mathematics and astronomy to make use of Maskelyne's tables effectively.

At the end of July, nearly two months after Cook first boarded his ship, the preparations for its voyage were completed. After a brief farewell visit to his family, Cook rejoined the *Endeavour* at

Deal, near the mouth of the Thames. It took a week to sail along the southern coast of England to Plymouth, where final supplies were loaded; here Banks and his entourage joined the ship. Finally, on August 26, 1768, the *Endeavour,* with ninety-four men and provisions for a year and a half, set its course for the Pacific.

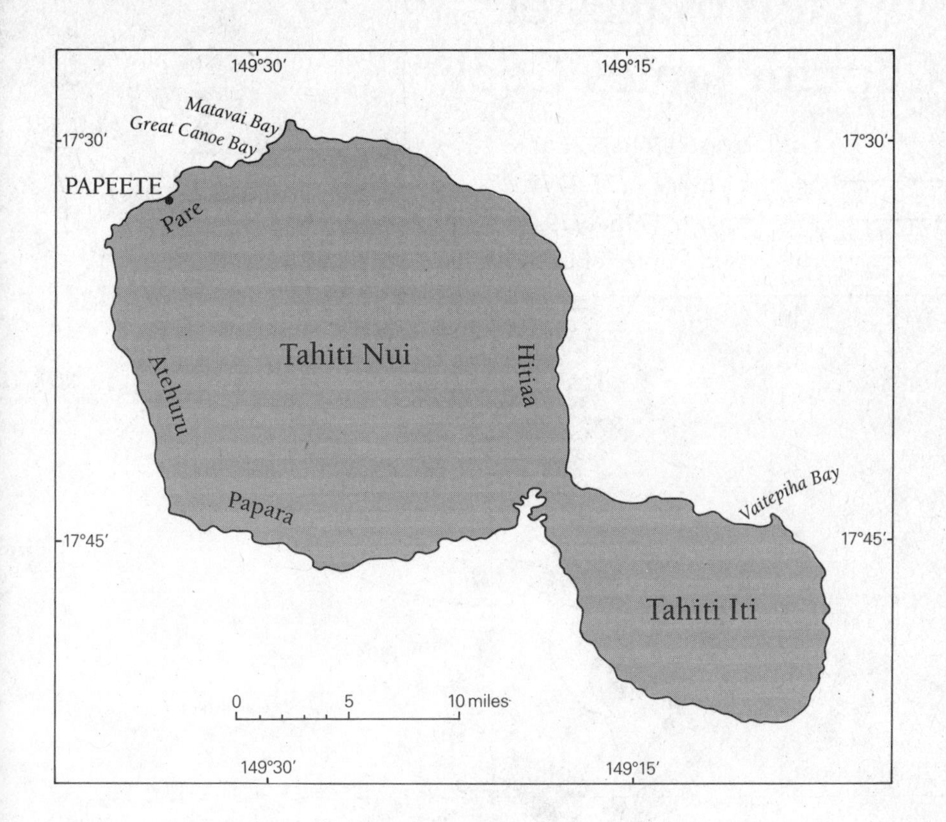
149°30′
149°15′
Matavai Bay
Great Canoe Bay
17°30′
17°30′
PAPEETE
Pare
Tahiti Nui
Hitiaa
Atehuru
Papara
Vaitepiha Bay
17°45′
17°45′
Tahiti Iti
0
5
10 miles
149°30′
149°15′

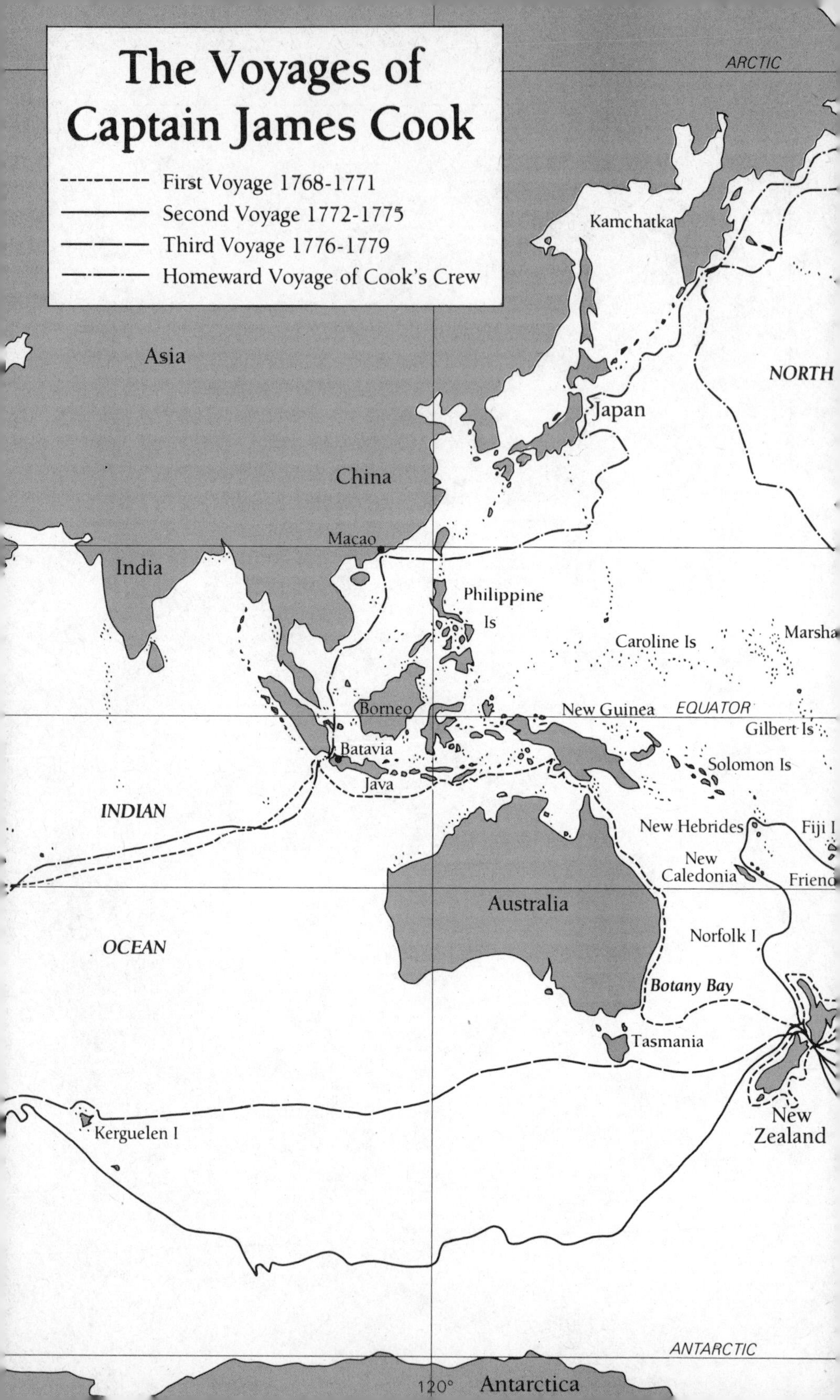
The Voyages of Captain James Cook
First Voyage 1768-1771
Second Voyage 1772-1775
Third Voyage 1776-1779
Homeward Voyage of Cook's Crew
ARCTIC
Kamchatka
Asia
NORTH
Japan
China
Macao
India
Philippine Is
Caroline Is
Marsha
Borneo
New Guinea
EQUATOR
Gilbert Is
Batavia
Solomon Is
Java
INDIAN
New Hebrides
Fiji I
New Caledonia
Friend
Australia
Norfolk I
OCEAN
Botany Bay
Tasmania
New Zealand
Kerguelen I
ANTARCTIC
120°
Antarctica

CIRCLE
g Strait
Alaska
Prince William Sound
laska
Nootka Sound
North America
C
OCEAN
Hawaiian Is
Christmas I
EQUATOR
Galápagos Is
Marquesas Is
Tuamotu Is
Tahiti
Is
South America
Easter I
OUTH
PACIFIC
OCEAN
Strait of Magellan
Cape Horn
CIRCLE
120°
60°

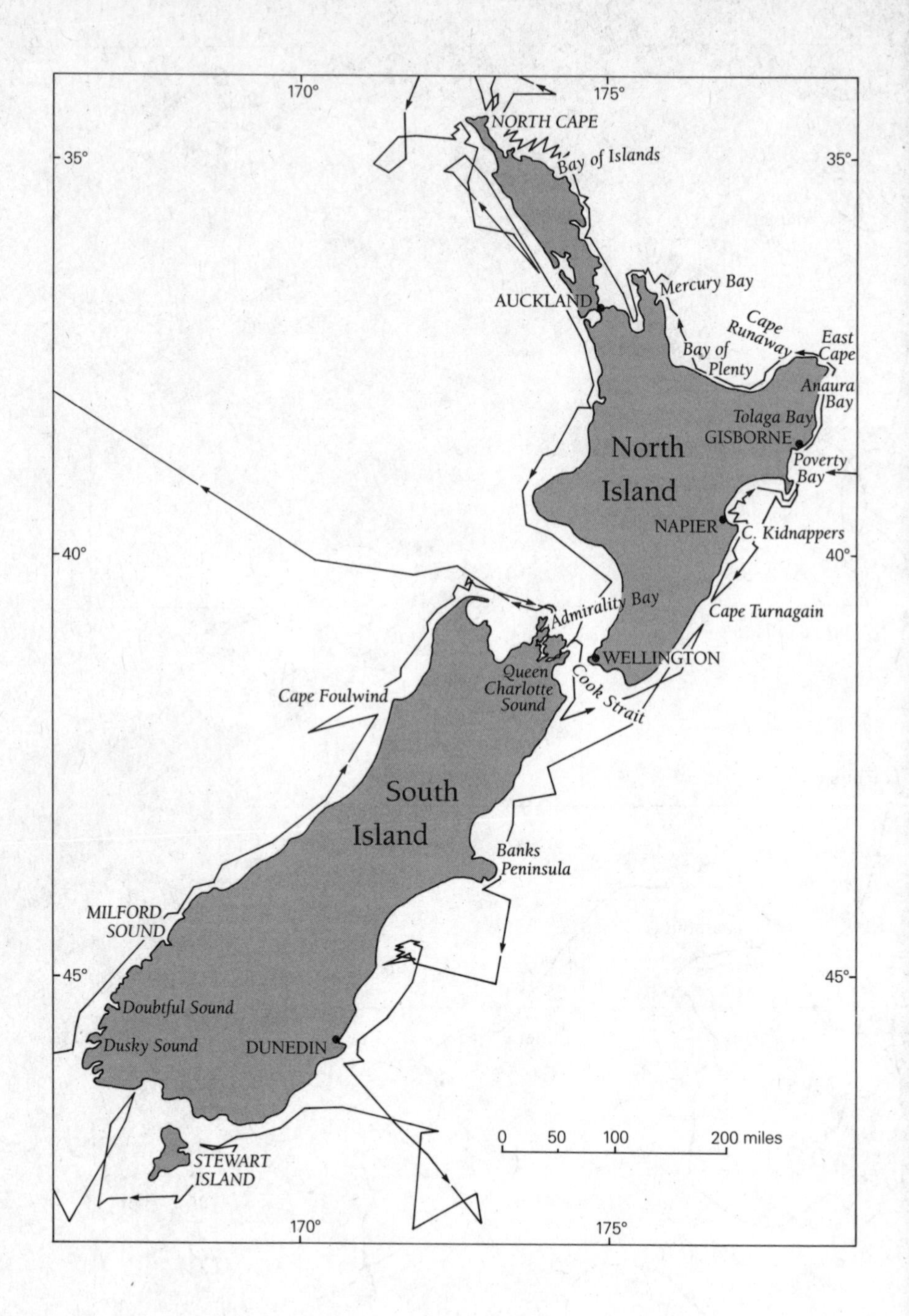
170°
175°
35°
40°
45°
NORTH CAPE
Bay of Islands
Mercury Bay
AUCKLAND
Cape Runaway
East Cape
Bay of Plenty
Anaura Bay
Tolaga Bay
GISBORNE
North Island
Poverty Bay
NAPIER
C. Kidnappers
Admirality Bay
Cape Turnagain
WELLINGTON
Queen Charlotte Sound
Cook Strait
Cape Foulwind
South Island
Banks Peninsula
MILFORD SOUND
Doubtful Sound
Dusky Sound
DUNEDIN
STEWART ISLAND
0
50
100
200 miles

II

THE DISCOVERY OF POLYNESIA

5. *Tahiti*

Once out of sight of land, the ship's company settled down to a familiar routine. Those who were not directly involved in sailing the ship occupied themselves mending and setting up the rigging, keeping the ropes in good repair, stretching the cables on deck to air, mending sails, and cleaning the ship. The last was a particular concern of Cook's, as it had been with Wallis, for although eighteenth-century ships were in general noted for their unsanitary conditions and noisome stench, Cook believed cleanliness would help maintain the good health of his crew.

The scientists were not idle either. Banks and his assistants began to collect and analyze specimens almost as soon as the ship cleared Plymouth harbor, even though their efforts were limited at this point to tiny sea creatures caught in a net dragged alongside the ship. Nothing was too insignificant for Banks, however, who was delighted to discover several new species and set Parkinson to sketching them.

Cook and Green made daily astronomical observations to determine the latitude and longitude of the ship's position. Green began to train the officers to assist him as a means of checking the accuracy of the observations. Although he often complained that he got little help from the younger officers, within six weeks of their departure Clerke, Spöring, Pickersgill, and several others were adept with telescope and quadrant; and Green began keeping a notebook with each man's observations, computing the average of all readings as the most accurate way to fix their location.

Slightly less than a month after leaving England, the *Endeavour* put in to Madeira, an island off the coast of Africa, to take on fresh provisions, including 270 barrels of beef, a six-hundred-pound live bull, and three thousand gallons of wine. Nevertheless, by the end of October provisions were running low, and Cook cut the daily ration of bread by one-third to preserve supplies. In the first of many experiments with food, Banks caught a shark and had it steamed for dinner, although not without studying and classifying it first. Quite pleased with himself, he announced that the animal had solved some confusion about the species and made tasty eating besides. Most of the seamen didn't agree, however, preferring their usual ration of salt beef and bread to shark's meat. Banks and his assistants also occasionally caught birds for study. One they kept as a pet for nearly a month, until one of the ship's cats decided that fresh bird made a better dinner than table scraps.

As a rule the crew did not take kindly either to the remains of Banks's zoological dissections or to Cook's experiments with scurvy preventives. They turned up their noses at sauerkraut, in particular, until Cook instituted "a Method I never once knew to fail with seamen. This was to have some of it dress'd every Day for the Cabbin Table, and permitted all the officers without exception to make use of it and left it to the option of the Men either to take as much as they pleased or none at all." Within a week he had to start rationing sauerkraut.

In November the expedition stopped at Rio de Janeiro for reprovisioning before continuing south toward Cape Horn. Based on his reading of other explorers' accounts, Cook decided not to attempt the Strait of Magellan but rather to sail around Cape Horn. Shortly after they passed the entrance to the Strait, Cook anchored the *Endeavour* along the coast of Tierra del Fuego to search for water and fresh greens. About thirty men greeted them. Astonished at their crude clothing and tools, minimal housing and apparent lack of boats despite their seaside habitation, Cook pronounced them "perhaps as miserable a set of People as are this day upon Earth."

As barren as this countryside was, Banks could not miss an opportunity to "botanize," in his words, and with Solander he organized an expedition inland. Monkhouse, Green, Buchan, two seamen, and two of Banks's servants accompanied them. Their aim was to get as far into the interior of the country as possible and return to the ship by nightfall, but what was intended as a simple day hike turned into disaster. Buchan had an epileptic seizure, forcing the group to slow its pace; Banks lost his bearings and wasn't sure which direction would take them back to the ship; by late afternoon the air had turned colder and snow began to fall. Banks decided they had no choice but to camp out and return to the ship the next morning. The group separated, some going ahead to find a suitable campsite, while Solander and the two servants, declaring they were too tired to walk another step, lagged behind. Banks eventually persuaded Solander to go ahead with him, leaving the servants to catch up later. The pair never made it to camp, but fell asleep where Banks had left them; men dispatched to rouse them could not, and covered them as best they could with leaves and branches. Such flimsy covering was poor protection against snow and cold, and the next morning the two men were dead. The survivors, themselves weakened from hunger and exposure, set off on what they thought would be a long, difficult trek back to the ship; but in their confusion of the day before they had followed a circular path and were much closer to the ship than they had realized.

The voyage around Cape Horn took thirty-three days, and by mid-February the *Endeavour* was in the Pacific, sailing directly northwest toward Tahiti. After two weeks of good progress through relatively calm seas, Cook noted that they had encountered no strong currents since entering the Pacific, which led him to believe that they had not sailed near any large bodies of land, since land masses and ocean currents were usually associated with each other. Cook held a skeptical view of the theories about a southern continent, and this observation only confirmed his suspicion that geographers had grossly exaggerated its size—if, indeed, it existed at all.

Banks, on the other hand, was more inclined to believe the theorists and was surprised when they had sailed for a month with no sign of a continent. Still, he took a certain pleasure in this negative discovery. A believer in the importance of detailed observation in the advancement of scientific knowledge, he scorned those who theorized from little evidence and felt vindicated at seeing them proved wrong. As for the notion that a southern continent must exist to balance the land in the northern hemisphere, Banks believed that "till we know how this globe is fixd in that place which has been since its creation assignd to it in the general system, we need not be anxious to give reasons how any one part of it counterbalances the rest."

On April 4 the Englishmen got their first glimpse of land in nearly two months: a tiny, flat island with a few coconut trees. The next several days revealed more islands of similar appearance, all part of the Tuamotu Archipelago discovered by the Spanish explorer Quiros in 1606 (as Cook knew from his reading of earlier explorers). Early on the morning of April 11 a mountainous island came into view, "as uneven as a piece of crumpled paper, being divided irregularly into hills and valleys," as Parkinson described it; "but a beautiful verdue covered both, even to the tops of the highest peaks." Gore, Wilkinson, and the others who had sailed in the *Dolphin* recognized it as Tahiti.

The *Endeavour* sailed into Matavai Bay the next day to a rather tentative welcome. Only a handful of canoes approached the ship, and all they had to offer was coconuts. Gore, expecting a joyous welcome from his old acquaintances, was disappointed and a bit puzzled at this subdued reception. He recognized just one familiar face among those who greeted them, and when he went ashore with Cook, Banks, Solander, and Green he saw that the lively village of two years earlier had all but vanished. The attractive houses, the large canoes, the plentiful supplies of hogs and fowl, were all gone, replaced by a handful of shabby huts. Gore spotted the stone foundations of some of the houses he remembered, but the spaces where others had been were overgrown with grass, and the few people in residence all seemed to be "of the inferior sort."

Cook and Banks, unburdened by memories of an earlier visit, were delighted with their first experience of a Polynesian-style welcome and with the lushness of the island's vegetation. Coconut and breadfruit trees, flowers, and a profusion of unfamiliar plants grew everywhere, conjuring up visions of paradise in Banks's mind—and also visions of dominion. Tahiti, he thought, "was the truest vision of an arcadia of which we were going to be kings that the imagination can form." That Gore thought the place much poorer than he remembered it did nothing to dampen Banks's enthusiasm.

Although no one rushed to the ship offering pigs or breadfruit, Cook was not concerned, since the *Endeavour*'s men were in good health, a fact he attributed to the beneficial effects of sauerkraut, "portable soup," and malt. Cook planned to remain at Tahiti for at least two months and would soon need to establish a regular supply of food and water, however, and to ensure an orderly trade for provisions and prevent the value of their trade goods from diminishing through unregulated trade—the problem encountered by the *Dolphin*'s crew—he immediately established rules to govern relations with the Tahitians. Everyone was to make his best effort to establish friendship with the islanders and to treat them well at all times. Trade would be carried on only by the individuals specifically appointed for that purpose. Aware of the problems that the *Dolphin*'s crew had with theft, Cook ordered everyone to pay close attention to work while onshore, guarding his tools carefully; anyone who allowed tools to be stolen through negligence or lack of attention would have the cost charged against his pay. Equally mindful of Wallis's problems keeping his own men under control, he ordered that anyone who stole ship's goods would be charged for them also, and that iron and cloth could be traded only for provisions.

A fleet of canoes visited the *Endeavour* on its second day at anchor. Bolder and of higher status than the men who had greeted the Englishmen the day before (or so Banks judged from their dress and general appearance), they went on board without hesitation. Cook invited two men who appeared to be the leaders of the group to join him on deck; many of the rest, without waiting for an invitation, scrambled nimbly up the ropes hanging over the sides of the ship. It was no use trying to keep them out, Cook remarked, because "they clime like Munkeys."

The two "chiefs" were tall, handsome men of middle age, with thick, black, curly hair, large eyes, and very white teeth. "I never beheld statlier men," Parkinson recalled later; they "had something of a natural majesty in them." Upon entering the ship's cabin, the two removed most of the cloth they wore draped around themselves and dressed Cook and Banks Tahitian-style. Cook gave them hatchets and beads in return. Then, in response to the chiefs' invitation, he had a boat hoisted out and, with Banks, Solander, and several of the officers, followed the pair and their entourage across the bay.

The group disembarked at a point about three miles from the ship, where scores of Tahitians greeted them. After more exchanging of gifts, the Englishmen walked among the houses to get a closer look at the way their hosts lived. The women proved as friendly as the *Dolphin*'s men had described them, "but as there were no places of retirement," Banks remarked, "the houses being intirely without walls, we had not an opportunity of putting their politeness to every test that maybe some of us would not have failed to have done had circumstances been more favourable." The women could not understand such bashful behavior, pointing at their mats and even going so far as to try to force the men down with them. As Banks remarked, "they were much less jealous of observation than we were."

The Englishmen took their leave of this group and continued walking along shore for another mile or so until they met another "chief" with a throng of followers. Well versed in Tahitian greeting ceremonies by now, they accepted the green branches that were always offered at new meetings, crossed their hands on their chests, and repeated with the Tahitians the word *taio*, which Banks correctly surmised meant "friend." Their new acquaintances offered more food, including raw fish, "which it seems they themselves eat," Banks noted with some surprise.

The chief's wife, an "ugly" woman, Banks thought, took an immediate fancy to him and stayed close by his side, but he spotted a pretty young girl "with a fire in her eyes" and beckoned to her to sit beside him. "I was then desirous of getting rid of my former companion so I ceas'd to attend to her and loaded my pretty girl with beads and every present I could think pleasing to her." The older woman "shewd much disgust," but would not leave, continuing instead to offer presents to Banks.

Before this tug-of-war could be resolved, Solander and William Monkhouse, the surgeon, discovered that some nimble-fingered Tahitians had picked their pockets, stealing a snuffbox and spyglass. One of the chiefs offered native cloth in return, but Banks, already proficient in sign language, told him they wanted only the return of their own property. The chief went off in search of the missing objects, returning triumphant within half an hour; but to his chagrin, the spyglass case was empty. He and Banks then went to the home of a woman who, the chief believed, had some knowledge of the missing glass. He gave her cloth, and Banks offered beads; she soon returned with the spyglass. Banks rejoined his companions, impressed with this method of practical justice, "exercised by people . . . uninstructed by the example of any civilizd countrey." The incident prompted him to name this chief

Lycurgus, after his form of justice, the second chief he dubbed Hercules because of his size and strength.

Banks continued for several days to endow his new acquaintances with Greek names. Two others who struck him as particularly thoughtful he named Solon and Mentor, another became Ajax because of his "grim countenance," and one "who eats most monstrously" he called Epicurus. Such intellectual nomenclature was lost on most of the crew, and Lycurgus and Hercules were the only names that stuck. After two weeks, Banks finally learned their real names: Tepau i Ahurai and Tuteha. It took him a month longer to learn the name of the island; until then, the English continued to use the name Wallis had given it, King George's Island.

The Tahitians were pleased to learn that their visitors also had names, although, Banks commented, "they make so poor a hand of pronouncing them that I fear we shall be obligd to take each of us a new one for the occasion." Banks they rendered as "Tapane," Cook as "Toote," Solander as "Torano," and Gore as "Toarro." Molyneux they gave up on entirely, sticking to his first name, Robert, which they pronounced Boba. The Englishmen didn't do too well with Tahitian names either; Banks, for example, called Tuteha "Doodahah," and Tahiti was always Otaheite. The latter was a common error among the English because they did not realize that the Tahitian language generally used the definite article *o* with proper names.

This first encounter with the Tahitian propensity for theft had ended innocently enough, but more serious events followed.

After scouting the whole length of Matavai Bay and the surrounding shoreline, Cook decided there was no better place for the *Endeavour* to anchor, so he and his men began making preparations to settle where they were and unloading their gear for the observation of the transit. Fixing on a small promontory, soon to be named Point Venus, Cook directed a team to set up their portable observatory and build a small fort around it. Fascinated, a large group of Tahitians watched the Englishmen at work, as Cook tried to make it clear to them that he and his men wished to use this land for a period of time and would then go away, leaving it as they had found it.

Cook left the construction crew under the direction of midshipman Jonathan Monkhouse while he and Banks continued their exploring. Shortly after they left, a band of Tahitians overpowered the sentry at the construction site and stole his musket. Monkhouse ordered his men to fire at the thieves, a directive they "obeyed with the greatest glee imaginable, as if they had been shooting at wild ducks," according to an outraged Parkinson. Some pursued the Tahitians into the woods, killing one and wounding several others.

Hearing the shots, Cook and his companions raced back to shore and tried to quell any further disturbances by explaining as well as they could manage that the man had been killed for stealing a gun and that the incident should not affect the friendship between the Tahitians and the English. The Tahitians were not persuaded by such arguments—understandably enough, since their previous experience with European weapons had taught them that the first shots only led to worse. But with a good deal of help from Banks, whose affinity for the islanders was already making itself felt, Cook finally calmed the terrified group. Although Banks tried to appease the Tahitians by telling them that the theft of the gun was a crime deserving of death (and in England at the time, minor theft could indeed bring sentence of capital punishment), in fact he thought it was a specious argument and blamed the English collectively for the death of an innocent man. "If we quarreled with these Indians," he told Parkinson, "we should not agree with angels."

The Tahitians stayed away from the ship and the fort for two days, but then resumed their visits as if nothing had happened. This showed "that their dispositions are very flexible; and that resentment, with them, is a short-lived passion," Parkinson thought. He and Banks were already developing a view of the Tahitians as generous and without guile, who stole out of curiosity rather than greed or ill will. It was an image that accorded well with eighteenth-century Europeans' propensity to idealize the lives of people unacquainted with industrial technology and other complexities of modern European culture.

Two days later tragedy struck again, when Alexander Buchan died after another epileptic seizure. Displaying less sympathy for Buchan than for the murdered Tahitian, Banks was concerned mostly about losing the young man's services. "My airy dreams of entertaining my friends in England with the scenes that I am to see here are vanishd," he wrote; "no account of the figures and dresses and men can be satisfactory unless illustrated with figures: had providence spard him a month longer what an advantage would it have been to my undertaking but I must submit." Fortunately for Banks's "airy dreams," Buchan had already been busy producing sketches of the voyage, and Parkinson, hired to do botanical and zoological drawings, broadened his scope to produce a record of people and landscapes as well.

Despite these rocky beginnings, within a week the *Endeavour*'s men were comfortably settled in their temporary camp. Fifty men spent most of their days on shore, some building the fort and others guarding it. On the ship, the boatswain supervised a crew overhauling the rigging; the quartermaster went through all the supplies in the hold, throwing out those that had gone bad; the carpenter and his mates examined the ship for any signs of damage; and the armorer set up his forge to make the fittings needed to repair the ship. Occasionally Cook declared a holiday and allowed all hands on shore, but under certain restrictions—forbidding them to go beyond Matavai Bay or molest the islanders.

Cook also worried about introducing venereal disease to the island, for he knew from Wallis's experiences that he would be powerless to prevent sexual encounters between his men and the Tahitian women. Before anyone left the ship at Tahiti, he had the surgeon check all hands for signs of disease. All but one got a clean bill of health, and he was prohibited from shore duty. As Cook expected, his crew lost no time in forming liaisons with the Tahitian women. Even Banks got over his initial reluctance, although Cook remained aloof and Parkinson eschewed what he considered to be immoral acts of fornication. But Cook's precautions were of little use, because symptoms of syphilis were already apparent among the islanders and a large number of the crew were soon afflicted as well. Suspicion immediately fell upon the men of the *Dolphin,* who had not stayed at the island long enough to see the effects of their contact. The Tahitians, however, blamed another group of Europeans who had anchored at the island about a year before Cook arrived, they claimed. The English were skeptical of this story, for as much as they would have liked to absolve the *Dolphin* of responsibility for bringing venereal disease to the island, the thought of competition in the Pacific was even less appealing.

The possibility gained more credence a few days into the Englishmen's stay, when Banks noticed an ax that was obviously not of English make. He dismissed the find on the assumption that the *Dolphin* had probably carried some old axes to use for trading purposes, but when he found an unusual type of adze several days later, Banks began to think the Tahitians' stories might be true. After painstaking questioning sprinkled with sign language, he learned that two ships had anchored off the island some months earlier, at a spot several miles east of Matavai Bay. A woman was among the crew, according to his informants, and when the ships left, a young Tahitian man went along. To learn the nationality of the ships, a matter of no small importance to the English, Banks showed the Tahitians a large sheet of paper with pictures of European flags; several picked out the Spanish flag.

For the rest of their voyage, the Englishmen were puzzled about the "Spanish" ships that had preceded them at Tahiti, but the islanders had unknowingly misled their visitors. In fact the ships were French, commanded by Bougainville. His purpose, like theirs, was to search for a southern continent. As a result of France's agreement with Spain to give up claim to the Falkland Islands, the French government sent Bougainville to dismantle the colony he had established there in 1764. To soften the blow and in some measure compensate him for his efforts, they commissioned him to continue his voyage by going around the world.

Like Cook, Bougainville traveled with an astronomer and a naturalist, the latter a distinguished member of the Académie des Sciences named Philibert Commerson. The woman on board was Jeanne Baré, Commerson's valet; she managed to keep her identity a secret until the ships reached Tahiti, where the islanders had no difficulty discerning her femininity. (Commerson, a prudish and rather self-righteous man, took a good deal of ribbing over this incident, although he claimed to have been taken in himself by Baré's disguise.)

Cook was not pleased to learn that a rival nation had visited Tahiti ahead of him, but he took some solace in this evidence that his men were not responsible for introducing venereal disease, although the question of who did in fact first infect the Tahitians remained a matter of debate. The English would later blame the French, while the French accused the English—syphilis was labeled the "French" or "English" disease depending on one's home country—and some eventually tried to absolve Europeans of blame altogether by arguing that venereal diseases were endemic to the Pacific islands. As Cook remarked, knowing the source was "little satisfaction to them who must suffer by it." He predicted that venereal disease "may in time spread it self over all the Islands in the South Seas, to the eternal reproach of those who first brought it among them."

In an attempt to ensure a steady supply of food and prevent any further violence, Cook appointed Banks as chief broker with the Tahitians. It was an ideal choice. Banks had already demonstrated his popularity with the islanders, and he was happy enough to turn over the labor of collecting botanical specimens to Solander and Spöring, devoting himself to playing trader and diplomat. He and his group set up camp in a cluster of tents on shore, which became the center of most English-Tahitian gatherings. Tepau i Ahurai, who quickly emerged as the Tahitians' chief emissary in dealing with the English, did the same, moving his family from their home at Pare, a district about ten miles away, to temporary quarters near the English encampment.

Both men threw themselves into their new roles with enthusiasm. Banks, who had never managed to master Latin or Greek during his years at Oxford, became quite proficient at Tahitian (as did several of the younger officers). And Tepau i Ahurai, who displayed a considerable talent for mimicry, picked up a number of English words and did his best to imitate English customs. He was, in particular, quite taken with the English habit of eating with knives and forks, managing them, according to Banks, "more handily than a Frenchman could learn to do in years." Sometimes his enthusiasm for trying English ways got him into trouble, however. Late one evening Tepau i Ahurai's wife came to Banks's tent in tears; her husband, she said, had vomited and now lay dying after eating something the English had given him. She had the remains of the offending substance wrapped in a leaf. It was chewing tobacco, Banks discovered, and Tepau i Ahurai, attempting to imitate the Englishmen who gave it to him, had swallowed it.

Under Banks's direction, trade with the Tahitians proceeded in an orderly if sluggish fashion. Breadfruit, coconuts, and plantains were offered in abundance, but it was nearly a week before the Englishmen had their first taste of native pork. The animals were scarce and expensive; in the interim since the *Dolphin*'s visit, the Tahitians had learned the value of iron tools, and hogs, which could formerly be bought with nails, now commanded hatchets when they were available at all. This escalation in prices reflected in part a shortage of meat, but it was also a consequence of the Tahitian chiefs' understanding that they controlled the supply of food and could charge accordingly. The Englishmen suspected as much after a few days of trade, and Molyneux, who had become Banks's chief assistant in procuring food, decided to take one of the boats and seek food at more distant parts of the island, where the inhabitants might be less inclined to charge whatever the market would bear. But although Molyneux, with Green for a companion, sailed twenty miles east of Matavai Bay, he found hogs scarce everywhere, and the Tahitians refused to sell what few they had, claiming that they all belonged to their chief, Tuteha.

The Englishmen ate well despite the scarcity of hogs, adopting the Tahitians' primarily vegetarian diet. Breadfruit, a starchy, fleshy fruit that grew abundantly on large trees, was the islanders' staple food, supplemented by sweet potatoes, taro root, coconuts, bananas—Banks counted thirteen varieties—plantains, an applelike fruit, sugar cane, several kinds of nuts, and other fruits less familiar to the English. They also ate wild fowl and fish, which they caught with bamboo poles and shell or mother-of-pearl hooks, speared with harpoons made of sugar cane, or trapped in nets of coconut fiber. Pork was an occasional delicacy, a point that had escaped the men of the *Dolphin* with their English taste for meat.

Dog meat was the rarest and most prized delicacy of all, much to the disgust of the English, whose appetites did not run to trussing and roasting man's best friend. But, curious to experience everything in this exotic place, the officers permitted one of their Tahitian acquaintances to prepare roast dog for them. "Few were there of us but what allowe'd that a South Sea Dog was next to an English Lamb," was Cook's judgment. Cook would eat almost anything, however—his taste was "surely, the coarsest that ever mortal was endowed with," one of his officers remarked—and most of the others did not share his enthusiasm. Parkinson thought the roast dog tasted like "coarse beef" and had a "strong, disagreeable smell." Only Cook, Banks, and Solander could stomach it, he claimed.

On the whole, relations between the Tahitians and the English proceeded smoothly, although the islanders' light-fingeredness became an increasing annoyance. "Great and small chiefs and common men all are firmly of opinion that if they can once get possession of any thing it immediately becomes their own," Banks complained. Only Tepau i Ahurai and Tuteha seemed immune to the temptations of theft—until one morning when Banks woke up to discover his knife missing. Since Tepau i Ahurai was the only Tahitian who always had free access to Banks's tent, he seemed the obvious culprit. Banks refused to believe his denials, and Tepau i Ahurai began searching for the knife, as if to prove his innocence. Within minutes one of Banks's servants produced the missing knife, and Tepau i Ahurai turned to Banks "with a countenance sufficiently upbraiding me for my suspicions; the scene was immediately changed, I became the guilty and he the innocent person."

Two days later, Tepau i Ahurai's ire was aroused again, this time when the tables were turned and a Tahitian was the victim of theft. Early in the morning the chief appeared at Banks's tent, grabbed his arm, and pulled him along to a spot where they found one of the ship's crewmen, a sailor named Henry Jeffs. Tepau i Ahurai told Banks that Jeffs had threatened his wife, and Banks responded that he would be punished, if only Tepau i Ahurai could explain the offense—for the Tahitian was so angry that Banks had difficulty understanding him. Jeffs, it developed, had visited Tepau i Ahurai's house and attempted to purchase a stone hatchet for a nail. When the Tahitian refused to sell it, Jeffs took the hatchet anyway and threatened to cut her throat if she attempted to stop him.

Upon hearing this story, Cook confined Jeffs on board ship. The next morning he invited Tepau i Ahurai and his wife on board, called all hands together, lectured the crew on the odiousness of the crime, and ordered Jeffs flogged. Despite his fury of the previous day, when confronted with the reality of a Navy flogging, Tepau i Ahurai tried to have the punishment stopped. But Cook, who wanted to make an example of Jeffs in order to preserve peaceful relations with the Tahitians, would not be deterred.

The next day, several canoes that the English had never seen before paddled into Matavai Bay and landed on the beach near the English encampment. The leader of the group was a stately woman, perhaps forty years old. Tall, heavy, light-skinned, with expressive eyes and a vigorous manner, "she might have been handsome when young," Banks wrote, "but now few or no traces of it were left." Cook thought she was "like most of the other women very Masculine," although Parkinson considered her attractive. She and a few of her companions went to Banks's tent. A few minutes later Molyneux joined the group and recognized their guest as Purea (or Obarea, as the English usually spelled her name), the "queen" of his former visit. Several of her companions were familiar to him too, notably a young man named Tupaia, a priest and confidant of Purea. (Banks later referred to him as "Obereas right hand man.") Banks and Molyneux escorted Purea and Tupaia on board the *Endeavour* to meet Cook and the other officers. Gore met them as they boarded, delighted at the reunion with his friends from the previous voyage.

There was the usual exchange of gifts and greetings. Cook charmed Purea with a doll, telling her, with uncharacteristic levity, that it was an image of his wife. When she returned to the beach Purea encountered Tuteha, who was quite put out at all the attention lavished on her—and especially at the doll, which was unlike anything the English had given him. Cook regained Tuteha's good humor by inviting him on board and giving him more presents, including a doll, which Banks noted with amusement had suddenly replaced hatchets as the most honored gift the English could offer.

In the meantime, some of the old hands from the *Dolphin*, wondering why it had taken over two weeks for Purea to make her appearance and what was behind the changes around Matavai Bay, questioned her about events since their last visit. Limited in their ability to communicate, they learned only that there had been a battle nearby and that she and her family had lost their land and moved to a more distant district as a result. Skeptical about the stories of Purea's former glory, Cook concluded that she was "chief of her family" (which included Tepau i Ahurai, her brother), but probably had no authority over other inhabitants. Tuteha was chief of the island, he thought, or at least of the area around Matavai Bay.

About three weeks after the *Endeavour* anchored at Tahiti, the shore party finished the fort that would enclose their observatory. Perched on the little peninsula that defined one side of Matavai Bay, it looked like a miniature castle, with a turretlike center tower, rounded towers on each side, and a British flag flying above. A wall of earth protected the structure from the water side, with a stockadelike row of wooden posts guarding the side exposed to land. It had a bank of dirt four and a half feet high, with a ditch ten feet wide and six feet deep on two sides. A third side had wooden palisades built on top of the dirt wall, and the fourth side was protected with a double row of casks. Eight cannon were mounted along the walls, and forty-five men with small arms guarded it during the day. A smaller group slept in the fort at night. "I now thought my self perfectly secure from any thing these people could attempt," Cook wrote.

He had not reckoned with Tahitian ingenuity. The morning after the fort was completed and the quadrant taken ashore, Cook and Green discovered it missing. "It was a matter of astonishment to us all how it could be taken away," Cook said, because the instrument was still inside its heavy packing case and a sentry had been posted within five yards of it all night. Nothing else had been taken; the Tahitians realized that the quadrant was the prize object for which all this construction had been undertaken.

The islanders generally had few qualms about offering information when one of their number had stolen English property, and Cook quickly learned who had stolen the quadrant and which direction he had taken. Finding the man and retrieving the quadrant were far more complicated matters than simply learning the identity of the thief, however. Cook decided to adopt a tactic that he would use increasingly in encounters with native thievery: he seized all the large canoes in the bay, refusing to release them until the quadrant was returned. In the meantime, he dispatched Banks, Green, and a midshipman to search for the instrument.

They soon encountered Tepau i Ahurai, who led them through the woods, stopping at every house to inquire about the thief's whereabouts. The Tahitians readily cooperated with information, but the thief moved faster than the three Englishmen. After half running, half walking about four miles in 90-degree heat, they stopped, exhausted and uncertain about what to do next. Banks sent the midshipman back to Matavai Bay to ask Cook for reinforcements, while he and Green pressed on with Tepau i Ahurai; three miles farther along, they met the thief with a piece of the quadrant in his hand. By the time Cook reached Banks and Green, the pair were already on their way back to Matavai Bay, carrying the broken quadrant.

Back at the fort, they learned that the men left behind, not content with holding the canoes hostage, had seized Tuteha against Cook's explicit orders. Several dozen frightened Tahitians waited outside the fort; Tuteha, equally frightened, thought he would be killed for the crime. Outraged, for he believed Tuteha had nothing to do with the theft, Cook ordered the chief's immediate release. Tuteha gave Cook two hogs in return, to the captain's chagrin, "for it is very certain that the treatment he had met with from us did not merit such a reward."

The gift was more an attempt to placate Cook than a sign of friendship, however. Tuteha felt insulted, and like Cook he had methods of showing his displeasure: no one appeared at the fort with provisions for trade the next day. Pursued long enough, this tactic could work serious hardship on the English by cutting off their supply of fresh food; the English needed the Tahitians, but the reverse was not true, no matter how much the islanders coveted nails and European cloth. When trade had still not resumed after three days, Banks got worried. "We are for the first time in distress for necessaries," he wrote. Worse still, word reached the English camp that Tuteha had no intention of visiting them for another week, and it was rapidly becoming clear that he controlled the flow of trade. (The Tahitians who had refused to sell hogs to Molyneux because they "belonged to Tuteha" were not just making excuses.) As long as Tuteha remained angry, his subjects would stay away from the Englishmen.

With no end to the stalemate in sight, Cook decided to visit Tuteha, and accompanied by Banks and Solander he set out for the chief's home at Pare. Almost as if he had been expecting them, Tuteha led his guests to a large courtyard where a group of young men entertained them with a wrestling match; when it was over the chief suggested that they all return to the *Endeavour*—which signaled his permission to resume trade. As soon as the Tahitians around Matavai Bay learned that Tuteha had returned, they flocked to the fort with food for barter. "We now begin to think that Dootahah is indeed a great king much greater than we have been used to imagine him," Banks concluded.

So a prank that had threatened to ruin the transit of Venus observations and English-Tahitian relations in the bargain ended with no apparent ill effects. Spöring was able to repair the quadrant, and trade with the Tahitians was soon as vigorous as ever.

By the end of May, all preparations were completed for the transit observations, exactly on schedule; and June 3 brought ideal weather for the event, clear and hot.

Cook and Green established several observation sites to increase their chances of getting accurate readings, stationing themselves and Solander at Point Venus and Banks, Spöring, Gore and Monkhouse at Imeo, a smaller island about twelve miles from Tahiti. (It is today called Moorea.) Clerke, Pickersgill, and Hicks went to a tiny islet just offshore. Each was to record the exact time that Venus appeared to touch the sun's surface and the moment it completed its passage across the sun. Precise timing was essential, since an error of even a few seconds in recording the transit would seriously undermine the accuracy of the observations. The astronomers planned to compare and average their results to increase their chances of achieving an accurate conclusion.

The observations could hardly have been more carefully planned, but no one had foreseen one serious difficulty: it was nearly impossible to distinguish the penumbra, a hazy glow surrounding the planet, from the edge of Venus itself, and consequently the exact time that Venus began its transit across the sun could not be determined accurately. Even Cook and Green, the most experienced astronomers in the group, differed considerably in their observations. Pickersgill, commenting on the ideal weather conditions for the observations, believed the observers alone would be to blame if their results were inadequate. But in fact the disappointing results at Tahiti were duplicated at other sites around the world; and although much more information was amassed in 1769 than in 1761, it was so inconclusive as to be nearly useless.

Cook was disappointed at the poor results of their observations, but he believed the most significant work of his voyage was just beginning: the close observation of natural phenomena and conditions of life in the South Pacific, the search for a southern continent, and the charting of whatever land he and his crew might find along the way.

6. *Amateur Ethnographers*

Cook and his men remained at Tahiti almost six weeks after the transit of Venus. Comfortably encamped and enjoying the islanders' goodwill, except for occasional misunderstandings, they began the second phase of their expedition with a close examination of Tahiti and its people. Cook believed the island was an ideal location for the base England needed to continue exploring and eventually expanding its trade in the Pacific, and he wanted to chart and describe it thoroughly for the benefit of those who would come after him. From a purely scientific point of view, he also wanted to take advantage of the Tahitians' friendliness to learn as much as possible about their way of life, for Cook interpreted his responsibilities as an explorer in the broadest sense of the word.

As a practical matter, however, Banks took on the major responsibility for collecting information about Tahitian culture, while Cook concentrated on describing the physical characteristics of the island and charting its coastline. It was a division of labor they would maintain throughout the voyage, and one that suited their different talents and temperaments. Cook was a literal-minded man, always hesitant to draw conclusions about anything he could not observe directly, and reserved, even aloof, in his dealings with people. Although Cook felt a deep affinity for the native peoples he encountered and generally enjoyed their respect, he could not shed his position as leader of the expedition to partake of native life, however temporarily, as Banks did. But his skills as a navigator and cartographer were unmatched, and lent themselves to close observation of natural phenomena as well. Banks's gregarious personality and his unquenchable curiosity about everything he saw—aided, no doubt, by his youth and the cocksure self-confidence born of great wealth—earned him the trust of the Tahitians and allowed him to become more fully absorbed in their lives than any of his fellow explorers. He was also more willing than Cook to ask questions about abstract issues like political organization, religion, and social structure, and more likely to get answers.

Banks continued to supervise the collection, classification, and drawing of botanical and zoological specimens, the task that had been his primary reason for joining the voyage. But Spöring and Parkinson could be trusted with most of the day-to-day work involved, and as the weeks in Tahiti went by Banks increasingly concentrated on observing people. The close study of man was gaining popularity among Europeans as men of philosophic and scientific bent believed it possible to come to a greater understanding of the fundamental nature of man through methods of scientific observation. Men untouched by the complexities of European civilization, and presumably therefore closer to the "state of nature" postulated by Enlightenment thinkers, were ideal subjects, since by observing them it should be possible to separate the influence of environment from innate qualities shared by all humans. In this spirit, Banks approached the Tahitians as he did any other species—as a phenomenon to be studied and described—but his personality and his genuine empathy for the islanders permitted him insights that the purely dispassionate scientist might well have missed.

Banks had a practical advantage in that he and his entourage set up permanent quarters on shore, unlike everyone else, who slept on board ship. In addition, his role as chief trader for the English put him in daily contact with the Tahitians, and his tent became a gathering place for the islanders to the extent that, he remarked, "I fear I shall scarce understand my own language when I read it again." He was especially popular with the women. Purea's "ladies in waiting," as he called them, fought among themselves for the privilege of spending nights in his tent, and Purea herself quite openly sought him for her lover, although without success. "Her majesties person is not the most desireable," Banks judged. The younger women, on the other hand, he thought strikingly attractive, and one of Purea's attendants, Tiatia, became his mistress. Banks likened the Tahitian women to Greek statues in their grace and bearing, and praised their simple, uncontrived dress and freedom of expression—a sharp contrast to the fussy garb and false modesty, which he considered prudery in disguise, of European women.

The Tahitians' affection and trust allowed Banks glimpses into native life that the other Englishmen were not privileged to share—as, for example, when he was invited to take part in a funeral ceremony. Upon learning that a funeral procession was planned, Banks asked permission to watch. The Tahitians agreed, but only on the condition that he would play a role. In anticipation, he wrote, "Tomorrow . . . I am to be smutted from head to foot and do whatever they desire me to do." The next day, with several Tahitians, he was stripped of his clothes except for a bit of cloth around his waist—"I had no pretensions to be ashamd of my nakedness for neither of the women were a bit more coverd than myself"—and covered from head to shoulders with a grimy charcoal-and-water mixture. Thus adorned, the group marched to Matavai Bay, as all the Tahitians they encountered ran away at their approach. The purpose, Banks learned, was to appear as frightful as possible, as if mad with grief for their departed friend; the people who ran away from them were part of the act.

On another occasion, Banks watched a young girl undergoing her first "tattoo" operation. The Tahitians did not have elaborate ornaments, but preferred instead to decorate their bodies directly, with designs made by injecting a black dye under the skin. It was an exceptionally painful process, as each design required repeated injections with a wooden or shell comb about two inches long lined with about thirty sharply pointed teeth. The islanders acquired the designs gradually, beginning in early adolescence; most eventually sported tattoos over much of their buttocks, thighs, and sometimes other parts of their bodies as well. (In the course of their stay at Tahiti, several sailors also submitted to the process, taking home a unique souvenir and inaugurating a long-standing custom among seafaring men.)

This sort of participant observation was an important part of Banks's method in studying the Tahitians, supplemented by direct questioning. He relied primarily on his closest acquaintances, like Tepau i Ahurai, Purea, and Tuteha, as informants. Banks became especially interested in the Tahitian social and political system, and his attempts to understand it involved both observing the way certain groups of Tahitians behaved toward each other and questioning them to learn the meaning of these relations. One peculiar incident, for example, offered some insight into the strict hereditary succession that dominated Tahitian politics. A chief named Amo, whom the Englishmen had not previously met, came to their camp with a young woman and a boy of about seven or eight, who was carried on another man's back despite his healthy appearance. Immediately Purea and the other Tahitians went to meet them, baring their heads and bodies to their waists. "We thought that this Oamo must be some extraordinary person, and wonder'd to see so little notice taken of him after the ceremony was over," Cook noted, but Banks learned by questioning his Tahitian friends that it was the child, not Amo, who commanded such respect. The son of Purea and Amo, who had once been married but no longer lived together, he was "heir apparent to the Sovereignty of the Island," a more important person even than Tuteha.

This notion that a child of seven might enjoy more political and social status than chiefs like Purea and Tuteha puzzled the English, as indeed the political and social system of Tahiti in general did. Cook and Banks had concluded earlier that Tuteha was the most powerful chief in the vicinity of Matavai Bay and that Purea, while influential, did not exercise political power to the same extent. In addition, they had spent enough time with Tahitians to understand that the islanders maintained clear-cut class distinctions, to the point of attaching names to certain status groups. But the Englishmen's knowledge was limited and in some instances faulty because of the difficulty of learning about complex, abstract subjects with their rudimentary knowledge of the Tahitian language and because their experience on the island was confined to the area around Matavai Bay.

Cook remedied the second deficiency toward the end of June, when he decided to journey around the entire island, primarily to make an accurate chart, but also to add to his store of knowledge about the island and its inhabitants. Banks and enough sailors to man one of the boats went with him. Traveling east from Matavai Bay, they first came to a district called Hitiaa, where the "Spanish" ships were supposed to have anchored. Here they met a chief named Ereti, whose brother Ahutoru had gone with the Europeans; his story corroborated the version Banks had heard from his informants at Matavai Bay.

The next day they reached a narrow isthmus connecting the larger part of Tahiti, which the islanders called Tahiti-nui (the name means "big Tahiti") to a much smaller peninsula called Tahiti-iti ("little Tahiti"). A local resident warned them not to cross over to Tahiti-iti, because it was beyond the bounds of Tuteha's influence and the people there, subject to a chief called Vehiatua, were enemies.

Unconcerned by the warning, Cook and Banks continued across the isthmus to Tahiti-iti, where they met Vehiatua, a frail old man with white hair and a beard, and toured his district, which was more elaborately cultivated than any part of Tahiti they had yet seen. Banks found the women here just as friendly as everywhere else in Tahiti. As night came on, he recalled, "I stuck close to the women hoping to get a snug lodging by that means as I had often done; they were very kind, too much so for they promised more than I ask'd"; and when they saw that he was really interested only in a bed for the night, they left him "jilted."

The trip so far led Banks to conclude that Tahiti was divided into two "kingdoms": Tahiti-nui, ruled by Tuteha, and Tahiti-iti, ruled by Vehiatua. This was a revision of his earlier opinion that Tuteha reigned over the entire island, but was still not an entirely accurate assessment. Not surprisingly, in trying to understand Tahitian culture, the Englishmen looked for parallels to their own, and they assumed that Tahiti, like European nations, must have a single ruler. So the men on the *Dolphin,* identifying Purea as the most influential individual they met, called her the queen; and when Banks learned that people of the highest social class were called *arii* and that the highest-ranking member of the group was called the *arii rahi,* he assumed the latter was the equivalent of a king. He pursued his analogy with European political systems still further, arguing that the Tahitian class system was similar to the feudal system of medieval Europe. The *arii,* he thought, were equivalent to noblemen; the next class, called *manahune* by the Tahitians, were like the vassals or knights of the nobles; and the lowest class, or *teuteu,* were the Tahitian equivalent of serfs.

Banks's analogy was flawed for several reasons. There was no single "king" on the island; the trip to Tahiti-iti and the meeting with Vehiatua demonstrated that. In fact there were three major districts on the island: Tahiti-iti, controlled by Vehiatua; the area from Pare, just west of Matavai Bay, eastward to Tahiti-iti, which was controlled by Tuteha; and the western coast of Tahiti-nui. These areas were subdivided into smaller districts, each under the control of lesser chiefs, Purea among them. Rivalries and shifting alliances among the local chiefs complicated the system.

In addition, social status and political influence were not so closely linked as he assumed. The *arii rahi*—there were actually three of them, one for each major district—enjoyed the highest social status in Tahiti, but they were not necessarily political leaders. Purea's seven-year-old son, who had been carried to the English encampment and commanded such respect from the Tahitians there, was an *arii rahi,* but Tuteha was not.

Such discrepancies confused the English, and especially Banks, who tried to analyze Tahitian ways in European terms. In fact there were three social classes in Tahitian society: the *arii* or chiefs; *ra'atira,* a class of landowners, lower in prestige than the *arii;* and *manahune,* who were nonlandowning commoners. The *arii rahi* were the highest-status chiefs. The *teuteu* of whom Banks wrote were not a separate class, but commoners who functioned as attendants to the *arii.* Membership in these classes and status within the *arii* group were determined entirely by family descent, with the highest-status individual in each major group designated as the *arii rahi.* Birth, not age or ability, was of paramount importance. The *arii rahi* of each generation assumed his title on birth; hence Purea's son enjoyed the title despite his youth.

The overwhelming importance of hereditary social status was apparent in the Tahitians' religious beliefs as well. Tupaia, in one of his attempts to explain Tahitian religion to Banks, told him that the islanders believed in an afterlife and described two different places where the souls of the dead were transported. Banks equated them with the Christian heaven and hell, even though Tupaia made it clear to him that the Tahitian version of "hell" was simply a place of fewer luxuries than their "heaven." The Tahitians believed the souls of their chiefs went to "heaven" and those of the inferior classes of people to "hell." The notion that one's status in the afterlife should depend on heredity rather than behavior on earth led Banks to conclude that Tahitian religion was unconnected with any ideas of morality, but he missed the point: the Tahitians' "heaven" and "hell" represented not judgment for earthly behavior, but simply a continuation of earthly status.

While social standing in Tahiti was rigidly hierarchical and entirely dependent on family succession, political influence was based on a combination of hereditary status and ability. For example, the *arii rahi* in Tuteha's district was his nephew, a young boy named Tu; Tuteha acted in effect as his regent. This arrangement continued for some years, even when Tu became old enough to assume chiefly duties—an indication of the young man's weakness and his uncle's strength as a political leader.

On their journey back from Tahiti-iti, Cook and Banks learned a bit more about Tahitian politics. On the west side of Tahiti-nui they passed through Purea's home territory, a district called Papara, where they walked along a stretch of road covered with human bones. Banks asked for an explanation and was told that warriors from Tahiti-iti had attacked the people of Purea's district several months earlier. She and her family fled to the mountains, but many of the district's residents were killed and nearly all their houses were burned. The invaders also took their hogs, fowls, and as many other possessions as they could carry.

Such warfare among Tahitians was not frequent, but it did occur, usually when one chief tried to extend his or her influence into the districts of others or made unreasonable demands on his subjects. In this case, Purea and her family had visions of glory that did not sit well with other chiefs on the island.

Purea's ambitions were clearly visible in an enormous *marae,* or temple, that she had built in the months before the *Endeavour* arrived. Banks reported it to be 267 feet long, 71 feet wide, and 44 feet high. It was constructed of white coral stones squared and polished, some of them as large as three or four by two feet. "It is almost beyond belief that Indians could raise so large a structure without the assistance of Iron tools to shape their stones or mortar to join them . . . it is done tho, and almost as firmly as a European workman would have done it," he remarked. (Banks could not resist adding that the construction, while remarkable, was not perfect—the steps weren't straight.) The *Dolphin*'s visit unwittingly contributed to Purea's ambitions and her subsequent downfall, as the attention and new possessions showered on Purea and her family only increased their already substantial sense of self-importance. Cook and Banks did not at this point understand the reasons for the attack on Purea, however; it would take more time for them, and others, to recognize the consequences of European contact with Tahitians.

Although the English achieved some understanding of the complexity of Tahitian culture, they didn't realize that the islanders were trying just as assiduously to understand their visitors and use them for their own purposes.

The memory of the *Dolphin* was still strong in the minds of the Tahitians. "They have often described to us the terrour which the Dolphin's guns put them into," Banks commented, "and when we ask how many people were killd they number names upon their fingers, some ten some twenty some thirty." As a result, they approached the men of the *Endeavour* warily at first, employing the combination of cautious and suppliant behavior that had worked well with their previous English visitors.

The Tahitians were concerned that the English, at best, might seriously deplete their food supplies and, worse, might try to take over the island. But they also had something to gain—iron, which they valued but had no other way of obtaining—and were prepared to treat the English well in order to get it. The Tahitians, like the English, attempted to maintain close control over the trade in foodstuffs, as Molyneux discovered in his search for hogs; and although the women who prostituted themselves for nails obviously enjoyed the novelty of the encounters, it was iron they were after, not the pleasure of sex. In this they were encouraged by the men of the island—to the amazement of the English, who believed that men should defend the honor of young women, not push them into prostitution.

Like the Englishmen, the Tahitians were fascinated with this strange new culture they had encountered. They constantly imitated English actions; Cook and Banks were particularly amused at Tepau i Ahurai's attempts to mimic English eating habits. Presumably he was equally amused at the contortions the English went through with knives and forks. Nor could the Tahitians understand why the English insisted on inviting women to eat with them, for they observed rigid segregation of the sexes at mealtime. This particular custom was incomprehensible to the English, "especially," Cook noted, "as they are a people in every other instance fond of Society and much so of their women." Only once could the English persuade a woman to eat with them, and then only after promising not to tell her countrymen. When Cook asked the reasons for this habit, the Tahitians replied only that it was right; and they registered "disgust" at being told that men and women ate together in England.

The Tahitians also tried to learn some of the English language, with mixed results. The notion of written language was entirely new to them, and they interpreted the English custom of letter-writing as a form of tattooing.

Inevitably, conflicts arose from a lack of understanding on both sides. Sometimes the confusion was humorous, as on one occasion when Gore challenged Tepau i Ahurai to an archery contest. Neither man realized at first that the Tahitians and the English had different goals in archery. Gore wanted to shoot at a fixed target, with accuracy the goal, while Tepau i Ahurai was only interested in seeing who could shoot the farthest, regardless of whether or not the arrow hit a bull's-eye. "Neither was at all practisd in what the other valued himself on," noted Banks.

More often the misunderstandings were serious, especially when they stemmed from the two cultures' different views about property. The Tahitians observed private property among themselves, but felt no compunctions about observing property rights among the English. Their frequent theft of English possessions was, in part, a way of expressing hostility; it was also a game, a fact that was apparent in some of the more spectacular thefts. The man who stole the quadrant from the portable observatory at Point Venus had no idea what this instrument was used for, but could easily tell that it was an object of great importance to the English. Taking it from under the noses of the men appointed to guard it on the very day it had been placed in the fort for safekeeping was the supreme challenge for a nimble-fingered Tahitian. Another man stole Cook's stockings from under his head while he was sleeping, a less dramatic but equally impressive accomplishment.

As the weeks wore on, the thefts became more frequent and shifted from the level of amusing pranks to harassment. The *Endeavour*'s long stay had seriously cut into the island's food supplies—a fact Cook and Banks discovered on their trip around the island, although they attributed the shortages to the recent battles and the end of the breadfruit season without realizing the extent to which they had contributed to the problem. Neither Purea nor Tuteha gave any indication that the English presence posed difficulties for the Tahitians, but their continued generosity only increased the hardship felt by their subjects. Ordinary Tahitians could cease trading with the English when supplies ran low, but they could not refuse their chiefs' demands for provisions, which were in effect taxes used to entertain the English. The chiefs enjoyed the prestige conferred by the foreigners' attentions, but the common people, once the novelty was over, would have preferred that their visitors go. They expressed their resentment through petty theft, which in turn angered Cook and pushed him to respond out of proportion to the crimes.

Matters came to a head when an iron rake was stolen from the fort after several days of small but annoying thefts. Cook decided to take canoes hostage, as he had when the quadrant was stolen, but on a much larger scale this time; he had twenty-five large canoes, just returning to Matavai Bay loaded with fish, captured and moored behind the fort. Cook believed the chiefs condoned or at least overlooked thievery against the English and hoped by this tactic to force them into taking responsibility for their subjects' behavior. The rake was returned quickly, but Cook saw his opportunity to get back many of the items stolen in the previous weeks. He had a list read to the Tahitians and told them he would burn their boats if the objects were not returned, hoping to scare them into returning the contraband and curtailing their theft.

This time Cook's strategy backfired. The canoes did not belong to the people who had stolen English property, nor to influential chiefs who might pressure the guilty to give up what they had stolen, but rather to ordinary people who depended on their cargo of fish to survive. Banks recognzied this fact, but, well aware of Cook's determination to impress upon the Tahitians the seriousness of their thefts, he did not attempt to persuade the captain to relent. The stalemate continued for three days, while the canoes' owners importuned Cook to let them go and the fish began to rot, "so as in some winds to render our situation in the tents rather disagreeable," as Banks put it. But Cook was too stubborn to admit that his plan had failed. Normally sympathetic toward the Tahitians and scrupulous about treating them fairly, he invoked his notions about English justice without realizing that they meant nothing to the islanders. Theft to them was not a crime, but a way of exercising power over the English, who otherwise held most of the odds in any struggle. Cook, who abhorred the idea of displaying European power by violent means, instead resorted to force in a different way, by seizing canoes. The situation was resolved only when Purea visited Cook and made a personal plea for the release of the canoes.

Banks was more tolerant of Tahitian foibles, but his very closeness to the islanders and his tendency to romanticize their way of life sometimes worked against him. When one of his favorites occasionally lapsed into behavior of which he didn't approve, he reacted with the air of an injured parent disciplining a wayward child. Curious about English weapons, Tepau i Ahurai once grabbed Banks's pistol, cocked it, held it up in the air, and pulled the trigger. The gun failed to fire and no harm was done, but Banks was furious. He had attempted—successfully, he thought—to make the Tahitians understand that they were not so much as to touch his gun; that Tepau i Ahurai, of all people, should be the one to violate this injunction angered and disappointed him. "I scolded him severely and even threatned to shoot him," Banks recalled. Tepau i Ahurai listened quietly to Banks's tirade, but then packed up his family and belongings and moved back to Pare. A subdued Banks, who genuinely liked Tepau i Ahurai and also recognized how important he was to the English, followed the chief to Pare and persuaded him to return to Matavai Bay.

On another occasion about a week later, one of Banks's servants noticed Tepau i Ahurai hiding a large nail under his clothes. When Banks confronted him with the theft, he admitted taking the nail but claimed he no longer had it. Certain the chief was lying, Banks threatened him and succeeded in getting him to turn over the nail, but then admitted, "I was more hurt at the discovery than he was. I firmly believe he was the only Indian I trusted and in him I had placed a mos' unbounded confidence." Unwilling to let the incident drop, Banks brought Tepau i Ahurai back to his tent for "judgement" by the chief's companions, who argued that the offense was understandable under the circumstances. Banks finally agreed. He had left the nails out in plain view, and the temptation was simply too great. Having convinced himself that he was partly to blame for the incident, Banks offered to forgive Tepau i Ahurai if he would return the nails. The chief promised to do so, but instead of complying, he and his family once again decamped to Pare, staying away nearly a week this time. Like Tuteha after his capture, Tepau i Ahurai did what he thought necessary to appease Banks, but then demonstrated his displeasure by his absence.

The English also violated Tahitian notions of property, despite Cook's attempts to enforce an attitude of respect toward the islanders. William Monkhouse was attacked apparently without provocation after picking a flower from a tree; Banks recognized that the tree was at the edge of a burial ground and therefore considered sacred. When a group sent to collect stones for ballast took several from a *marae,* a Tahitian messenger sent to the English camp demanded that the desecration stop, and again Banks intervened. But while Banks and Cook understood the importance of the Tahitians' sacred sites, the men under their command did not, nor did they necessarily share their leaders' belief that the English were obliged to respect native customs; only Cook's watchfulness and his ability to control his men prevented more serious incidents.

None of the English fully comprehended the level of importance the Tahitians attached to their religion, an intricate system with a complicated hierarchy of gods and lesser spirits who manifested themselves in many forms on earth. The Tahitians believed that everything, including people, plants, animals, even some inanimate objects, was invested with a kind of spiritual aura called *mana,* which flowed from the gods and was stronger in some people and objects than others—stronger in men than in women, for example, and strongest of all in chiefs. Too much exposure to *mana* could be harmful to ordinary people; this was the reason men and women did not eat together, for *mana* was strongly associated with food, and it was therefore considered dangerous for men to consume food in the presence of women. For the same reason, common people dared not approach too closely to *arii* or touch anything they had used. Beliefs surrounding *mana* led to an elaborate system of rules known as *tabu,* which governed many aspects of Tahitian behavior, including contact between *arii* and commoners and use of sacred sites. Picking flowers and taking stones from a *marae* were not merely breaches of etiquette, but violations of *tabu* which, according to Tahitian belief, could bring strong repercussions.

Outwardly a simple society with little in the way of material culture, the Tahitians placed small value on possession of physical objects but valued instead matters of the spirit. The principles of *tabu* were as important and as inviolable to them as the principles of private property were to the English. Of course they had no possible way of knowing that they were in effect violating the Englishmen's *tabu* system by stealing their belongings.

After completing his journey around the island at the end of June, Cook decided it was time to move on, especially given the increasing difficulty of obtaining provisions.

The Englishmen contemplated their departure from Tahiti with reluctance, despite the mounting tensions of recent weeks, for the island was the closest thing imaginable to a paradise on earth and months at sea in close quarters were not an inviting prospect by comparison. Tahiti was a sailor's paradise, with its beautiful and willing women, abundant fresh food, and balmy weather, and the slower pace of life in port gave the *Endeavour*'s crew ample time to enjoy the island's attractions.

In the years after Cook's expedition, descriptions of Tahiti published in Europe would elicit condemnation of the islanders' uninhibited sexual behavior, but the sailors aboard the *Endeavour* were not troubled by such scruples, and Cook and Banks, approaching Tahitian culture in the spirit of scientific observation, were able to view the islanders as people altogether different from themselves and therefore not necessarily subject to the same moral standards. Cook, for example, after witnessing a public entertainment that included a ritual act of sexual intercourse, wrote dispassionately that "it appear'd to be done more from Custom than Lewdness, for there were several women present particularly Obarea and several others of the better sort and these were so far from shewing the least disaprobation that they instructed the girl how she should act her part, who young as she was, did not seem to want it." Only Parkinson condemned the practice of free love at Tahiti, and even he blamed the Englishmen rather than the islanders, lamenting that his shipmates behaved "as if a change of place altered the moral turpitude of fornication."

But for Europeans in general, a society so devoted to pleasure, where work was scarcely necessary, raised profound moral questions. In some ways the fact that the Tahitians could devote most of their time to play and still live comfortably was even more disturbing than their sexual license. As Banks explained, a Tahitian who planted ten breadfruit trees, a task requiring about an hour's labor, did as much for his family's food supply as an English farmer who labored the year round planting and harvesting crops. Houses were little more than shelters open to the air, devoid of partitions or furniture. Cloth draped around the body in various ways sufficed for clothing; tools were made of bone or shell; eating utensils were nonexistent. Cook recognized that the Tahitians' modest possessions were sufficient for their needs and well suited to their environment; he was especially impressed by the way in which Tahitians had adapted canoes, the simplest of all boats, into vessels that would serve every need, from offshore fishing to journeys to islands hundreds of miles away. Banks, however, less capable of shedding European preconceptions than Cook, could not resist pointing out how much more the Tahitians might have accomplished with their natural wealth had they been inclined to work beyond what was necessary for subsistence. "The great facility with which these people have always procured the necessaries of life," he thought, "may very reasonably be thought to have originally sunk them into a kind of indolence which has as it were benumbnd their inventions, and prevented their producing such a variety of Arts as might reasonably be expected from the aproaches they have made in their manners to the politeness of the Europeans."

Again Banks was judging Tahitians by European standards. For all his skill as an ethnographer and his empathy with the Tahitians, he failed to understand that they had no concept of "progress" and saw no reason to change. Unlike Europeans, they made no connection between work and the accumulation of material goods, or between wealth and moral worth. As Parkinson put it, "They seem . . . as contented with what is spontaneously produced, as if they had attained to the ne plus ultra, and are therefore happier than Europeans in general are, whose desires are unbounded." Such a statement ignored the less attractive side of Tahitian life (largely hidden from the English anyway), such as Purea's and Tuteha's greed for European goods and attention, which had created jealousy among the other chiefs on the island and contributed to Purea's downfall. Indeed, in two visits to Tahiti, the English had already instilled in the islanders a desire for exotic goods that would contradict Parkinson's statement about their contentment and upset the equilibrium of their self-contained society. For the moment, however, and for years into the future, Tahiti was for Europeans a living example of a society neither savage nor "civilized," whose people lived happily and comfortably, content with what they had, free of the poverty, social conflict, and crime that plagued European nations.

It was no wonder that Tahiti looked especially attractive to English sailors, who were generally drawn from the lowest ranks of society; and just days before the *Endeavour*'s planned departure, two men deserted. One was known to be strongly attached to a woman on the island, and the other, in Banks's estimation at least, was a "wild young man" drawn into deserting by the prospect of an easy life in a tropical paradise. According to midshipman James Magra, the two deserters were merely the remnants of what started out to be a much more serious mutiny. Several sailors and two or three officers were prepared to stay at Tahiti, and only the virtual certainly of contracting venereal disease, without hope of medical attention, deterred the others.

Although Cook suspected nothing about this plot, he did fear the effect of desertion on ship discipline and decided to get the men back at all costs. Realizing they had left with the cooperation of the islanders, he seized Purea and three other chiefs as hostages for the sailors' return. This tactic had the desired effect, as the chiefs took Cook where they had gone, and Cook sent one of his lieutenants, Zachary Hicks, with a group of Tahitians as guides to capture them. When the search party reached the deserters' hiding place, their Tahitian companions seized one of the group, apparently in retaliation for Cook's kidnapping of their chiefs. Hicks chased them, warning the Tahitians not to harm their captive on pain of retaliation against their chiefs. This strategy succeeded, and the deserters were finally returned to the ship.

Cook regretted the entire incident. He had not wanted to seize the native chiefs, but thought the deserting marines would not have been captured without extreme measures. Unfortunately, he concluded, "we are likly to leave these people in disgust with our behaviour towards them, owing wholy to the folly of two of our own people."

Meanwhile, Tupaia had decided that he wanted to go to England, much to Banks's delight. The young man would be a great help in their further travels in the South Pacific, and with Tupaia along he could continue his study of Tahitian culture. Cook was less pleased, not wanting the responsibility of taking a Tahitian from his native land with no certain prospect of sending him back. Banks interceded, offering to pay the young man's expenses and take care of him after they returned to England. "I do not know why I may not keep him as a curiosity, as well as some of my neighbours do lions and tygers at a larger expence than he will probably ever put me to," he remarked. "The amusement I shall have in his future conversation and the benefit he will be of to this ship . . . will I think fully repay me." Thus he characterized one of the most respected priests of Tahiti.

Early on the morning that Cook planned to sail, Purea and several of her friends came to Matavai Bay to wish the English an emotional farewell. Banks, to the end still trying to convert his favorite Tahitians into people of European sensibilities, told Purea and his other favorites that the "clamourous weeping" of some of their compatriots appeared to be more "affected than real greif."

On July 13, three months after anchoring at Matavai Bay, the Englishmen left Tahiti, and, with Tupaia as their guide, turned toward the other islands of the South Pacific.

7. *In Search of the Southern Continent*

Before taking up his secret charge to search for a southern continent, Cook wanted to pursue the Tahitians' reports of other islands some distance west, both to help fill in the map of the South Pacific and to add to the *Endeavour*'s store of provisions for the long journey south. Tupaia dictated the names of seventy-four islands a few days' sail from Tahiti, including some as far away as the Marquesas, Tonga, Samoa, and Fiji, and drew a rough map for Cook showing their approximate locations. Cook decided to seek four, which according to Tupaia lay only a day or two's sail from Tahiti: Huahine, Raiatea, Tahaa, and Bora Bora.

Tupaia's skill as a navigator proved as good as he had promised. A day after leaving Tahiti, the Englishmen anchored at Huahine, the largest of the four islands. Cook and the island's chief, an elderly man named Ori, took to each other at once, exchanging names as a sign of friendship. The people of Huahine were obviously culturally related to the Tahitians, although Cook thought they were lighter in color and noted with approval that "they are not addicted to stealing." He did not stay long enough to test this observation, however, but after two days sailed on to Raiatea, Tahaa, Bora Bora, and the tiny islets of Tubuai and Maupiti.

Cook was especially curious about Bora Bora, because the people of both Huahine and Raiatea considered the Bora Borans enemies. The residents of Huahine, in fact, were most interested in the English as potential allies against their neighboring islanders, who, they claimed, invaded Huahine every month or so, stealing what they pleased and killing anyone who opposed them. Many years earlier, Tupaia explained, the chiefs of Tahiti and the neighboring islands had banished criminals to Bora Bora, until there were so many that the island could no longer support them. Then the Bora Borans became pirates, attacking and stealing from neighboring islands. Reefs and contrary winds made landing there impossible, however, and after claiming for England all the islands he had discovered—naming them the Society Islands because of the friendly reception the English had enjoyed there—Cook continued south, into the region where Wallis believed he would have found terra australis, given enough time to look for it.

By now Cook doubted that he would find a continent in temperate latitudes. Like earlier explorers, he was puzzled at encountering variable winds in an area known to be dominated by easterly trade winds, but unlike his predecessors, he did not assume that a large land mass nearby was the obvious explanation. Strong ocean swells from the southwest argued against the continent theory, he believed, and the variable winds might be explained by their presence close to the fringes of the trade wind belt. Only one bit of evidence from earlier voyages held out even a remote possibility that a southern continent might exist, and that was an isolated and not very well documented discovery made by a Dutch explorer, Abel Tasman, more than a century earlier.

Tasman had sailed from Amsterdam to the Pacific in 1642 by way of the Cape of Good Hope with instructions to find the "remaining unknown part of the terrestrial globe." He sailed south of New Holland, continuing farther east than any explorer before him and discovering a large island off the southeast coast that he named Van Dieman's Land after the governor-general of the Dutch East Indies (today known as Tasmania); then he proceeded north until contrary winds forced him to turn east. Hundreds of miles later, in a part of the Pacific never before explored by Europeans, Tasman approached a mountainous and forbidding coastline. Thinking perhaps he had found the long-sought-after southern continent, he anchored in a sheltered bay with the intention of exploring further, but when he sent out a scouting party in one of the ship's boats, hostile inhabitants attacked and killed four men with heavy, blunt clubs. Three others escaped by jumping overboard and swimming back to the ship. Given this inauspicious beginning, Tasman decided to move on, naming his landfall Murderers' Bay. Soon he approached a long, narrow cape marking the northern limit of the coastline, but bad weather prevented him from rounding the peninsula to determine the extent of the land; instead he headed northwest, discovering several of the Tongan and Fiji Islands before ending his voyage at Batavia.

Tasman thought he had discovered the western edge of a continent extending east to Tierra del Fuego, but the year after he completed his voyage another Dutch explorer destroyed that theory by proving conclusively that Tierra del Fuego was surrounded by water. (Drake had established the same point sixty years earlier, but his discovery of Cape Horn had been conveniently ignored.) Dutch officials, never much interested in exploration unless it had immediate commercial benefit, concluded that they were better off concentrating on their possessions in the East Indies, and Tasman's discovery, named Nieuw Zeeland by the Dutch government, remained a tantalizing but uncertain bit of evidence for the possible existence of a southern continent.

Cook had a copy of the English translation of Tasman's journal along with Alexander Dalrymple's volume of Pacific voyages, which discussed Tasman's voyage at some length. He thought it possible that New Zealand might be part of a continent; and then, too, there was the unresolved mystery of New Holland—Dutch explorers had mapped much of its northern, western, and southern coasts, but no one knew how far east it extended or whether it was a single land mass or a group of islands. Conceivably, New Zealand and New Holland might be connected, and both might be part of a larger continent. In any case, that part of the Pacific seemed to offer the most fruitful possibilities for further exploration.

The next several weeks brought long, monotonous days of open sea. The naturalists used the time to advantage, spending hours around the table in the ship's large cabin—Parkinson drawing specimens, Solander filling notebooks describing them, and Banks "journalizing." On August 25, 1769, they celebrated one year at sea by breaking out a Cheshire cheese and a cask of porter. For a day, Banks noted, "we livd like English men and drank the healths of our freinds in England." A comet enlivened the night sky through much of September, but otherwise there were few diversions until the first week in October, when Nicholas Young, one of the youngest sailors on board, spotted land from the masthead. As the ship drew closer, it became obvious that his discovery was not simply another tiny Pacific island, but something much more substantial; a wide, curving bay stretched out ahead, bounded on the left by high cliffs extending well out into the ocean—named by Cook and still called today Young Nick's Head—and on the right by a river and steep hills. Mountains were visible some distance inland. Since mountains and rivers were commonly believed to be characteristic features of large land masses, Banks concluded confidently that this must be the "Continent we are in search of." Cook, more cautious, reserved judgment until he could explore further.

As Cook suspected, they had reached New Zealand. Their landfall was along the east coast of the North Island, near the present-day town of Gisborne.

A few small canoes paddling across the bay were evidence that the region was inhabited, but as soon as Cook had boats hoisted out and readied for a shore excursion, the canoes disappeared. Cook had one of the boats hauled up close to the mouth of the river while he, Banks, and Solander landed to have a look around, but they hadn't gone far when they heard shots. Racing back to the water's edge, they learned that one of the crew left behind had shot and killed a New Zealander who had attempted to attack the boat, or so his companions claimed. At this inauspicious beginning, Cook ordered everyone back to the ship.

When the New Zealanders first saw the *Endeavour* on the horizon, they thought it was an enormous bird, and they talked with wonder about its huge, beautiful wings. As it came closer, they saw a smaller bird without wings descend to the water, and "a number of party-coloured beings, but apparently in the human shape" descended into it. The amazed New Zealanders thought the figures in human shape were gods capable of unleashing thunderbolts at will. Some were frightened and ran away when Cook first landed, but at least fifty men, armed with pikes and clubs of polished stone, were gathered at the riverbank when Cook returned the next morning, accompanied by a party of armed marines.

As Cook disembarked with Banks, Solander, and Tupaia, the New Zealanders, arrayed on the opposite side of the river, waved their weapons in the air and broke into a violent dance, sticking out their tongues, rolling their eyes back into their heads until only the whites showed, and singing in harsh, guttural tones. Assuming this display was a war dance, Cook fired a musket over their heads and retreated to await the marines' landing. When Cook approached the New Zealanders again, Tupaia called out to them and, to the Englishmen's amazement, got a response in a language that sounded much like Tahitian. The men continued their conversation in shouts across the river, as the New Zealanders asked Tupaia where the strangers had come from and berated them for the previous day's killing. Tupaia tried to convince them that the Englishmen intended no harm and eventually the two sides agreed to exchange food for iron, but still the New Zealanders refused to put down their weapons or venture across the river. Not until Cook handed his gun to a companion and walked to the water's edge alone would any of them budge. Then one man swam across, hestitating for a time on a rock in midstream; Cook offered him small gifts, and soon another crossed, although Monkhouse noticed that he carried his weapon with him, concealing it under the water. Now "the ice was broken," in Monkhouse's phrase, and the rest of the group followed. After Cook distributed more presents, they danced again, just as violently, leading Monkhouse to conclude that this was not necessarily a war dance, but rather an expression of strong passion of whatever kind. His judgment was close to the mark, as the New Zealanders' vigorous dances, *haka* in their language, could be a form of greeting as well as a prelude to war.

Gifts of small trinkets did not satisfy the men for long, however. "Active and alert to the highest degree, . . . they were incessantly upon the catch at every thing they saw," Monkhouse noted; "every moment jumping from one foot to the other; and their eyes and hands as quick as those of the most accomplished pickpocket." When Monkhouse tried to barter for a canoe paddle, one of the men readily agreed to trade his for a musket, and reproached Monkhouse when he refused. Failing barter, some attempted to steal what they wanted, and the English, overly sensitive on the matter of theft after their experience in Tahiti, threatened to kill anyone who stole weapons. One man, undaunted, grabbed a knife that Green wore suspended from his belt and "set up a cry of exultation and waving it round his head retreated." Annoyed at so blatant an act of defiance, Cook ordered his men to fire, and Monkhouse felled the thief with a single shot. Others wounded three more. The toll was now two New Zealanders dead—Cook learned later that the man Monkhouse had shot died of his wounds—with the Englishmen no closer to establishing friendly contact.

Later in the day two canoes approached the *Endeavour*. Curious to learn more about these people and convinced by now that his usual methods of establishing trade and friendly relations would not work here, Cook decided to capture one of them and take its occupants on board, assuming that because the men were unarmed, they would surrender without resistance. He hoped that after talking with the men and treating them well, they would spread the word that the English meant no harm. This proved to be a serious miscalculation, which only emphasized how little Cook understood about the nature of the people he was dealing with, despite their outward similarities to the Tahitians.

Tupaia called to the men, inviting them on board, but instead they paddled furiously away from the ship. Next Cook ordered a gun fired over their heads, thinking that they would either surrender or jump overboard, in which case they could be easily captured. Again he was wrong. The men responded by throwing stones, and even their canoe paddles, at the ship. Such behavior induced Cook to fire directly at the canoes, killing three or four men immediately and wounding at least one more. Three who were unhurt jumped into the water, and the *Endeavour*'s crew took them on board.

The captives were boys ranging in age from about ten to eighteen, two of them brothers. Terrified at first, they cheered up as it became apparent that the English did not intend to harm them. The ship, the men, and especially the food fascinated the boys, and they quite willingly shared the crew's dinner and slept on board that night, to the amazement of the English, who wondered at how quickly they seemed to recover from the deaths of their companions.

Cook and Banks, however, did not spend so easy a night, for they were deeply disturbed at the death toll from their first forty-eight hours on the New Zealand coast: between four and six men, depending on who was doing the counting, were now dead, with little progress toward establishing cordial relations. Cook believed that most "humane men" would censure him for firing on the canoes, and that his reason for approaching the canoes in the first place could not justify his conduct. Had he thought the men would resist, Cook wrote, he would never have gone near them, "but as they did I was not to stand still and suffer either my self or those that were with me to be knocked on the head." Banks could write only that it had been the most "disagreeable" day of his life. "Black be the mark for it," he concluded, "and heaven send that such may never return to embitter future reflection."

The next morning Cook planned to take the boys ashore near river where he had landed the day before, but the three resisted the prospect strenuously, insisting that the people living there were enemies who would kill and eat them. Cook didn't take their protests seriously, preferring to view this talk of cannibalism as youthful imagination, but he agreed to deliver them to a spot they pointed out on the opposite side of the bay instead. Convinced by now that further attempts to make contact with the New Zealanders were pointless, Cook decided to move on. He called their first landfall Poverty Bay, "because it afforded us no one thing we wanted." Today a productive agricultural area, it is still known by Cook's name.

Shortly after the *Endeavour* got underway, several canoes paddled out from shore, and the men in them signaled their desire to come on board. In the end, it turned out, Cook's strategy had worked. The three boys, recounting their stories of good treatment from the Englishmen, persuaded some of their listeners to satisfy their curiosity about the strangers. Once on board, the New Zealanders were eager to trade, especially for Tahitian cloth. They offered their clothes, their clubs, and their canoe paddles, and even a canoe in return. Three men who stayed overnight on board urged yet more friends to board the *Endeavour* the next morning, reassuring those who hesitated that the English did not eat men—the second indication that the New Zealanders might be cannibals and a clue about why they had been so aggressive at first.

The New Zealanders' initial hostility stood in sharp contrast to the Tahitians' exuberant friendliness, but in other respects the two groups were remarkably similar, especially in light of the distance that separated them. The New Zealanders' language was not precisely the same as Tahitian, but close enough that Tupaia could understand them and make himself understood without difficulty. Nearly all the New Zealand men sported tattoos, if in different places and different patterns than the Tahitians. They oiled their hair in the same manner as the Tahitians, although most tied it into topknots adorned with feathers and small combs instead of letting it flow loosely, and they placed great value on

Tahitian cloth despite the fact that their own fabric was quite different. The New Zealanders were skillful boatmen and swimmers, and their canoes, though small, were equally well constructed and decorated with intricate carving in patterns not unlike the tattoos on their faces. The boats' prows were adorned with the carved images of men's heads, tongues protruding, as if in imitation of their war dance.

These similarities between the New Zealanders and the Tahitians mystified the English. The only logical explanation was a migration from Tahiti to New Zealand at some point in the distant past, but the thought of making such a long ocean voyage in canoes, even the Tahitians' huge sailing canoes, seemed preposterous. And yet that is exactly what had happened. Sometime between the end of the eighth century and the beginning of the eleventh, a group of men and women from central Polynesia, most likely either the Society Islands or the Marquesas, made their way to New Zealand on a planned colonization expedition. They settled originally on the North Island, and by the time of Cook's visit had grown to a population of 100,000 to 150,000, divided among about forty tribes. According to native tradition, a second and larger group of settlers came to New Zealand around the middle of the fourteenth century, and while this is possible, it is equally likely that the story refers to a major population shift within New Zealand itself.

The migrants attempted to transplant their culture, but differences in climate required some modifications. New Zealanders practiced more elaborate agriculture and built sturdier houses than the people of central Polynesia. They also learned new techniques of cloth manufacture, because the trees from which Tahitian cloth was made—the migrants had brought some in their canoes—would not grow in New Zealand's cooler climate. But historical memory was so strong among these people that they knew about the ancient art of making cloth from bark, and consequently, as Cook discovered, Tahitian fabric was the most valuable item the Englishmen could offer in trade.

The New Zealanders' behavior intrigued the men of the *Endeavour* even more than their appearance. Although they all knew the stories about the Tahitians' resistance to the men of the *Dolphin* (and several of the *Endeavour*'s crew had firsthand experience of that encounter), for most of this group, Poverty Bay was their first experience with native hostility toward Europeans. For the New Zealanders, of course, it was their first encounter of any sort with Europeans, which accounted for much of the difference between the receptions Cook met at New Zealand and at Tahiti. The Tahitians had learned something about Europeans: they would not offer unprovoked violence, they had useful products to give if provided with food and women, and, most important, they would in due time go away. As a consequence, the Tahitians were prepared to humor the Englishmen and use their visit to best advantage. The New Zealanders knew none of this. Even though Tupaia spoke their language and assured them of the Englishmen's good intentions, he was a stranger and New Zealanders did not trust strangers; they lived in small tribes and frequently warred with their neighbors, a fact the English did not yet understand.

The English reacted to the New Zealanders' hostility with mixed emotions. Some viewed them as insolent savages who had to be taught European superiority through a judicious display of firepower. Cook and Banks were inclined to be more sympathetic and regretted the necessity to display force, although they believed that under the circumstances, force was necessary. But the New Zealanders did not react to the *Endeavour*'s guns as the Tahitians had reacted to the *Dolphin*'s; instead of recognizing the superiority of the English weapons and surrendering, they continued to resist. Some of the *Endeavour*'s crew took this as a sign of extreme insolence. Banks reacted with despair, convinced that they would never make friends with men who were not frightened of their guns, while others displayed a certain grudging admiration. Pickersgill thought that "this . . . courage . . . is greatly to be admir'd for it has allways been remark'd amongst Savages lett them be ever so much us'd to fire arms that as soon as they see a man or two fall that they immeddeately fall in to disorder and give way yet these People was so far from shewing any kind of fear that when they saw the man fall they immediately had ye Presence of mind to attempt it a second time." Not until the New Zealanders had firsthand testimony of friendly behavior from the English, in the accounts of the three captured boys, would they put down their arms and venture to the ship.

From Poverty Bay Cook headed south, staying close to the coast in search of a suitable place to anchor and replenish provisions. The rugged coastline offered no shelter, however, and the boatloads of New Zealanders who paddled out to satisfy their curiosity about the ship were just as hostile as the inhabitants of Poverty Bay, although none attempted a direct attack on the *Endeavour*. Rather, they seemed to be trying to establish their authority and scare the intruders away by making threats while prudently avoiding open hostilities. The English, sobered by their experiences at Poverty Bay, confined their responses to grapeshot fired over their adversaries' heads.

In one disturbing incident reminiscent of the Englishmen's capture of the three boys at Poverty Bay, a group of New Zealanders, having been persuaded to come close enough to the ship to trade, seized Tupaia's servant boy Tayeto when he leaned too far over the side and paddled away quickly amid shots from the *Endeavour*. In the ensuing confusion Tayeto jumped overboard and swam back to the ship, frightened but unharmed. Cook named a nearby peninsula Cape Kidnappers to commemorate the incident.

When several days went by with no sign of a decent harbor, Cook decided to reverse direction, at a point that he named Cape Turnagain. (Cook's names for the places he discovered seemed to change with his mood; he variously adopted native names, honored patrons at home or members of his crew, or took inspiration from physical features or events associated with particular places. Along the coast of New Zealand, he was clearly in one of his periods of literal-mindedness.) Heading north, the Englishmen discovered that their reputation had spread and that the New Zealanders were now quite willing to trade and converse with them. One afternoon five men actually ventured on board, ate dinner with the officers, and stayed overnight. Another group guided them to a small harbor just north of Poverty Bay where, they said, the Englishmen would find fresh water.

It was a beautiful little bay, a shallow crescent with steep, jagged hills rising sharply from a narrow strip of level, fertile land. The New Zealanders called the bay Anaura, but through a misunderstanding Cook thought its name was Tegadoo, and he so labeled it on his charts. The people who lived there had heard about the Englishmen from their acquaintances at Poverty Bay, and the men of the *Endeavour* found a cordial reception. Finally they were able to get ashore and explore the country at leisure, a task made much easier by the presence of Tupaia, who continued to act as interpreter and proved popular with the New Zealanders.

The Englishmen approved of what they saw at Anaura. Unlike

the Tahitians, these men and women cultivated their land with obvious care and skill, clearing it of all weeds, planting yams in neat hills, and fencing their fields with closely spaced reeds. Houses were neatly constructed, with walls made of thick reeds and roofs of grass covered with bark. The interiors were clean and well kept. Banks was especially impressed by the presence of outhouses—one for every three or four houses—a convenience he had not yet encountered in the Pacific.

The residents at Anaura, like those at Poverty Bay, were tattooed in intricate, spiral patterns, and with more opportunity for observation the Englishmen began to detect some relationship between the extent of tattooing and social status. Chiefs had the most elaborate designs, while very young men had little tattooing and women usually none, except for a few with designs on their lips. Most women instead painted their faces with a thick red paint that never completely dried. The English thought them rather unattractive even without the paint, and worse with it. Appearances were not a great deterrent to the sailors, however, judging from the number of men who sported some red paint of their own after a day ashore. Banks thought the young women were "as great coquetts as any Europeans could be," and one seaman observed that many of them went regularly to the spot where the crew collected water, granting their favors "on very reasonable conditions" to anyone who asked.

Although the people of Anaura Bay displayed some curiosity about their visitors and readily engaged in trade, for the most part they kept to themselves, going about their daily business as if nothing unusual were going on. "Such fair appearances made Dr. Solander and myself almost trust them," Banks commented; and the two scientists took advantage of the New Zealanders' trust by visiting their homes, looking at everything, and asking questions freely of their hosts, who were not at all reluctant to show them whatever they wanted to see.

As productive as this stay was in some ways—Banks and Solander eagerly made up for lost time in collecting specimens of plants and birds—the supply of water was not all that Cook had hoped for, and after two days he decided to move on. The local residents told him about another bay a bit farther south, where water was more plentiful, and the *Endeavour* anchored there the following day. Again Cook chose the native name for his chart and again he was wrong, probably because he misinterpreted a word descriptive of some natural feature for the name of the entire area; but the name he fixed upon, Tolaga Bay, has stuck.

The stay at Tolaga Bay afforded time for a closer look at certain elements of native life. Parkinson showed an artist's appreciation of New Zealand carving, which ornamented boats, houses, even canoe paddles and walking sticks. The residents of Tolaga Bay possessed one of the largest and most elaborately carved canoes the Englishmen had yet seen; Banks measured it at sixty-eight and a half feet long, five feet broad, and three and a half feet high. The carvings were all built upon a spiral pattern, Parkinson observed, but with infinite variations and executed as precisely "as if done from mathematical draughts," despite the fact that the New Zealanders' only carving tools were chisels and axes made of stone. (There are, in fact, about fifty different spiral forms used in Maori carving.) The designs were "wild and extravagant," showing no imitation of nature—"unless," he added, "the head, and the heart-shaped tongue hanging out of the mouth of it, may be called natural." On this point Parkinson's impressions were accurate. The New Zealanders' art was almost entirely symbolic, with stylized images of men, birds, and fish entwined with abstract forms. Such abstraction seemed bizarre to Parkinson, reared in a culture that valued representational art and himself a master at meticulously accurate drawing.

Tupaia occupied much of his time talking with the local priests, and concluded that they agreed with the Tahitians on most religious matters. The hints he picked up about the New Zealanders' cannibalism, however, disturbed him deeply, and he asked his informants at Tolaga Bay whether in fact they ate human flesh. The answer was yes, although only enemies captured in war.

After a week at Tolaga Bay, the *Endeavour* continued a northeasterly course, rounding a long, wide peninsula known today as the East Cape. From there the coastline trended northwest, and the rugged, mountainous terrain of the cape gave way to a wide, crescent-shaped bay with sandy beaches and fertile, well-cultivated land stretching miles inland. The next few days revealed it to be a heavily populated area, abundantly supplied with food—Cook would later name it the Bay of Plenty—but the residents were as mistrustful as any they had encountered at Poverty Bay, and sporadic attempts at trade usually ended with New Zealanders refusing to give anything in return and the English shooting at them.

Sailing along this stretch of coastline over the next two or three days, Cook and Banks were struck by the number of villages perched on tops of hills and surrounded by ditches and wooden palisades. Tupaia, detecting a certain resemblance to the *marae* of his home island, thought these were the New Zealanders' places of worship, but Cook and Banks more accurately concluded that they must be intended for defense, and Banks took them as evidence that the residents must be "much given to war."

Impressed with signs of a large population and material abundance, Banks also believed that some of the New Zealanders' "princes" might live in this region. He was, in general, inclined to impose European terms and views about government on the new cultures he encountered, as was obvious in his analysis of Tahitian government; in New Zealand, his attempts to question the inhabitants about their government had so far elicited the name of just one chief, "Teratu." From these conversations Banks assumed that the entire area they had seen fell under the jurisdiction of a single man, and he theorized that the Bay of Plenty might be his seat of government. Unfortunately for Banks's theory, there was no such man as Teratu, and the New Zealanders, far from being united in a large kingdom, lived in small, highly localized tribes. He was correct, however, in describing the Bay of Plenty as one of the richest and most populous parts of New Zealand and in judging the New Zealanders to be a warlike people. Soon they would make friends with a group of men who explained much of this to them.

Cook knew that a transit of Mercury was scheduled to occur on November 9, and he wanted to find a suitable place to observe the phenomenon, since a good set of observations would allow him to fix the longitude of New Zealand accurately. He found such a place at the beginning of November, guiding the *Endeavour* to anchor in a narrow, sheltered harbor just beyond the Bay of Plenty. Cook would later name it Mercury Bay.

When canoes carrying more than a hundred armed men approached the ship, Cook feared an attack, but the men wanted only to talk and get a good look at the ship; they paddled around and around for nearly three hours. Years later an elderly chief, who had been a child at the time of the *Endeavour*'s visit, recalled that his compatriots had at first thought the English were goblins and their ship a god. It was the English who resorted to weapons, when some of the New Zealanders offered to trade and then refused to pay for what they were given.

Despite this inauspicious beginning, the next day the Englishmen and New Zealanders established a friendly trade. Two men boarded the *Endeavour* and explained that other New Zealanders occasionally attacked them, stealing their possessions and capturing their wives and children. This was the reason they had been so suspicious of the English at first, and it also explained the fortified villages along the coast. Some days later Cook and Banks visited two of the nearby villages, called *pa* by the New Zealanders. (Banks called them heppahs, or eppahs, confusing articles and nouns as he had with the Tahitian language.) One had only five or six houses, built on a small rock that was completely inaccessible at high tide. At other times, it could be reached only by a narrow, steep path. The second *pa* was much larger, surrounded by a double row of pilings and two ditches. More fences, with narrow passages and gates inside the fort, made it possible to defend just part of it if necessary, and the whole was accessible only by a narrow passage up the side of a steep hill.

The New Zealanders did not live in these villages normally, they explained, but retreated to them in the event of attack. A supply of fern roots and dried fish ensured their sustenance during a siege. Such care given to fortification must mean that the New Zealanders frequently warred with each other, Cook thought; "otherwise they never would have invented such strong holds as these, the errecting of which must cost them immence labour considering the tools they have to work with which are only made of wood & stone." Their informants also asserted, as others had earlier, that they ate the bodies of enemies killed in war. Cook was by now persuaded of the New Zealanders' cannibalism, but Banks remained doubtful, "loth . . . to believe that any human beings could have among them so brutal a custom."

The Englishmen remained at Mercury Bay almost two weeks without incident, until the day of the transit. While Cook, Green, and Hicks were occupied with their observations and Banks went off searching for plants, a new group of New Zealanders came to the ship to trade. When one man accepted a piece of Tahitian cloth but refused to hand over the fabric he had offered in return, Gore, in charge of the ship in Cook's absence, became so infuriated that he shot and killed the man. Though there had been much shooting over the past several days, this was the first fatality since Poverty Bay, and it served as a reminder that Cook's attitudes toward native peoples were a good deal more tolerant than those of his men and that his control was the only thing preventing more frequent violence. The New Zealanders themselves perceived a difference between Cook and the others; an elderly chief later said of Cook, "We knew that he was lord of the whole, by his perfect gentlemanly and noble demeanour. He seldom spoke, but some of the goblins spoke much. But this man did not utter many words; all that he did was to handle our mats and hold our *mere,* spears, and *wahaika* [clubs] and touch the hair of our heads. He was a very good man, and came to us—the children—and patted our cheeks, and gently touched our heads."

Although Cook was troubled by the incident, he decided not to reprimand Gore, saying only that he thought the punishment "a little too severe for the Crime, . . . we had now been long enough acquainted with these People to know how to chastise trifling faults like this without taking away their lives." His lenience on this occasion was an indication of his own ambivalence toward the aggressive New Zealanders. While Cook was by nature respectful of native peoples in a way that most Europeans were not, he also demanded in return a level of respect that was not always forthcoming. He interpreted theft, especially repeated theft, as a challenge to his authority, and indeed it often was. The Tahitians' ingratiating behavior made them appear to adopt the proper degree of deference toward European superiority, and the length of the Englishmen's stay in Tahiti allowed Cook to get to know them well enough to view them as individual personalities. But the New Zealanders were distant, haughty, and often violent; and as a result Cook's anger at their behavior often came into conflict with his principles.

The *Endeavour* left Mercury Bay on November 15, stopping briefly to explore a huge bay on its journey north. Although he did not take time to inspect every cove and inlet, Cook, with his unerring instinct about the natural features of sea and coastlines, judged that the western side of the bay would offer good harbors. The area he described is today the location of Auckland, New Zealand's largest city. Six weeks later the expedition reached the end of the long, narrow peninsula that marks the northernmost limit of New Zealand. Cook then turned south, intending to sail along the west coast, but strong winds drove the ship some distance offshore for several days, and Christmas found the men of the *Endeavour* in the open sea again, out of sight of land. Shipboard routine came to a halt on the holiday while the crew celebrated with a goose pie and liberal allowance of grog. By evening, Banks reported, "all hands were as Drunk as our forefathers used to be upon the like occasion."

At the end of December the *Endeavour* was once again back within sight of land, and for the next three weeks the Englishmen sailed south along a mountainous, forbidding coast until finally they reached a small, sheltered cove and anchored the ship for the first time in nearly a month. At this point they were on the northern tip of New Zealand's South Island, although they did not realize it yet because the passage between the two islands was not fully visible. Comparing his reading of the latitude and longitude with those recorded in Tasman's Journal, Cook concluded that the cove was quite close to Tasman's "Murderers' Bay." (In fact, "Murderers' Bay," now called Golden Bay, was about seventy miles from the *Endeavor*'s anchorage.) Cook decided to spend several days there, since the ship needed repairs and a fresh supply of wood and water, and he wanted to explore the area in some detail to see how it was connected with the land they had already seen. The stop would also give Green the opportunity to make detailed astronomical observations and establish an accurate longitude. Together with the readings taken at Poverty and Mercury Bays, these observations would supplement the shipboard charting techniques Cook had been using to create a detailed map of New Zealand.

Shortly after the *Endeavour* anchored, a group of men gathered on a small, steep island just outside the cove while several more approached the ship in four canoes. At first they appeared threatening, but then one elderly man went on board peaceably enough, and the next day about a hundred men and women visited the *Endeavour*—the presence of women, the English had found, was generally a sign of peaceful intentions—offered fish for trade, and talked to Tupaia about their customs and the legends of their ancestors.

These men and women were quite different in appearance from those the English had encountered on the North Island—shorter, wearing simpler clothing, and seldom tattooed; nor did they ornament their faces with paint or tie up their hair in neat topknots as the northern tribes did. Their small size and lack of ornamentation led the English to believe they must be much poorer than the North Islanders, a belief confirmed in Cook's mind later by the absence of the carefully cultivated gardens so prevalent farther north. Parkinson thought they lacked the "spirit and sprightliness" of their earlier acquaintances, while Pickersgill labeled them "the Poorest and most mizerable sett we saw on New Zealand." They understood the value of iron, however, unlike the men of the North Island, and preferred English cloth to Tahitian, which Cook thought "shew'd them to be a more sensible people than Many of their Neighbours."

In fact the South Island tribes were smaller in number and more widely scattered than those farther north. Their physical isolation insulated them from the cultural changes that had evolved among the northern tribes over the centuries, which helps account for the contrasts the Englishmen observed on the two islands. At the time of the *Endeavour*'s visit, the population of the South Island amounted to less than 20 percent of New Zealand's total, most of it concentrated in the northern coastal region. Forced to cope with a harsher climate, the people of the South Island seemed less vigorous and less interesting to the English; but Cook and his companions learned a good deal from them nonetheless, living in close proximity on friendly terms over a period of about three weeks, the longest time the English spent in any one place in New Zealand.

The day after anchoring, Cook and Banks took a boat to another cove nearby, where they were surprised to see the dead body of a woman floating in the water. On the beach they met a family gathered around a fire preparing dinner and noticed human bones, with shreds of meat still clinging to them, scattered on the ground nearby. Here, finally, was clear proof of cannibalism that even Banks could not deny; but Cook, unwilling to draw conclusions without further proof, asked the group if they were eating roast dog. In reply one man "with great fervency took hold of his fore-arm and told us again that it was that bone and to convence us that they had eat the flesh he took hold of the flesh of his own arm with his teeth and made shew of eating." He went on to explain what Cook and Banks had heard before, that the New Zealanders ate only their enemies; a few days earlier he and his companions had captured a boatload of people from a hostile tribe and were now enjoying the remnants of their feast.

Over the next several days the evidence of cannibalism mounted. Realizing how much it shocked and fascinated the Englishmen, the New Zealanders visited the *Endeavour* with half-eaten human bones and demonstrated how they killed their enemies by knocking them on the head with a small, heavy club called a *patu* and then tearing their bodies open with darts or spears. "The natives seemed to take pride in their cruelty," Parkinson observed, "as if it was the most laudable virtue, instead of one of the worst of moral vices."

The horrified sailors labeled the New Zealanders "Barbourous" and "Hatefull," but Cook and Banks approached cannibalism as one more phenomenon to be observed among strange cultures. Despite his earlier reluctance to believe that anyone could be capable of eating his fellow man, Banks now claimed to be "well pleasd at having so strong a proof of a custom which human nature holds in too great abhorrence to give easy credit to." Within a few days even the sailors let curiosity get the better of horror, as preserved human heads became the hottest new item of trade.

During the expedition's three-week stay in this cove, Banks had his first significant opportunity since Tahiti to exercise his charm on the local inhabitants, who became so friendly that they greeted him with "numberless huggs and kisses." His new friends took him to a nearby *pa,* where he toured all the houses, and he became conversant enough with one old man to ask him if his tribe had ever seen Europeans before or had any tradition of Tasman's visit. (Neither Banks's informant nor his father or grandfather had ever seen ships as large as the *Endeavour,* but according to tradition, two large vessels had visited long ago, and the local tribe had killed all the men in them.) Cook, typically, was more interested in exploring the natural features of the area and particularly in determining if there was a passage eastward into the open sea. Twice he climbed steep hills in an attempt to learn the answer, but the little cove where the *Endeavour* lay at anchor was part of a large sound dotted with islands and harbors, which made it impossible to see clearly where land ended and sea began. The local residents were quite firm in their assertions that such a passage existed, however, and Cook began to suspect that New Zealand was in fact two islands and not part of a continent, as Tasman and so many others had believed.

At the end of January, as the Englishmen began making preparations to move on, Cook put up markers commemorating their visit and claiming the land for England. He named the inlet Queen Charlotte Sound after the wife of George III (although at least one of the sailors preferred to call it Cannibal Harbour) and the little bay where they had anchored Ship Cove. It was in his opinion the best of all the places they had visited in New Zealand for a ship's crew to rest and reprovision, and Queen Charlotte Sound, like Tahiti, would become one of his regular stopping points on future voyages.

In the last days at Ship Cove, Cook thought the New Zealanders seemed glad to hear that their visitors would soon leave. Banks learned more: the Englishmen had consumed so much food during their stay that their hosts were in danger of running out of provisions themselves.

From Queen Charlotte Sound the *Endeavour* sailed east into the Pacific through the passage that the New Zealanders had described to Cook. (It is today called Cook Strait.) Although Cook was convinced by now that New Zealand was not a continent, some of his officers still insisted otherwise, and hazy weather, obscuring the extent of the land, allowed them to cling to their wishful belief. To prove them wrong, Cook decided to head north to Cape Turnagain, a journey that required just one day. When the Cape became visible on the horizon, Cook called his officers together on deck and asked them if they were finally satisfied that they had in fact sailed around an island. They could hardly disagree. Having made his point, Cook ordered the ship to reverse direction and head south.

But for the partisans of a southern continent there was still hope. The land south of the strait might be part of a great land mass, notwithstanding the local residents' talk of two islands; and although Cook did not agree, he was methodical and thorough, and would not leave New Zealand before establishing its exact dimensions. He now proceeded to do exactly as he had been doing for the past four months: follow the coastline as closely as possible, charting it in detail.

The weather was against the Englishmen, however, and a combination of winds and hazy skies kept them out of sight of land for much of the next month. When they saw high land at some distance, about a week after leaving Cape Turnagain, "We once more cherishd strong hopes, that we had at last compleated our wishes and that this was absolutely a part of the southern continent," Banks wrote. But Cook thought this was probably just an offshore island, which he named after Banks. (In reality it was a peninsula, still known as Banks Peninsula.) The *Endeavour* continued an erratic, zigzag course, often out of sight of land, for several days. At one point, when the ship rounded a peninsula and the crew could see nothing beyond it, the officers got into a lively discussion about whether they had just reached the end of another island. According to Banks, land visible in the distance "was supposd by the no Continents the end of the land; towards even however it cleard up and we continents had the pleasure to see more land to the Southward." The next day, it became obvious that the coast stretched some distance south, and the optimistic Banks wrote that "our unbelievers are almost inclind to think that continental measures will at last prevail."

Banks and his supporters were doomed to disappointment when, a few days later, their ship rounded another peninsula and sailed into the open ocean, with no more land in sight. Heavy swells from the south confirmed Cook's belief that they had reached the southernmost tip of New Zealand, and even Banks had to admit that the weight of evidence was against him. Strong winds, he wrote, "carried us round the Point to the total demolition of our aerial fabric calld continent."

Intending to complete his circumnavigation of New Zealand, Cook directed his course north, staying close to the coast but unable to find a safe harbor where he might anchor the ship. He and his men passed a beautiful bay carved out between steep mountains, but a strong wind blowing offshore made it too risky to negotiate a passage into it, so Cook sailed on, naming the harbor Dusky Bay. The next morning the ship approached a similar bay and Cook again rejected the idea of anchoring, fearing that if they got into the bay, they would have great difficulty getting back out. For this reason he named the bay Doutbful Harbor. Banks, who had not been ashore to "botanize" in more than a month, complained loudly, but Cook prevailed.

From Doubtful Harbor north the coastline was spectacularly beautiful, but no more hospitable for ships, an unbroken stretch of "mountains piled on mountains to an amazing height," in Banks's words. Their forbidding appearance and the total absence of any signs of habitation near the coast made him conclude that this part of New Zealand must be uninhabited. Cook made no judgments about population—in fact, the land was inhabited, although sparsely—but he did make some perceptive guesses about the terrain. Near Dusky Bay and for some distance north, he wrote, mountains "of a prodigious height," many of them snow-covered, extended as far inland as one could see; they were so massive that no valleys appeared to separate them. Farther north the mountains receded inland with a lower ridge of hills closer to the ocean and wooded valleys in between. In these valleys, Cook thought, there were probably many lakes and ponds "as is very common in such like places."

Two weeks' sail from Dusky Bay brought the Englishmen back to Queen Charlotte Sound. It was the end of March 1770, almost six months after they had first approached New Zealand. The "continent mongers," as Banks once called them, had to admit that if terra australis existed, New Zealand was not part of it. For Cook, the circumnavigation of New Zealand only confirmed his strong suspicions. It was typical of his dogged persistence and attention to detail that he did not leave when it became obvious that he had found two remote islands rather than a continent, but instead devoted weeks more to completing a chart of their entire coastline. The result was one of Cook's most accomplished maps. There were a few blank spaces for those stretches where the *Endeavour* had been forced out of sight of land—Cook indicated them with dotted lines—but otherwise it was a remarkably accurate map that was not superseded until the nineteenth century. A French explorer visiting New Zealand two years later made his own chart, but after comparing his work with Cook's he scrapped what he had done, saying "I cannot do better than to lay down our track off New Zealand on the chart prepared by this celebrated navigator."

8. *Homeward*

Leaving New Zealand, Cook deliberated about what to do next. The voyage had already lasted nearly two years and fulfilled all its instructions, so it was reasonable to think about returning to England; but Cook was not yet ready to give up exploring, and he wanted to choose a route home that would offer the prospect of more discoveries. The ideal course would be east across the Pacific to Cape Horn at an extreme southern latitude, continuing the search for a continent, but with winter approaching in the southern hemisphere such a route would be too risky, especially given the *Endeavour*'s weakened condition. Weather also ruled out a southern track to the westward. As an alternative Cook decided to sail due west toward New Holland and then home by way of the East Indies and the Cape of Good Hope.

Banks was disappointed at the decision to give up the search for a southern continent. Unlike Cook, he believed firmly that a continent existed, even though he admitted that his reasons were "weak" and that it must be smaller than geographical theorists supposed. The *Endeavour*'s voyage had proved as much, especially by delineating the actual boundaries of New Zealand, and it seemed unfortunate to Banks that he and his companions could not finish the task.

Instead he had to content himself with working out a plan for another voyage that would solve the continent riddle. If a ship sailed to New Zealand by way of the Cape of Good Hope and then across the Pacific, timing its voyage to make the Pacific crossing in midsummer, it would in theory be a simple matter to discover the continent, if one existed. The Royal Society should sponsor such an expedition in a ship provided by the government, Banks thought, as it would be a voyage of "Mere Curiosity." The expense would be "trifling" considering "the Praise which is never denied to countries who in this publick spirited manner promote the increase of knowledge." Indeed, "the Smallest Station Sloop in his majesties service is every year more expensive than this ship where every rope, every sail, every rope yarn even, is obligd to do its duty most thoroughly before it can be dismissd." Banks and Cook discussed the scheme, and by the time the *Endeavour* reached England, Cook had a well-developed plan in mind for another voyage.

For the moment, however, Cook was more concerned with New Holland, which, like New Zealand, had been discovered in the seventeenth century but then largely ignored. Spanish and Dutch explorers, as well as William Dampier, had visited widely scattered patches of New Holland's northern, western, and southern coasts, but its eastern coast—the *Endeavour*'s next destination—was totally unknown.

The first European to see New Holland was probably the Spanish explorer Luis Vaez de Torres, who sailed through the narrow strait between New Guinea and the northeastern tip of New Holland in 1606. His discovery was kept secret from the rest of Europe to prevent other nations from taking advantage of new knowledge, and although Alexander Dalrymple had learned of Torres's voyage while working in India in the 1750s, he did not publish his find until 1767. (Cook did not see Dalrymple's volume on Pacific voyages until he returned to England.)

In 1616 the Dutch ship captain Dirck Hartog discovered the west coast of New Holland while trying a new route from Amsterdam to the East Indies. Spurred by the belief that this area might be a good place to reprovision, other Dutch navigators stopped at points along the continent's west coast over the next few years. More incentive to explore the region came in 1622, when an English ship was wrecked on hidden rocks just off the coast. Most of the crew survived the shipwreck, but the ship's boats would hold only about a third of them. Those fortunate enough to get into the boats made it to Batavia, but the rest, stranded on a barren shore, were never heard from again. This disaster, which occurred in good weather within sight of land, made clear the importance of getting better charts of the area, and Dutch officials in the East Indies initiated a drive to explore the coasts of New Holland. As a result, by the end of the 1620s the entire west coast as well as the western parts of the northern and southern coasts had been charted with reasonable accuracy. But the eastern half of the continent remained unknown and the question of Australia's connection with New Guinea and New Zealand unanswered.

In 1642 the Dutch government commissioned Abel Tasman to solve these mysteries. It was on this voyage that he made his two great discoveries: Van Diemen's Land and New Zealand. Tasman and those who studied his results were uncertain about whether these lands were islands, part of New Holland, or perhaps part of another continent. On a second voyage, in 1644, Tasman was charged with determining if New Guinea was separate from New Holland and how his previous discoveries were connected to the continent, if at all. Again his results were inconclusive; he failed to answer the two major questions posed by his instructions, but did complete an accurate chart of New Holland's northern coast.

The Dutch government and the directors of the East India Company were not impressed with Tasman's results and decided that further exploration was unwarranted. The land was uniformly barren and the inhabitants hostile; there was nothing in New Holland, Dutch officials thought, that could benefit them in trade, not even sufficient food and water to reprovision ships. Since the Dutch had exclusive control over that part of the world—they had succeeded in pushing the Portuguese out of the East Indies altogether and the English were busy creating their own trading empire on the subcontinent of India—their decision put a stop to exploration in the region for more than a century. Ships continued to stop occasionally at various parts of New Holland's coast (Dampier's visits in 1688 and 1699, for example), but their crews made no new discoveries.

One of the best maps of New Holland available in Cook's time, published by the French cartographer Thévenot in 1663, showed the western three-quarters of Australia in reasonably accurate outline, with its northeast coast extending up to New Guinea. A small gap indicated the uncertainty about whether the two land masses formed a continuous coastline. Far to the south was the outline of Van Diemen's Land, shown as an island with its northern coast blank. And in the corner of the map, well east of Van Diemen's Land, was New Zealand—also shown as a fragment of coastline, its boundaries unclear. It was tempting to think of New Holland as a huge unbroken continent, and the American mapmaker Emanuel Bowen suggested as much in a chart published in 1747; he based his work on Thévenot's map, but drew dotted lines connecting New Guinea, New Holland, and Van Diemen's Land.

Such a map was no more than a hypothesis, however, for none of the early explorers had seen anything that might remotely be considered an eastern boundary. Filling this gap was the task Cook set himself as he left New Zealand. If Thévenot's map was even close to accurate, such a course should bring him to Van Diemen's Land, and from there he would sail north along the coast until he found Torres's strait or reached New Guinea. If he succeeded, he would complete the map of New Holland and solve the question of its connection with New Guinea.

Two weeks after leaving New Zealand, Zachary Hicks saw land in the distance. Judging from his maps and from Tasman's description, Cook thought they were probably just north of Van Diemen's Land. The appearance of the sea and the trend of the coastline made him think there was land to the south, but that it was probably not connected to the coastline ahead of them. In all respects Cook was correct. This first landfall, which Cook named Point Hicks after his lieutenant, was on the extreme southeastern coast of New Holland, just north of the strait separating the continent from the island of Tasmania.

Following the coast north, Cook repeated the procedures he had used in making his New Zealand chart while Banks chafed at being confined on board so close to land but unable to explore it. Wisps of smoke indicated that the land was inhabited, but only once were people visible on shore. Banks thought them "enormously black," but then admitted that imagination rather than observation influenced his impressions. He had read Dampier's account of his experiences on the coast of New Holland, which described the native men as black, much like Africans in physical features, and exceedingly primitive; "the Inhabitants of this

Country are the miserablest People in the world," he had written. Dampier's account had made such an impression on him and on others in the ship, Banks said, "that we fancied we could see their Colour when we could scarce distinguish whether or not they were men."

About a week later Banks finally got his first chance to explore the land more closely, when Cook located a narrow break in the coastline. Beyond it was an enormous bay. Naked men painted with bands of white in geometric patterns across their chests and legs stood on the rocks near the entrance, shaking their weapons—long pikes and short, curved pieces of wood that Banks compared to scimitars—at the *Endeavour*. Other men fishing in canoes nearby ignored the intruders, however; they didn't even look up as the ship passed. Cook chose to anchor near a small village, where a woman surrounded by children took note of the *Endeavour*'s presence without interest and proceeded to prepare her family's dinner. (Banks was shocked to note that even the women wore no clothing.) In time the fishermen paddled ashore, beached their boats, and sat down to eat, still paying no attention to their visitors.

Later in the day Cook and his usual emissaries—Banks, Solander, and Tupaia—went ashore, with Banks optimistically predicting that the villagers' apparent lack of concern about the *Endeavour*'s presence boded well for establishing friendly relations. As the boats drew close to shore most of the inhabitants ran off, however, while two men faced the Englishmen and waved their spears menacingly. Cook tossed nails and beads, momentarily distracting them, but as soon as he moved the boats closer the men reverted to their threatening pose. When Tupaia spoke, they replied in a language altogether unlike the Tahitian and New Zealand tongues, and Cook's attempt to signal his friendly intentions with sign language elicited no response.

Friendly gestures having failed, Cook tried a show of force, first shooting a musket over the men's heads and then firing small shot directly at them. His shots hit one of the men, who ran to a hut nearby and returned with a shield. Finally the Englishmen pushed their boats ashore and landed. At this the native men threw their spears, landing one squarely between Parkinson's feet. The English responded with more small shot; their adversaries threw more spears and then ran off into the woods. Cook thought momentarily of trying to capture one of the men as they ran away, but Banks was afraid their darts might be poisoned, and in any case the last experiment in capturing native inhabitants had produced unfortunate consequences. Instead the Englishmen followed the two men into the woods, where they discovered a deserted village. Cook left gifts of beads, ribbons, and cloth in the houses and returned to the ship, collecting as many spears as he could find on the way back. The next day he and a few others returned to the village, but again the residents had anticipated their arrival and disappeared. The gifts remained untouched where Cook had left them.

The *Endeavour*'s crew stayed a little over a week, but could get no closer to the inhabitants than on their first day. Clearly the villagers were curious about their visitors, for they frequently collected near the place where the crew gathered water for the ship; but they would not get too close, and whenever the English attempted to approach, the men ran away. Hicks, in charge of the watering party one morning, tried to entice them closer with presents, and Cook, alone and unarmed, once followed a group along the shore some distance hoping to initiate a conversation, but neither succeeded. "All they seem'd to want was for us to be gone," Cook said.

The Englishmen had to content themselves with observing the men at a distance. After spending time with two groups of people who were culturally very similar despite living more than twenty-five hundred miles apart, to encounter a third, very different group was curious, to say the least, and Banks was frustrated at his inability to study the New Holland men more closely. Based on superficial observation, it seemed obvious that these people were not only physically different from the Tahitians and New Zealanders—the New Hollanders were much smaller in stature and blacker in color, although Banks disputed Dampier's assertion that they resembled Africans—but also far more primitive in their way of life. They lacked clothes, although they painted their bodies in elaborate designs; they did not cultivate the land at all, but subsisted on fish and shellfish; both their houses and their canoes were small and crudely built. (Cook thought the Australian canoes were the worst-constructed boats he had ever seen.) Pickersgill, for one, agreed with Dampier that the New Holland men were "the most wretched sett I ever beheld or heard of."

The English were much more successful in learning about plant life than human. The bay where they anchored was a botanist's paradise, and Banks collected so many new species of plants in three or four days' work that he was afraid they would spoil before they could be properly preserved. He spent one entire day laying out his collections on a sail in the sunshine, while other plants that he wanted to preserve in their fresh state were wrapped in wet cloths and stored in tin chests until Parkinson had time to sketch them. The artist worked frantically for the next two weeks, completing ninety-four drawings. Cook joined Banks on some of his walks into the countryside, and was so impressed with the enormous variety of plants that he named the place Botany Bay. Both he and Banks were also struck by the richness of the soil, and commented that the area should be well suited for a variety of crops.

The *Endeavour* left Botany Bay, water casks full and the officers' cabin stuffed with plants, on May 6. The weather was good, and Cook was able to keep the ship close to the coast. North of Botany Bay the land became hillier, with sandy beaches broken by occasional rocky promontories. It was lush, tropical terrain, thick with vegetation, in contrast to the barren desolation reported by other explorers of Australia, but for miles it offered no shelter for ships.

After nearly three weeks of following the coastline north, with just one brief stop to replenish water supplies, the Englishmen suddenly found themselves in shallow water surrounded by shoals. Cook had the ship's boats sail ahead to find a way through the treacherous waters, but it soon became apparent that this was not a temporary problem; all that day and the next, the crews manning the boats reported an unbroken string of small islands ahead as far as they could see. Determined to continue charting the coast, Cook chose not to get clear of the islands by heading out to sea but directed his men to continue threading their way among the islands, staying about ten miles offshore. After two weeks of this, they reached a wide, shallow bay. Cook named its northern point Cape Tribulation "because here begun all our troubles."

After another tedious day of navigating around islands and hidden shoals, Cook decided to shorten sail, reduce the ship's speed, and move farther away from the coast to avoid the danger of grounding on some unexpected reef in the dark. It was a clear night with plenty of moonlight and a good breeze for sailing. The depth of the ocean varied radically; one moment it was twenty-one fathoms, then eight, and then just as suddenly back to deep water. About 11:00 P.M., the sailor taking depth soundings reported seventeen fathoms and then, before he could cast his lead to get another reading, the *Endeavour* smashed against a hidden reef and began taking on water rapidly.

Cook, who had gone to bed just a few minutes earlier, was on deck in his nightshirt in moments. Men began working the ship's pumps while others lowered the sails, hoisted out the boats, and tried to tow the *Endeavour* off the reef. But it was low tide, the ship would not budge, and the men at the pumps could not keep up with the water flowing into the hold. To lighten the ship as much as possible, the crew transferred the anchors into the boats and threw all nonessential cargo overboard, including six of the ship's cannon. They attached buoys to the guns, hoping, if they survived, to hoist them back on board.

At high tide that evening, water began to accumulate in the hold more rapidly. Banks despaired of saving the *Endeavour* and with it his precious cargo of specimens, and started packing what he thought he might save if they had to abandon the ship. But he realized, as everyone else did, that the *Endeavour*'s boats would not hold the entire crew, and land was twenty miles away. Even those who survived a shipwreck, Banks reflected, would be abandoned on a barren coast, at the mercy of presumably hostile inhabitants, with no hope of rescue.

The crew remained calm despite the obvious dangers, somewhat to the surprise of Banks, who had heard too many stories about panic and rebellion on ships in trouble. The "Seamen worked with surprizing chearfullness and alacrity," he wrote; "no grumbling or growling was to be heard . . . no not even an oath," a situation that he attributed to the cool and calm manner of the officers. Cook commented later that every man had a clear understanding of their danger and realized the importance of concentrating on saving the ship. Calm weather, at least, worked in their favor; there was no wind or surf to grind the ship's bottom against the reef and worsen the damage to the hull. At the next high tide, a little after 10:00 the next morning, the *Endeavour* floated free.

Unfortunately the danger only increased as water flowed into the damaged hull. Cook ordered a third pump into operation, but a fourth was broken, and the water level rose steadily despite the crew's constant efforts. Then midshipman Jonathan Monkhouse proposed an obscure procedure known as "fothering" the ship. The crew chopped oakum (loose fibers picked from old ropes) into small pieces, mixed it with bits of wool and animal dung, and spead fist-sized bunches of the mixture on a sail in rows three or four inches apart. Attaching ropes to the sail, they dragged it under the ship, where the pressure of the water forced it against the hull, plugging the leak. The effect was apparent immediately. As soon as the sail was in place, the men at the pumps began making headway against the rising water. For the moment, at least, the ship was saved, and Cook turned his attention to finding a safe place to anchor while they repaired its hull.

Even the short journey to shore was a precarious one, as the waters were studded with shoals. After two days of cautious sailing, anchoring at night to guard against any further mishaps, the crew thought they had found a suitable harbor, but closer inspection showed it to be too shallow for the ship. Cook named the place Weary Bay and continued on. Later that evening, the boats' crews reported another harbor, which was actually a wide river with broad, sandy banks where the ship could be beached for repair. The channel of the river was quite narrow, however, and strong winds made it difficult to hold a course, so Cook anchored just outside the mouth of the river and went himself in one of the boats to mark the channel with buoys.

Bad weather kept them at anchor for three more days, but finally the crew succeeded in getting the ship up on the beach where they could inspect the bottom. Only then did they discover what had really saved them; a large chunk of coral wedged in the hole in the *Endeavour*'s hull. Without it, the ship's pumps would never have been able to stem the flow of water into the hold.

The land where the Englishmen found themselves was barren and dry compared with the region around Botany Bay. The river—Cook later named it the Endeavour River—was broad and shallow, lined on both sides with short, stubby trees; beyond them the land rose in low, rocky hills sparsely covered with brush. Cook climbed one of the hills to get a better view of the countryside and was not impressed by what he saw. But fresh water was available, and some fresh game to supplement the men's diet; and so the crew settled down for seven weeks, their only extended stay on any part of the New Holland coast.

It took about ten days for the carpenters to repair the damage to the *Endeavour*. In the meantime, the officers and most of the crew went searching for food and explored the countryside. Banks and Solander were off gathering plants as soon as the ship anchored. Tupaia, who was suffering from a severe case of scurvy, took to fishing and ate only the food he caught or gathered himself. He recovered fully in a matter of days. Several of the sailors were assigned to fish, harvest the shellfish that grew in abundance along the reefs offshore, and shoot birds for the crew's mess; Cook found greens that made a tasty vegetable dish.

One of the hunting parties discovered an unusual animal, about the size of a greyhound and "very swift," according to Banks, but could not get close enough to shoot it. Over the next few days the animals became one of the major topics of conversation, for they resembled nothing that the Englishmen had ever seen before. They had long tails and short front legs, and most curious of all, they did not walk or run but jumped on their hind legs only—and at incredible speed. Even Banks's dogs were not fast enough to catch them. Some days later Gore succeeded in killing one of the animals, which the Englishmen eventually learned was called a *kangaroo.* It was one of the few words of the local dialect they remembered.

Once the *Endeavour* was repaired, Cook and his men faced a new set of problems. They had done such a good job of getting the ship out of the water and up on the beach, a very steep one, that they couldn't get it back into the water. Cook began to fear that they would be stuck until the next spring tides in September, three months away. But within a week the crew managed to float the ship by shifting the weight of the cargo and lashing empty barrels to the hull. After testsing the carpenters' work, Cook decided to make a few more repairs, but on the whole he felt reasonably confident that the *Endeavour* would carry them safely home.

The next problem was more difficult: finding a safe passage away from the coast and beyond the reefs where they were trapped. Cook climbed to the top of a hill near the spot where they had beached the *Endeavour,* only to discover shoals and sandbanks all along the coast, from three or four miles offshore out into the ocean as far as the eye could see. The only hope of a way out appeared to be off to the north, where he thought he could make out a clear passage. The next day Cook sent Molyneux in one of the boats to investigate; he returned to report that there was indeed a passage, but that it would be difficult to negotiate. On the positive side, Molyneux and his crew brought back several enormous shellfish, each one a generous meal for two men, from the more distant reefs.

After three weeks at the Endeavour River the Englishmen still had met none of the native people despite signs that the area was inhabited. Finally Tupaia spotted two men at a distance, but they ran away as soon as they saw him. The next day Banks, Gore, and three companions took one of the smaller boats to explore upriver, hiking overland during the day in search of animals. On their third day out they saw smoke a little distance away, and Banks, hoping the inhabitants would not be afraid of a small group, headed toward it, but again the men fled as the English approached.

The next day, the *Endeavour*'s crew chanced upon several men fishing in a canoe. Some of the officers wanted to take a boat and row out to them, but Cook decided on a more subtle strategy: he ignored the fishermen. In time, two of them approached the ship. The crew working on board tossed them pieces of cloth, nails, and paper, which they took without any sign of interest; but when someone threw a small fish overboard, the pair grabbed it with obvious signs of pleasure and signaled that they would bring over their companions. A few minutes later they beached their canoes near the *Endeavour*, where Tupaia persuaded them to put aside their spears and sit down. Nothing would make them venture on board the *Endeavour*, but the following day three of the men returned to the beach near the ship with two others, bringing a fish in exchange for the one given them the day before. Their visit was brief—when the men observed some of the sailors inspecting their canoe, they quickly walked over to it and left—but they came back the next day with more friends, including a woman and child. This time they stayed most of the morning, but never strayed more than twenty yards from their canoe.

Like the people at Botany Bay, these men were small by English standards, most about five feet six and very slender, and much darker than the Polynesians. They were so thin, Parkinson wrote later, that he could span their ankles and upper arms with his hand. They too painted their bodies with white and red paint in lieu of clothing, although Banks was persuaded that the men, when standing still, usually covered their genitals with their hands or some object, "as if by instinct." In addition to the paint, they ornamented themselves with shell necklaces, fiber bracelets, and pieces of bark tied around their foreheads; most also wore a small bone through their noses, a decorative touch Banks found "preposterous."

The Englishmen's clothing was a great puzzle to the Endeavour River inhabitants, and after a few days of friendly acquaintance they asked some of the sailors to take off their clothes. The men examined the garments closely, talking animatedly among themselves, seemingly trying to make some sense of them. The English, thinking they had finally found something that the local residents would appreciate as gifts, offered them articles of clothing; but later Banks, on a plant-hunting expedition, found most of their presents lying in a heap "doubtless as lumber not worth carriage."

After about a week of friendly interchange, the Englishmen caught several turtles and spread them out on deck. Some of the local inhabitants boarded the ship and asked for one, but Cook refused, as he intended to have the animals prepared for the crew's dinner. One man attempted to seize a turtle and others grabbed whatever they could get their hands on, until the crew forcibly subdued them. Cook tried to placate them by offering bread, the only food readily available, but the men were interested only in turtles. Finally they jumped in their canoes and left. Shortly afterward, they set fire to the grass at a spot where the Englishmen had left nets and clothing to dry, although with little damage, since the crew had been loading the ship preparatory to departure and little of value remained on shore. Cook fired a musket at the men, the first time any of the English had used gunfire against them, which drove them away. Then Cook and Banks, with a few others, followed the men, who "after some little unintelligible conversation had pass'd . . . lay down their darts and came to us in a very friendly manner." The two groups made peace and the men returned to the ship, but this time they refused to go on board.

Later one of the sailors found a wooden harpoon in a turtle he had caught, indicating that the local residents hunted them. Cook surmised that they caught turtles only when the animals came to the mainland to lay eggs, as their boats were too flimsy to travel out to the distant reefs where the turtles normally lived. Consequently the animals were a valuable commodity, much more so than the cloth and nails that the English had offered as gifts.

It was by now the third week in July, the ship was repaired and its provisions stowed, and Cook was eager to move on. For days, however, it was too windy to make the precarious passage through the shoals, and so they remained landbound for another two weeks. Even Banks, having exhausted the natural history of the area, grew bored. Finally, on August 4, the *Endeavour* ventured to sea.

The expedition made slow progress north. Shoals stretched endlessly north and east, forcing Cook to send a boat ahead of the ship at all times to search for a safe passage. After five days of this tedious sailing, Cook faced a serious dilemma. The only certain way out of the shoals was to turn south and go back the way they had come, but this would be a long, arduous journey, complicated by the strong winds that were now blowing continuously from the south. Sailing north would allow Cook to continue charting the New Holland coast, but he had no way of knowing whether a passage back to the open ocean existed at all in that direction. If there was no passage, they would eventually have to backtrack anyway. Molyneux, the ship's master, recommended the southern course, but Cook overruled him, preferring to risk the northerly route.

When the *Endeavour* approached a group of small, hilly islands the next day, Cook decided to anchor and climb to the top of the largest island to get a better view of the area. With Banks he hiked to the island's peak, only to see a reef stretching north as far as the eye could see. There were some breaks in the reef, however, and Cook thought it would be possible to get the ship through one of these openings and out to the open sea. He stayed overnight, hoping to get a better view the next morning, while Banks turned his attention to the island's natural history. Its distinguishing feature was an enormous number of lizards, which inspired Cook to name the place Lizard Island. (Its highest point is today memorialized as Cook's Look.)

The passage Cook saw from Lizard Island was a narrow one, but he managed to get the ship safely through it on August 13. The next day the Englishmen were out of sight of land for the first time in almost three months. It was a sunny day with a brisk wind, and the deep, rolling swells of the sea were a clear sign that they were well away from reefs and islands. For once, Banks noted with amusement, everyone rejoiced at being far from land. As relieved as the others at being finally out of danger, Cook nevertheless was disappointed at losing his opportunity to explore the rest of this coastline. He believed that New Holland and New Guinea were separate, and had been close to proving it; now it looked as if he would be unable to do so, although he still hoped to get back to the coastline at some point farther north.

He had not long to wait. Two days later, the *Endeavour* was about a mile outside the reef in water too deep to anchor, with ocean swells pushing it rapidly toward what Banks described as "a wall of coral rock rising almost perpendicularly out of the unfathomable ocean." A dead calm made getting away impossible. Cook ordered two boats manned with rowers to tow the ship farther out to sea, but it was a futile effort, and the *Endeavour* drifted inexorably toward the reef. "All the dangers we had escaped were little in comparison of being thrown upon this Reef where the Ship must be dashed to peices in a Moment," Cook wrote.

When the ship was within two hundred yards of the reef, a light breeze combined with the towing of the boats briefly pushed it away. Then some of the crew noticed a narrow opening. After waiting several hours until the tide changed, rowing constantly to keep the ship in place, they managed to steer it through the channel. (Later they realized it was not the breeze, but an ebb tide flowing out through the opening in the reef that had saved them.) Cook called the passage Providential Channel and observed with some irony that he and his crew were "happy once more to incounter those shoals which but two days ago our utmost wishes were crowned by getting clear of."

For another four days the *Endeavour* and its crew slowly made their way among more tiny islands and shoals (as Banks explained, "so much do great dangers swallow up lesser ones that these once so much dreaded shoals were now look[ed] at with much less concern than formerly") until they reached the tip of a narrow peninsula surrounded by open water. Cook called the peninsula Cape York and claimed the entire area he had discovered for England, naming it New South Wales. Two days later he sailed through the passage between Cape York and New Guinea, now known as Torres Strait after its original discoverer, into the Indian Ocean. Although Cook thought there must be an easier way through this strait, he decided not to search for it himself, "having been already sufficiently harrass'd with dangers without going to look for more."

The dangers they had been through caused Cook a great deal of soul-searching as he pondered the proper balance between the explorer's daring and the responsible ship captain's concern for the safety of his crew. Previous explorers had often failed to accomplish as much as they might because of their unwillingness to push on in the face of hardship or uncertainty—both Byron and Wallis had suffered from this failing—and yet Cook's stubborn persistence, his decision to continue north along the Australian coast when Molyneux had advised the more cautious policy of turning back, had almost cost the lives of his men and himself. "The world will hardly admit of an excuse for a man leaving a Coast unexplored he has once discover'd," Cook wrote. "If dangers are his excuse he is than charged with *Timorousness* and want of Perseverance and at once pronounced the unfitest man in the world to be employ'd as a discoverer; if on the other hand he boldly incounters all the dangers and obstacles he meets and is unfortunate enough not to succeed he is than charged with *Temerity* and want of conduct." Clearly Cook came down on the side of temerity, admitting that "I have ingaged more among the Islands and shoals upon this coast than may be thought with prudence I ought to have done with a single Ship and every other thing considered"; but he justified his conduct in this instance by the importance of the knowledge gained.

This would not be the last time Cook risked his life and those of his men in the pursuit of knowledge, but the intrepid side of him, essential in any good explorer, was tempered by intelligence, a clear head, and a genuine concern for the welfare of those under his command. At bottom was a curiosity about the unknown and a pride in his accomplishments that, in his mind, justified the risks: "was it not for the pleasure which naturly results to a Man from being the first discoverer," he wrote, ". . . this service would be insuportable."

Cook had ample reason to take pleasure in his discoveries of the previous eleven months. In that time he had knocked several new holes in the theory of a southern continent; essentially completed the map of New Holland, sailing inside the fifteen-hundred-mile Great Barrier Reef in the process; and proved beyond a doubt that New Holland and New Guinea were separate land masses. His maps of New Holland and New Zealand were strikingly accurate—although he warned anyone who might come after him that he was not confident he had charted every shoal and reef along New Holland's coast—despite the tedious method of determining longitude and the great possibility of error. Green was "indefatigable" in making the astronomical observations required, according to Cook, with the cooperation of several junior officers. This teamwork ensured that multiple observations could be made, increasing their level of accuracy. It was the only way that the lunar-distances method of determining longitude could be made reliable and practical, Cook thought; "would Sea officers once apply themselves to the making and calculating these observations they would not find them so very difficult as they at first imagine."

In addition to completing the map of New Holland, Cook and his men made the first detailed and thoughtful observations about its natural history and people. What little had been written on New Holland by the mid-eighteenth century portrayed it as a dry, barren, ugly country and its people as exceedingly primitive and hostile. The men of the *Endeavour* helped change this image, as the discoverers of the more fertile east coast and the only explorers up to that time who had stayed long enough to learn something about the inhabitants beyond the merely superficial.

Both Cook and Banks commented that much of the land along the coast was arid and sandy, except for the area around Botany Bay, and that food crops did not grow there naturally as they did on the South Pacific islands they had visited. "However," Cook added, "this Eastern side is not that barren and Miserable Country that *Dampier* and others have discribed the western side to be." He pointed out that the inhabitants did nothing to cultivate their land; it was, in short, "in the pure state of Nature" untouched by "the Industry of Man." With some effort, he thought, most kinds of grain and fruit would grow well, but even so, in his opinion New Holland did not offer any useful products for trade and would not be a good site either as a base for shipping—the Endeavour River was suitable only for emergency shelter—or as a potential location for European settlement. Banks agreed, noting that the New Holland coast "could not be supposed to yeild much towards the support of man."

Cook and Banks's assessment of the continent's people, like their assessment of its land, was more positive than that of earlier explorers but still, on the whole, negative. The men they had met did not compare favorably with the Polynesians, although they were not the hostile savages depicted by Dampier and the Dutch explorers either. Their manner of living was primitive to be sure, but supplied all necessities. Their dwellings, which Cook described as "mean small hovels not much bigger than an oven," were sufficient for the warm, dry climate—in the warmer northern areas, Banks observed, the houses were even smaller and simpler than at Botany Bay—and the hot weather made clothing superfluous. Even the crude canoes, as Cook pointed out, were useful in their simplicity because they were so small that they could be taken right up onto mudbanks to harvest shellfish. The New Hollanders cared nothing for objects they could not use; they spurned English gifts and carried their few possessions in small woven bags on their backs. Everything they had, Banks observed, would fit in the crown of a hat: a bit of paint, fish hooks and lines, some shells, points for their darts, and a few ornaments. Nevertheless, their society was not a merely functional one; the English found some evidence of artistry in the designs the New Hollanders painted on their bodies and the ornaments they fashioned of wood and shells. But this evidence of creativity did not impress the English, for they considered it, in Banks's phrase, "useless." That these people would devote energy to turning shells into necklaces or decorating their bodies with paint when they obviously invested no more than the minimum effort necessary in building their houses and gathering food was beyond English understanding.

In New Holland as in Tahiti, the English were struck most forcefully by the residents' contentment with their simple style of life. In the case of the New Holland tribes, however, such contentment was even more remarkable because their culture appeared to the English far more primitive than that of the Tahitians and because they displayed no interest whatever in getting anything more than what they already had, in contrast to the Tahitians, who craved English goods and could drive a hard bargain to get them. The native men were curious about their visitors—the incident in which some of them asked the English to disrobe made that clear—but they did not want the gifts the English offered and never stole anything (a definite point in their favor compared with the Polynesians). The only thing the English possessed that they wanted was turtle, and this, of course, was useful to them as food. Ironically, as Banks put it, turtle "of all things we were the least able to spare them."

The New Hollanders' lack of interest in material things provoked Cook to much thought and a certain admiration. Recognizing that his description of them was not a particularly flattering one, he went on to express his opinion that "in reality they are far more happier than we Europeans; being wholy unacquainted not only with the superfluous but the necessary Conveniencies so much sought after in Europe, they are happy in not knowing the use of them. They live in a Tranquillity which is not disturb'd by the Inequality of Condition: the Earth and sea of their own accord furnishes them with all things necessary for life."

Banks was less sympathetic, describing the New Hollanders as "but one degree removd from Brutes," although at the same time he professed to envy their freedom from the "anxieties attending upon riches, or even what we Europeans call common necessaries." But just as Banks had employed European concepts to make sense of his observations about Tahitian society, so too his musings about New Holland's people were inspired more by his reflections on life in England than by what he had seen in the wilderness around the Endeavour River. Recognizing that the more money Europeans had the more they seemed to want, he took inspiration in the New Hollanders' total lack of materialistic desires to suggest that the rich of Europe were really no more content than the poor; the "anxieties" attached to wealth might be seen as a way of balancing the pleasures of material possessions for those who could afford them with the wants of those who could not. "Providence seems to act the part of a leveler," he explained, "doing much towards putting all ranks into an equal state of wants and consequently of real poverty: the Great and Magnificent want as much and maybe more than the midling." In drawing this analogy between the New Hollanders and the poor of Europe, Banks ignored the fact that the former did not live surrounded by the fruits of prosperity that they could never hope to attain—demonstrating once again that, although unmatched as an ethnographic observer, he had little talent for analyzing what he saw; his comparisons between Pacific and European cultures more often obscured than contributed to understanding them. Cook, on the other hand, took a more dispassionate and therefore less culture-bound view of the people he met, perhaps because he felt less compelled to explain what he saw than Banks did.

After rounding the Cape York peninsula, Cook directed his course to Batavia, skirting the southern coast of New Guinea along the way. For the first time since rounding Cape Horn, he was in well-known waters, extensively charted by previous navigators, and yet he found his maps unreliable. Islands were shown in places where Cook found nothing but open ocean, an annoyance that led him to blast unscrupulous captains who drew coastlines they had never seen for the sake of filling in a map and careless publishers who printed the results or failed to include on a published map the mapmaker's caveats about potential flaws for fear of hurting sales of the work. The consequence, Cook complained, was that "we can hardly tell when we are possessed of a good Sea Chart untill we our selves have proved it."

The limitations of their maps notwithstanding, the *Endeavour*'s crew reached Batavia in the second week of October, about six weeks after leaving the Australian coast. For the first time in more than two years, the Englishmen had the company of other Europeans, in a seaport where they could find the amenities of civilization once again.

While Banks and Tupaia were walking through town on their second day ashore, a stranger approached them and asked Tupaia if he had visited Batavia before. Further conversation revealed that the French explorer Bougainville had stopped at the port a year and a half before; among his companions was a Tahitian who looked—in the Dutchman's eyes at least—just like Tupaia. Banks questioned the man further about Bougainville's voyage, finally solving the mystery of the European ships that had visited Tahiti between the *Dolphin* and the *Endeavour*.

There were certain formalities to be observed with the local Dutch government, and it took several days to get permission to use the facilities of Batavia's docks to repair the *Endeavour*. Cook, meanwhile, began writing letters to send back to England by ships that would be ready to sail long before the *Endeavour*'s repairs were completed. He praised the courage and hard work of his men to Admiralty Secretary Phillip Stephens; "they have gone through the fatigues and dangers of the whole voyage with that cheerfulness and allertness that will always do honour to British Seamen," he wrote, and then added, "I have the satisfaction to say that I have not lost one man by sickness during the whole Voyage."

That was the third week in October, twelve days after the *Endeavour* anchored at Batavia. Within days, more than half the crew were seriously ill. The first week in November, William

Monkhouse died, followed a few days later by Tayeto and Tupaia. Banks, Solander, and two of Banks's servants were all sick, Solander so seriously that Banks feared he would not survive more than another day or two. On the advice of a local doctor, Banks rented a house in the country and took Solander and his servants to convalesce there. Spöring, still in good health, went with them as nurse and general helper, and Cook sent his personal servant and a seaman as well. Both Banks and Solander recovered, but the ship's crew did not have the benefit of country air, and the illness continued to spread until most of them, including Cook, were sick. Some days only about a dozen were capable of working.

The diseases that afflicted the men were malaria and dysentery. Batavia's low-lying location and wet climate made it an ideal breeding ground for malaria-bearing mosquitoes, and poor sanitation contributed to the prevalence of disease. It was widely known as an unhealthy place to live—more than a million deaths were recorded there between 1730 and 1752—but the city remained a popular port of call for trading and exploring ships because it was one of the few places in that part of the world where they could make repairs and get provisions.

The general sickliness delayed the *Endeavour*'s departure until the day after Christmas. Seven men were dead, and every member of the crew had been sick for some period of time except a sailmaker who was the oldest member of the group and usually drunk. Although many were still sick, Cook hired nineteen additional seamen and decided to leave before the health of his men deteriorated further. Few recovered, however, and still more fell ill on the voyage to the Cape of Good Hope. Within a six-day period at the end of January, Green, Spöring, and Parkinson died, and Banks became ill again. Barely enough men were healthy enough to sail the ship. By the time they anchored at Table Bay on the Cape of Good Hope in mid-March, the disease had fairly well run its course, but at the cost of twenty-nine lives. Another four men died at the Cape.

In talking with other ship captains—Table Bay, like Batavia, was a major stopping point for trading ships—Cook learned that their ships had been even more seriously afflicted. Yet he felt the deaths among his own crew as both a professional and personal tragedy. He also believed that the illness aboard the *Endeavour* would be widely publicized because of the unusual nature of the voyage. This prospect disturbed him, partly because he was annoyed that hardships and disasters rather than solid accomplishments were the stuff of fame with the public, but also because he had taken pride in maintaining the health of his men. Before anchoring at Batavia, only seven men had died, but by now he had lost more than four times that number, including almost all the scientific groups, of which only Banks, Solander, and two of Banks's servants survived.

At Table Bay Cook learned more about Bougainville's voyage, including the fact that his Tahitian companion Ahutoru had made it safely to France and was to be returned to Tahiti on another voyage. What he did not learn was that Bougainville too had discovered the Great Barrier Reef. Like Cook, he had decided to search for the east coast of New Holland, but he had approached the continent much farther north than Cook and so encountered the Great Barrier Reef head-on. For three days the French explorers sailed along an almost unbroken line of sandbanks, shoals, and rocks, until Bougainville decided that it was too dangerous to go on and altered his course for New Guinea.

Banks feared that Bougainville would publish a description of his voyage before Wallis and claim Tahiti as his own discovery, which made it imperative to get an account of the *Endeavour*'s voyage into print as quickly as possible. Cook also worried that Tahiti, although it was an island of "little value" in his opinion, would become a source of conflict between England and France, particularly if the French attempted to establish a settlement there. (They misjudged Bougainville, who, in a spirit of scientific cooperation, had crossed the English Channel to report on his voyage at a meeting of the Royal Society in May 1770.)

Cook was more seriously disturbed about the possibility that the French would stumble across New Zealand on the voyage to return Ahutoru to Tahiti, but his concern was wasted, since the French had already discovered New Zealand for themselves at almost exactly the same time he had. In mid-December 1769, Jean François de Surville anchored in a bay (Cook had named it Doubtless Bay) on New Zealand's North Cape. At that moment Cook was several miles offshore, rounding the North Cape and beginning his journey south along the west coast of the North Island. He and de Surville had missed seeing each other by a day.

In mid-April the *Endeavour* left Table Bay on the last leg of its journey home. Two months later the crew met three ships from New England, bound on a whaling voyage, and got their first recent news of England; and on July 12, they anchored in the Downs. Their voyage around the world had taken almost exactly three years.

9. Mr. Banks's Voyage

When the *Endeavour* reached the Downs, a pilot came on board to guide it up the Thames, and Cook, his work finished, went ashore with Banks and Solander to go home. Constant companions for three years, Cook and Banks went their separate ways, Cook to his modest house in Stepney on the northeast edge of London, and Banks to his home in the fashionable center of the city.

When he sailed on the *Endeavour,* Cook left behind two young boys, an infant daughter, and a pregnant wife. He returned to find his sons flourishing, now aged six and eight; but his daughter had died three months before he came home, and the son born shortly after his departure in 1768 had survived only a few weeks.

Within a month the Admiralty promoted Cook from lieutenant to captain and appointed him commander of the *Scorpion,* which had been commissioned to sail around the British Isles correcting the charts of the English coast, an appropriate if rather mundane assignment. But it was clear almost from the moment the *Endeavour* returned to England that any such assignment was simply temporary employment, for Cook and the Lords of the Admiralty agreed that another voyage should be outfitted soon to capitalize on the discoveries of the *Endeavour*'s expedition.

* * *

The newspapers were filled with stories about the voyage during July and August, and, as Cook had predicted, they were highly selective in what they chose to report. The near-disaster on the Great Barrier Reef made good copy, and one story reported that the *Endeavour* had sailed "many hundred Leagues with a large Piece of Rock sticking in her Bottom, which had it fallen out must have occasioned inevitable Destruction." Another writer embroidered his account of the crew's stay at the Endeavour River, informing his readers that "the savages were very troublesome" and had often attacked the Englishmen. An account of Banks's ill-fated expedition at Tierra del Fuego was even more excessively dramatic: he and his companions "undertook to climb to the summit of a prodigious mountain," and when they did not return by nightfall, their shipmates feared they "must be either cut off by the natives, or devoured by the wild beasts."

Most of the stories, however, focused not on spectacular dangers and near-shipwrecks, but on the explorers' experiences on the exotic islands of the South Pacific. Few articles failed to report that Mr. Banks and Dr. Solander had returned with chests full of "curiosities"—the eighteenth century's all-purpose term for any unusual objects. (Plant and animal specimens were labeled "natural curiosities" and man-made objects "artificial curiosities.") Estimates of Banks's collection of plant specimens ranged from one thousand to seventeen thousand items.

The longer newspaper pieces, which purported to be letters written by men who had sailed on the *Endeavour,* regaled their readers with tales of the South Pacific islanders, especially the women. The Tahitian women were the most beautiful in the world, according to newspaper accounts, notwithstanding their rather bizarre custom of tattooing their buttocks black; their dances were extraordinarily lascivious, and their sexual favors easily bought. "They marry at nine and ten. . . . A virgin is to be purchased here, with the unanimous consent of the parents, for three nails and a knife," one writer claimed. The authors of these titillating stories freely admitted that they had enjoyed the charms of the Tahitian women and had been reluctant to leave their island paradise, although they invariably blamed Bougainville's men for introducing the "French disease."

Newspaper writers entertained the English public with more

serious fare as well, attempting descriptions of Tahitian government and religion. Banks's analogy between the Tahitian political system and European feudalism was convenient for the purpose, and newspaper readers learned that Tahitians lived under a chief who was "despotic without controul," who could inflict punishment without cause and force his men to the battlefield upon whim. A description of the Tahitian *marae* emphasized the enormous size of the stones used in their construction—the writer had apparently heard about Purea's giant *marae* and extended the description to all such structures—and compared them to Stonehenge. Another article, reminiscent of the stories about Patagonian giants, claimed that the Englishmen had visited a shrine in which they found a box containing a human skeleton "of an enormous size."

The old geographical myths got their share of attention too. In a remarkable collage of incidents from various parts of the voyage, one writer announced that the explorers had discovered a southern continent in the latitude of the Dutch East Indies where the people were "hospitable, ingenious, and . . . politely civilized" although they observed no kind of religion, and that two of the native inhabitants had joined the expedition but died at Batavia. Another reported with disappointment that the *Endeavour*'s men had discovered no gold or silver, while a third writer asserted that the crew members would make their fortunes, having come home well supplied with "some of the richest Goods made in the East, which they are suffered to dispose of without the Inspection of Custom-house Officers." Still another, closer to the truth, confidently stated, " 'tis expected that the Territories of Great Britain will be widely extended in Consequence of those Discoveries."

The more learned members of London society also pondered the marvelous stories brought back by the men of the *Endeavour* and tried to analyze them in accordance with their own views of the world. Benjamin Franklin, working in London as an agent for several American colonies, attended a dinner party in honor of Banks and Solander at the home of Sir John Pringle, president of the Royal Society. After listening to Banks on Tahitian society, Franklin reported to Bishop Jonathan Shipley that the people of Tahiti "live under a regular feudal Government, a supreme Lord or King, Barons holding Districts under him, . . . Farmholders under the Barons; and an Order of Working People Servants to the Farmholders." Tahitian morals, according to Franklin, were "very imperfect," since they did not consider chastity a virtue or theft a vice. On a lighter note, Franklin—who had a reputation for being a bit of a rake—was amused to learn that the Tahitians had no notion of kissing with the lips, "tho' they lik'd it when they were taught it." Franklin was surprised to learn that the Tahitians preferred their own way of life to that of their visitors, despite the Englishmen's technological advantages. Nevertheless, when told that the *Endeavour*'s men considered the New Holland natives "a stupid race" because they refused to accept English presents, he mused, as Cook had, that perhaps they were happier for their lack of wants.

Beautiful, promiscuous women, a primitive version of European government in Tahiti, dangerous savages in New Holland: these were the themes that dominated talk of the voyage in London, whether among newspaper readers in the coffeehouses or Banks's socially distinguished friends. Not everyone took the voyage seriously, however. Samuel Johnson, who like the rest of London's literati dined with Banks on at least one occasion, thought collecting plants and animals was hardly an intellectual pursuit and all the fuss over the *Endeavour* was a bore. When Boswell pointed out that Banks had discovered many new species of insects, Johnson replied, "Ray reckons of british insects 20,000 species. Banks might have staid at home and discovered enough in that way." Cartoonists had fun with Banks—one caricatured him with one foot on each of two globes, a net in each hand, trying to catch a butterfly. The caption read: "I rove from Pole to Pole, you ask me why, I tell you Truth, to catch a ——Fly." The butterfly motif was popular. When Banks was awarded the Order of the Bath by George III, another cartoonist pictured him with butterfly wings and the caption "The Great South Sea Caterpillar, transform'd into a Bath Butterfly."

As the weeks went by, most attention focused on Banks and Solander, while Cook was almost a forgotten man. Both Cook and Banks had audiences—separately—with the King, but it was Banks who was invited for a second meeting to show the royal family the artifacts he had collected. Newspaper accounts of the voyage invariably talked about the two naturalists; occasionally someone acknowledged the importance of Green's astronomical observations, although others gave Banks credit for that too. Mr. Banks had gone to the South Seas to "discover the transit of Venus"; "Mr. Banks and Dr. Solander's Voyage" would prove "extremely subservient to the Purposes of Navigation"; the *Endeavour* was the ship that had "carried Mr. Banks and Dr. Solander round the World." Banks and Solander became London's leading social attractions, and both were awarded honorary degrees by Oxford.

If Cook resented the attention shown to Banks, he did not show it. He was well pleased with his accomplishments, and was the sort of man who derived most satisfaction from achieving his own very exacting goals without much concern about what others thought. He was also perfectly well aware of the social gulf that separated him and Banks, meaningless on board a ship in a remote sea but important in London, and it is doubtful that the popular attention lavished on the young socialite came as a surprise to Cook. The captain was more interested in the reactions of the Admiralty and the Royal Society than in those of the newspapers or Samuel Johnson.

Cook's first tasks when he returned home were reporting to the Admiralty and the Royal Society on the scientific results of his voyage. He had written to the Royal Society from Batavia summarizing the transit of Venus observations—the letter was read to the membership at a meeting in May, two months before the *Endeavour*'s return—and he sent copies of Green's papers ahead to Nevil Maskelyne. Shortly after landing in England, Cook wrote again to the Royal Society with a detailed description of the techniques used in observing the transit, and two weeks later he attended a council meeting to discuss his results, which were published in the Society's *Philosophical Transactions* for 1771. Cook was equally prompt in writing to the Admiralty Secretary and the Navy's Victualling Board about his experiments with a new type of compass and with scurvy preventives. The compass had not worked well in turbulent seas, but the remarkable lack of illness among the *Endeavour*'s crew, until the ship reached Batavia, was proof of the antiscorbutics' efficacy. Although it was impossible to determine which of the several concoctions was responsible, Cook emphasized the virtues of sauerkraut. (In fact, sauerkraut is not especially effective as an antiscorbutic; it was Cook's insistence that his men eat only fresh local produce when the *Endeavour* was in port that kept his men in good health.)

Later, when the first rush of postvoyage activity was over, Cook sent his collection of "curiosities" to the Admiralty Secretary. Unlike Banks, who had traveled as a private citizen and could do as he pleased with the objects he had collected, Cook was obligated to turn over what he had accumulated on the voyage as the property of the Navy. His collection included Tahitian *tapa* cloth and weapons; a headdress from Raiatea; a drum, five pillows, stone and wooden axes, implements used to make tapa cloth, and three carved images, all from the Society Islands; a carved box and several weapons from New Zealand; and fish hooks from Australia. Cook also spent some time writing more detailed explanations of his various scientific experiments; his observations on the magnetic variation of the compass and an explanation of tides in the South Pacific were both published in the *Philosophical Transactions*.

With his obligations to the government satisfied, Cook put down some of his more personal thoughts about the voyage in two long letters to John Walker, his old friend and employer in Whitby. In the first he told Walker about King George's and the Admiralty officials' praise for his accomplishments, adding, "I however have made no very great Discoveries, yet I have explor'd more of the Great South Sea than all that have gone before me so much that little remains now to be done to have a thorough knowledge of that part of the Globe." This kind of comment was typical of Cook; publicly he played down the significance of his achievements, yet privately he understood the importance of what he had done. He believed that the voyage's contribution lay not in new discoveries, but in its confirmation of and additions to old ones. And there was the one major negative discovery: the area in which a southern continent might lie was significantly reduced. This gradual piecing together of the map of the South Pacific was not so glamorous as finding new lands would have been, nor so exciting as tales of exotic peoples, but for Cook the satisfaction of bringing home an accurate chart of vast areas of the Pacific was sufficient.

His summary of the voyage eschewed the dramatic and shocking elements so popular in the newspapers in favor of describing the lands he had visited and his thoughts about the inhabitants' way of life. The simplicity and comfort of Tahitian life and the apparent contentment of the New Holland men despite their primitive existence were the points Cook emphasized in his narrative, rather than bizarre customs or sexual promiscuity. The treacherous passage inside the Great Barrier Reef sounded almost tame in Cook's calm, reasoned prose: "once we lay 23 hours upon a Ledge of Rocks . . . received very much damage in her bottom but by a fortunate circumstance got her into Port and repair'd her."

The major task to be accomplished after the voyage was the publication of its results. In accordance with Admiralty procedures, Cook had collected the journals kept by his officers and turned them over to his superiors along with his own journal and the official ship's log; Admiralty officials then hired John Hawkesworth, a free-lance writer and self-taught man of letters, to prepare Cook's journal for publication along with those of Byron, Carteret, and Wallis—for no official record of the three earlier voyages had yet been published, and with interest in the South Pacific increasing throughout Europe, government officials wanted to put the English discoveries on record as quickly as possible. Banks and Cook were not the only ones who worried that the French might try to claim the discovery of Tahiti or New Zealand as their own. These political considerations were the primary motive for the confiscation of the officers' journals and the publication of a single, official account of the voyage.

It was a lucrative assignment for Hawkesworth. Travel literature continued to be very popular, and Cook's voyage had captured even more than the usual amount of public attention. A London printer, recognizing a potential best-seller, paid him £6,000—an astounding sum at the time—for the rights to his account.

Hawkesworth's charge from the Admiralty was not merely to edit the captains' journals for publication in the modern sense of the term, but to transform them into a narrative that would hold the interest of the general reader, a task assumed to be beyond the capability of naval officers. He ignored most nautical and scientific matters, focusing instead on the expeditions' landfalls in the South Pacific and the people living there. In preparing his work, Hawkesworth drew on literary traditions ranging from classical mythology to the tales of previous explorers as a means of describing the voyages in a framework understandable to the European reader. He exaggerated Byron's account of the Patagonian giants, for example, giving new life to the popular legend; romanticized Wallis's encounter with the Tahitians; and turned Cook's experiences with the New Zealanders at Tolaga Bay into a discourse on the similarities between Pacific primitives and the mythical figures of ancient Greece.

The engravings illustrating the volumes also took the explorers' raw material and shaped them into images familiar to Europeans. A drawing of one of Byron's sailors standing waist-high next to a Patagonian was a blatant case of playing to public perceptions; a more subtle example was the transformation of Alexander Buchan's sketch of a family in Tierra del Fuego into an engraving of classically inspired figures clustered around a rustic hut surrounded by trees of a sort that did not grow within hundreds of miles of Tierra del Fuego. Parkinson's details were all there, but framed in a bucolic setting typical of English landscape drawing of the period. In both text and illustrations, Hawkesworth attempted to describe the exotic qualities of the people the explorers encountered while placing them in the context of the familiar, emphasizing those qualities that he believed all mankind had in common.

The volumes were an instant success. By the end of 1773, just a few months after publication, a second edition had to be printed to keep up with the demand, and the work was published in New York and in French and German translations in 1774. Soon afterward it also appeared in a cheap serial edition—sixty weekly parts at a shilling apiece—making it available to the many who could not afford the elegantly printed original.

Despite its popularity, the book drew heavy criticism from some quarters. The captains themselves were disturbed at seeing their words changed in unexpected ways and at the emphasis on what they considered the more sensational features of their voyages. Although Hawkesworth claimed that the captains had read and approved the work before publication, both Cook and Carteret denied ever seeing the manuscript; when Cook first saw the published volumes he found them "mortifying." (It is probable that Hawkesworth turned over his manuscript to Lord Sandwich, the First Lord of the Admiralty, on the mistaken assumpton that he would show it to Cook and the others.) Literary critics, on the other hand, thought Hawkesworth had included too much nautical detail. "The entertaining matter would not fill half a volume," Horace Walpole complained. Impatient with Hawkesworth's classical allusions, he made fun of the story about Purea and Wallis's encounter on Tahiti: "An old black gentlewoman of forty carries Captain Wallis across a river, when he was too weak to walk, and the man represents them as a new version of Dido and Aeneas."

In the interests of realism, Hawkesworth included descriptions

Captain James Cook. Engraving by J. Chapman. The scene below the portrait is modeled on John Webber's painting of Cook's death.

Joseph Banks. Oil by Joshua Reynolds. *National Portrait Gallery, London*

Charles Clerke. Oil by Nathaniel Dance. *His Excellency the Governor-General of New Zealand*

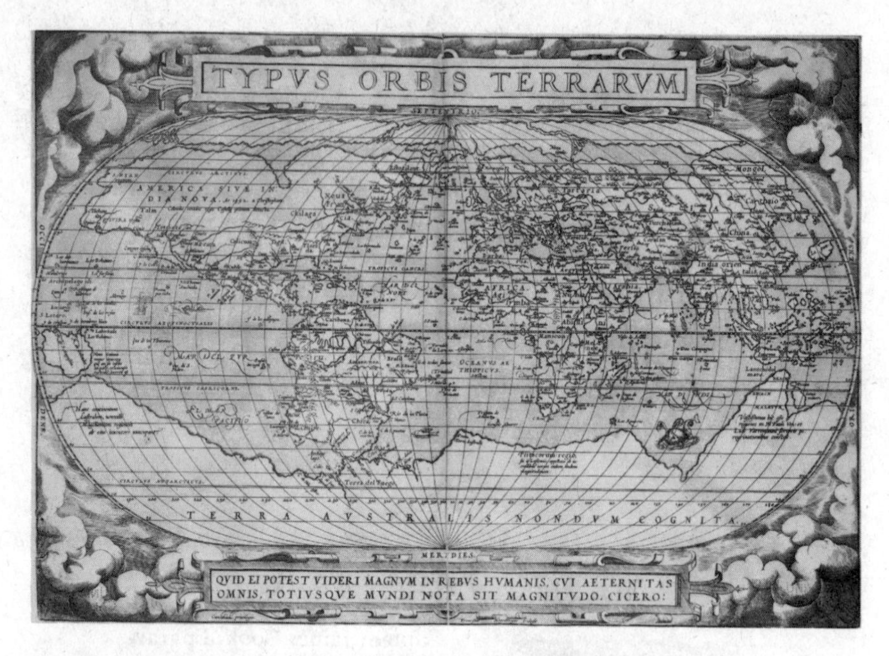

World map by Abraham Ortelius showing the continent presumed to exist in the southern hemisphere. From the atlas *Theatrum Orbis Terrarum*, first published in 1570. *Bancroft Library, University of California, Berkeley*

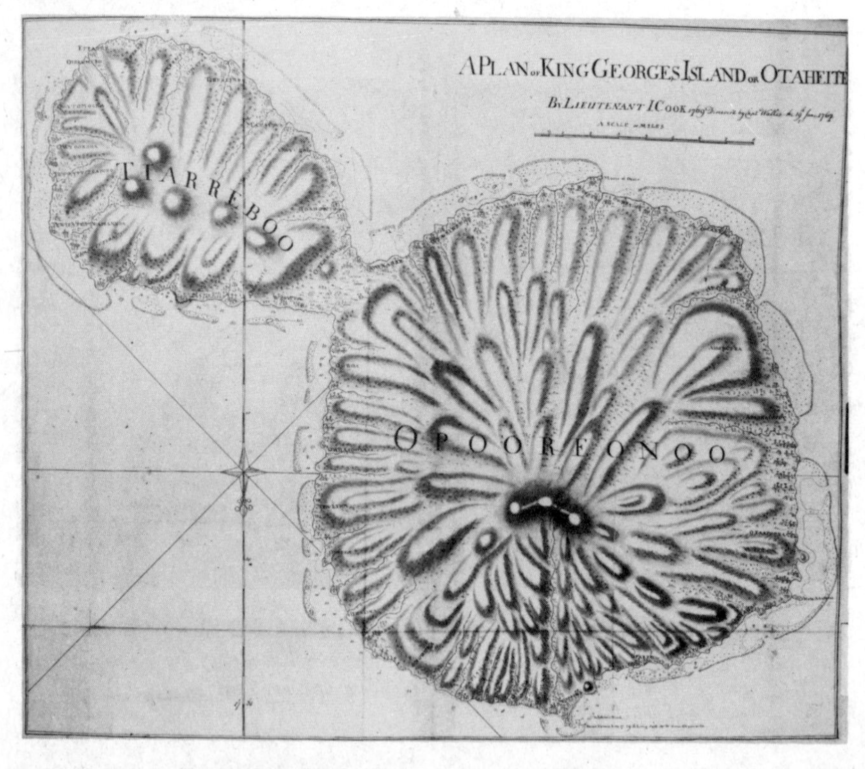

Cook's map of Tahiti. *British Library*

A Tahitian woman and child. Engraving based on a drawing by Sydney Parkinson. *Bancroft Library, University of California, Berkeley*

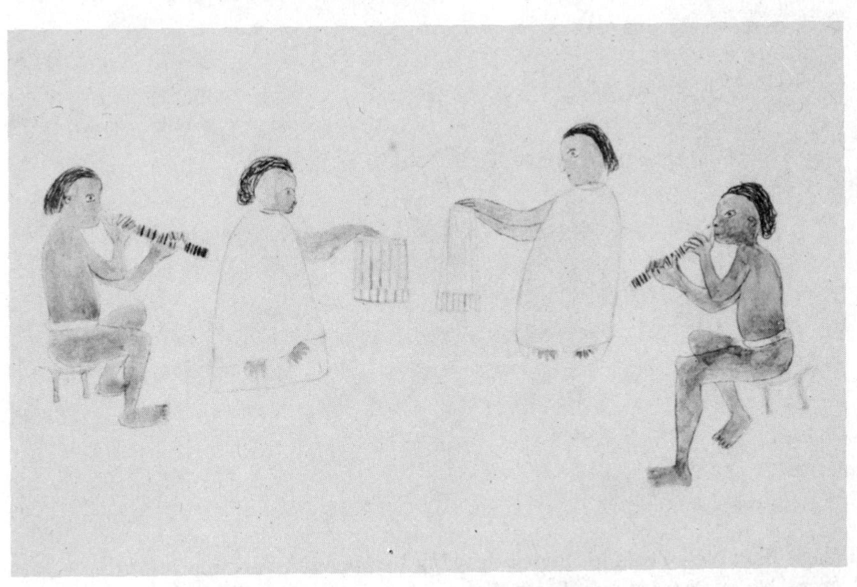

Watercolor drawing of a group of Tahitian musicians by an anonymous artist known as "the artist of the chief mourner," most likely Joseph Banks. *British Library*

New Zealand war canoe. Engraving based on a drawing by Sydney Parkinson. *Bancroft Library, University of California, Berkeley*

Tolaga Bay, New Zealand. Pencil drawing by Herman Spöring. *Mitchell Library, State Library of New South Wales*

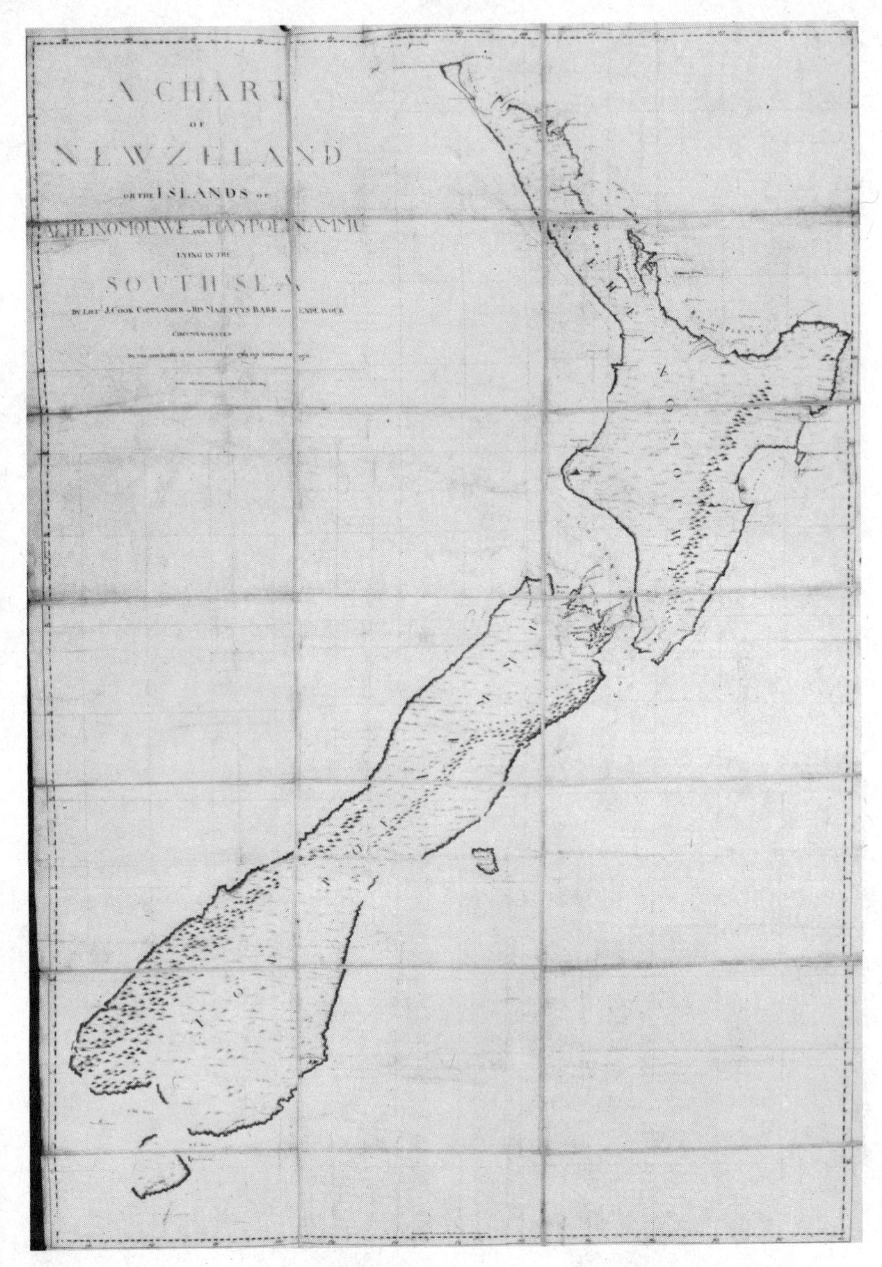

Cook's map of New Zealand. *British Library*

A New Zealand chief. Engraving based on a drawing by Sydney Parkinson. *Bancroft Library, University of California, Berkeley*

Two New Holland warriors. Engraving based on a drawing by Sydney Parkinson. *Bancroft Library, University of California, Berkeley*

A New Zealand chief. Chalk drawing by William Hodges. *National Library of Australia*

A Tannese woman and child. Engraving based on a drawing by William Hodges. *Bancroft Library, University of California, Berkeley*

The *Resolution*'s crew collecting ice to be melted down for water. Engraving based on a drawing by William Hodges. *Bancroft Library, University of California, Berkeley*

"View in Pickersgill Harbour, Dusky Bay, New Zealand." Oil by William Hodges. The fallen tree forms a bridge between an island in the harbor, on the left, and the *Resolution,* on the right. *National Maritime Museum, Greenwich, London*

A native family at Dusky Bay. Engraving based on a drawing by William Hodges. *Bancroft Library, University of California, Berkeley*

of the Tahitians' sexual practices, and these bits drew fire from moralists. The frank accounts of Tahitian customs, some thought, should have been toned down out of deference to the sensibilities of polite society, and it was considered bad form for ladies to admit they had read the book. John Wesley, the Methodist leader, found some of the incidents described in the book so appalling that he simply refused to believe them. "Men and women coupling together in the face of the sun, and in the sight of scores of people!" he considered "absolutely incredible." (Hawkesworth was tame, however, compared to the anonymous writers who published a series of pamphlets in 1773 and 1774 satirizing the Tahitians' uninhibited sexuality and Banks's fondness for Tahitian women in crude verse that left little to the imagination.) Wesley also refused to believe that people living as far apart as the Tahitians and New Zealanders could speak the same language. The book, he concluded, must be classified as fantasy along with Robinson Crusoe, with Tupaia "akin to his man Friday."

Alexander Dalrymple, the geographer who had been the Royal Society's initial choice to lead the *Endeavour*'s voyage, attacked Hawkesworth's book for what he considered its faulty geography. He refused to believe that no southern continent existed in the temperate latitudes, all but accused Cook of incompetence, and implied that if *he* had been given command of the voyage a southern continent would have been found. Dalrymple's attacks were leveled at Cook, of course, but Hawkesworth, as the author of the official account of the voyage, bore the brunt of his wrath. In the preface to the second edition of his work, Hawkesworth wrote, "I am very sorry for the discontented state of this good Gentleman's mind, and most sincerely wish that a southern continent may be found, as I am confident nothing else can make him happy and good-humoured." The attempt at humor only angered Dalrymple and provoked him to new attacks.

If the ship captains, the literary critics, John Wesley, and Alexander Dalrymple were not enough, there were those who criticized Hawkesworth simply for getting rich from his work. The £6,000 figure he had received for his volumes was widely publicized, and many thought it was absurdly high. He did not live to enjoy the fruits of his success, however, but died six months after the book was published—in part a victim of the heavy criticism of his work, in the opinion of his friends.

The range of reaction to Hawkesworth's volumes reflected the large and diverse audience for exotic travel literature and the conflicting opinions within European society about the significance of Pacific voyages. Cook's adventures and those of other explorers were at once entertainment for the masses, topics for polite conversation in upper-class circles, and raw material for serious scientific debate and social commentary. Pacific peoples were the topic of most enduring interest, both for those who cared only to be entertained and for those who sought deeper understanding of mankind in general by studying previously undiscovered cultures. Cook sailed at a time when the study of man was assuming greater importance among scientists and other intellectuals and the notion of a world centered on Europe was breaking down in the face of greater knowledge of other parts of the globe. Many intellectuals, critical of distinctions of wealth and social class in European society as well as the hardships created by economic change, questioned whether the achievements of modern civilization were worth its cost.

Belief in the corrupting influence of civilization and in the virtue of man in his natural state was not new in the eighteenth century, but it drew strength from increasing knowledge about cultures beyond Europe. Rousseau's "Discourse on Inequality," published in 1755, was especially influential in developing the concept of the "noble savage"—the primitive man who in his simple life was happier and more virtuous than his European contemporary. Rousseau's model was the North American Indian, but the Seven Years War had given Europeans greater exposure to the North American tribes and disillusioned them in the process. With convenient timing, however, the Tahitians emerged as even better candidates to make Rousseau's point about the virtues of a simple life. Philibert Commerson, the naturalist on Bougainville's voyage and a disciple of Rousseau, published a description of Tahiti as a utopian paradise, which in turn inspired Denis Diderot to write a "Supplement" to Bougainville's voyage contrasting the freedom and openness of Tahitian society with the corruptions and hypocrisies of modern Europe. (Bougainville's own version of his expedition, published after Commerson's, was more balanced in its description of Tahiti, but Commerson's account appeared first and remained more influential.)

Similar ideas were popular in England as well, and were reflected to some extent in Cook's and Banks's comments about the Tahitians and the men of New Holland. But fascination with the "noble savage" was, on the whole, less widespread than in France. Samuel Johnson, for example, besides poking fun at Banks over his passion for collecting plants and insects, wrote a novel called *Rasselas* that criticized the equation between a simple life and happiness as naive. The book is set in "Happy Valley," an impenetrable cleft in the mountains where everyone is content except the main characters, who are bored and want nothing more than to escape. In addition, as some of the criticism of Hawkesworth indicated, praise for Tahiti's sexual freedom, a major point of Diderot's work, drew less favor in Protestant England than in France; as the evangelical revival sparked by Wesley and the Methodist movement took hold toward the end of the eighteenth century, Tahiti was viewed increasingly as a source of moral corruption rather than virtue. Eventually the evangelical point of view would predominate and the Pacific would be seen as a new arena for Christian (and therefore European) influence; but in the early 1770s, English reaction to the discovery of Pacific cultures was a complex mix of admiration, titillation, skepticism, and moral indignation.

By the time Hawkesworth's volumes appeared and unleashed their controversies, Cook and many of his men were already back in the South Pacific.

There had never really been any doubt, from the time the *Endeavour* returned to England, that the Admiralty would sponsor another voyage and that Cook would be asked to command it. His first venture to the South Pacific had been far more successful than any of his predecessors', and, more important still, he came home with a plan for another voyage that should finally settle the southern continent question. Within two months of Cook's return, the Admiralty was shopping for a suitable ship. (The *Endeavour,* worn but still seaworthy, was refitted and sent as a storeship to the Falkland Islands.)

Given the general public reaction to the first voyage, it was not surprising that the proposed new expedition was billed in the popular press as Banks's voyage. At least three newspapers confidently reported that "Mr. Banks is to have two ships from government to pursue his discoveries in the South Seas." Others went so far as to say that "the celebrated Mr. Banks" would soon make another voyage to Tahiti with three ships "in order to plant and settle a colony there." Banks was inundated with letters from men who wanted to sail with him or offer advice on potential experiments. The letters came by the dozens, from France, Switzerland, and Germany; most were from individuals with scientific interests, but even sailors wrote to ask for his help in getting a berth on the next voyage.

At the Admiralty, of course, it was clear that the new expedition would be Captain Cook's voyage and not Mr. Banks's, and indeed Cook had a good deal more to say about the organization of this undertaking than he had in 1768. The Navy Board consulted him about the purchase of ships and selected two vessels originally built for the North Sea coal trade, similar to the *Endeavour*. The decision to send two ships instead of one was also based on Cook's advice; his close call off the coast of Australia had shown the wisdom of traveling with an escort. The two ships were purchased in mid-November and renamed the *Drake* and the *Raleigh*. At 450 tons, the *Drake* was somewhat larger than the *Endeavour;* the *Raleigh,* at 336 tons, was slightly smaller. In December, one of the Lords of the Admiralty pointed out to Sandwich that naming the ships after England's famous buccaneer-explorers would be an affront to Spain and suggested, in the interests of diplomatic relations, changing the names to something less provocative. Sandwich agreed, and the ships were renamed the *Resolution* and the *Adventure*.

Cook also had more control this time over the selection of his crew, and he saw that several men from the *Endeavour* were hired again, notably Charles Clerke as second lieutenant of the *Resolution* and Richard Pickersgill as third lieutenant. Three midshipmen—Isaac Manley, William Harvey, and Isaac Smith—were back, along with thirteen seamen, six of them promoted to petty-officer status. The top-ranking marine, Second Lieutenant John Edgcumbe, was a veteran of the *Endeavour;* ironically, the other returning marine was Samuel Gibson, one of the attempted deserters at Tahiti, now promoted to corporal. For the *Resolution*'s master Cook selected Joseph Gilbert, who had worked with him on the Newfoundland survey.

In choosing the rest of his crew, Cook had no shortage of applicants. After the fame of the *Endeavour*'s voyage, young men clamored for berths with Cook, using whatever influence they could muster. The *Resolution*'s first lieutenant, Robert Palliser Cooper, was a relative of Sir Hugh Palliser, comptroller of the navy and an early supporter of Cook's. Thirteen-year-old midshipman John Elliott was also appointed to the *Resolution* through Palliser's influence, and was quite frank about it: "It was thought, it would be quite a great feather, in a young Man's cap, to go with Capn Cook, and it required much Interest to get out with him," he recalled. Able seaman James Burney, at age twenty-one an eleven-year veteran of the Navy, was the son of the prominent musician and musical historian Charles Burney, a friend of Lord Sandwich. Elliott thought Burney was "clever & Excentric." Burney had a long and distinguished naval career, and eventually published a five-volume history of Pacific exploration. Among the other able seamen was George Vancouver, age fourteen, described by Elliott as a "quiet inoffensive young Man." The *Resolution* was his first Navy assignment. The *Adventure*'s crew was new to Pacific exploration, except for its captain: Tobias Furneaux, the competent and well-liked second lieutenant of the *Dolphin* under Samuel Wallis.

The two ships, following the example of the *Endeavour*, would carry their complement of scientists. Although the expedition had no specific astronomical assignment, it was expected to continue the work of perfecting methods of determining longitude at sea, and each ship had its astronomer, selected by the Board of Longitude. William Wales, assigned to the *Resolution*, and William Bayley, who sailed in the *Adventure*, had both observed the transit of Venus in 1769, Wales at Hudson Bay and Bayley in northern Europe. The thirty-eight-year-old Wales, a native of Yorkshire, was married to Charles Green's sister; he had worked with Nevil Maskelyne on the *Nautical Almanac*. Bayley, the son of a farmer, had taught himself astronomy and worked as an assistant at the Royal Observatory.

The natural history of the expedition was to be once again in the hands of Joseph Banks, who planned to take an even larger collection of helpers with him this time. Solander would join him again, and the noted painter Johann Zoffany would supervise the artistic tasks, assisted by three draftsmen. In addition, the group would have two secretaries, six servants, and for entertainment, two horn players. Banks also wanted the company of a distinguished astronomer, notwithstanding the Admiralty's appointment of Wales and Bayley, and he tried to get the Board of Longitude to hire Joseph Priestley, one of England's most distinguished scientists. Priestley was best known for his work in chemistry, but was a competent astronomer as well. Unfortunately for his prospects as a member of the *Resolution*'s crew, he was also a Unitarian, and the Board of Longitude, which numbered among its members several priests of the Church of England, thought Priestley's religion made him an inappropriate choice. Next Banks turned to James Lind, physician, astronomer, and author of a largely neglected treatise on scurvy. Banks convinced Parliament to make Lind a special grant of £4,000 to persuade him to join the expedition.

Since his group was so large, and very likely because his ego had been inflated with all the attention lavished on him, Banks decided the *Resolution* in its present form simply would not do to carry him and his companions to the South Pacific. He would have preferred a different sort of ship altogether, ideally a large man-of-war or merchant ship of the sort used in the East Indies trade. When it became clear that Cook and the Admiralty were set on the slow, squat, ungainly former coal ship, Banks asked them to alter the *Resolution* to allow his group more room by raising the deck several inches and building a cabin on the quarterdeck. Cook, who liked Banks and wanted him along, agreed to the modifications.

Meanwhile, during the late fall and winter of 1771-72, Cook spent much of his time overseeing the provisioning of his ships, a task he supervised closely himself rather than delegate it to his junior officers as most ship captains did. One successful voyage around the world had given him definite ideas about fitting out ships for long voyages, and the Admiralty was disposed to let him have what he wanted. "Portable soup," sauerkraut, malt, and a syrup made from oranges and lemons were provided in large quantities—three thousand pounds of portable soup and over thirty thousand of sauerkraut. In addition, the Navy's Office for Sick and Hurt Seamen wanted Cook to test carrot marmalade, reputedly an effective scurvy preventive, and Cook agreed to carry enough to give two hundred men a teaspoonful daily for six weeks. There would be other experiments with food too; the *Resolution* was to carry beef preserved by different methods, to determine how well it kept over long periods of time.

Mindful of the many times his crew had repaired the *Endeavour*, Cook asked for three extra carpenter's mates beyond the usual allotment, two for the *Resolution* and one for the *Adventure*, and for a complete supply of cooper's tools, a total of 182 different objects. He also knew exactly what sorts of goods would be most useful in trading with South Pacific islanders, and on Cook's advice the Admiralty provided each ship with a generous supply of axes, hatchets, spike nails, smaller nails, chisels, saws, augurs, knives, scissors, tweezers, combs, mirrors, beads, used clothing, and red woolen fabric. Cook also requested a supply of garden seeds and live animals to leave at some of the places where they landed, partly in the belief that English-style food supplies would benefit the local inhabitants and partly out of a practical desire to ensure adequate food supplies for subsequent ships stopping at these ports.

This experiment had been first proposed by members of the Royal Society during their discussions of the *Endeavour*'s voyage. Benjamin Franklin and Alexander Dalrymple, in particular, urged the Society to sponsor a voyage to New Zealand to stock the place with livestock, grain, and iron tools. Franklin pushed the idea as one that would benefit England's commerce in the long run. "A commercial nation particularly should wish for a general civilization of mankind," he wrote, "since trade is always carried on to much greater extent with people who have the arts and conveniences of life, than it can be with native savages." Unfortunately for Franklin's plan, the Royal Society proposed that Dalrymple should lead this voyage, and the Admiralty was no more willing to consider Dalrymple as captain of one of its ships than it had been in 1768.

Potential scientific experiments also occupied a good deal of attention. The Commissioners of the Navy asked Cook to test a new type of azimuth compass and to attempt two different methods of distilling fresh water from seawater; one involved boiling salt water in huge copper kettles, while the other, proposed by Priestley, injected "fixed air" (carbon dioxide) into salt water. Most significant, Cook was to continue his efforts to find a reliable method for determining longitude at sea. The *Endeavour*'s voyage had demonstrated that the lunar-distances method could yield consistently accurate results, although it required officers trained in astronomical observation. Cook planned to continue using these methods to establish the location of his landfalls, but he was also directed to test four different chronometers which, if they worked properly, would vastly simplify the determination of longitude. One was designed by John Harrison, a Yorkshire carpenter and self-taught clockmaker. Earlier versions of his chronometer had been tested successfully on voyages to the West Indies in 1762 and 1764, earning for Harrison half the £20,000 prize designated by Parliament for the inventor of the first workable method of determining longitude at sea. But the Board, and especially the Royal Astronomer, Nevil Maskelyne, remained skeptical of Harrison's machine and wanted more elaborate tests. Cook's expedition was also to test three other chronometers, made by John Arnold along somewhat different principles; the *Resolution* carried one and the *Adventure* the other two.

By mid-April 1772, the two ships were nearly ready for their voyage. The *Resolution* was so full of provisions, according to Cook, that there was no room for the men's sea chests; he asked the Navy Board to have canvas bags made for the sailors' possessions in place of the chests, since they would take up less room. When the *Resolution* took on its last provisions on April 25, it became apparent to Cook that the structural modifications made the ship extremely difficult to maneuver. Nevertheless, he thought some excess items could be eliminated to allow the ship to proceed safely. On May 2, Banks and his party came on board and celebrated their arrival by giving a party for Sandwich, the French ambassador, and a few others, complete with servants in scarlet and silver livery and a small orchestra. Then they left the ship, intending to rejoin it at Plymouth. Cook also left the ship for a few days to attend to personal business, leaving Cooper, his first lieutenant, to command the *Resolution* as it proceeded from the shipyard at Woolwich to the Downs. The *Adventure,* meanwhile, was already on its way to Plymouth.

But the *Resolution* made such slow progress downriver and was clearly so cumbersome in the water that after four days, the pilot brought on board to guide the ship to the Downs refused to go any farther. When Cook returned to the ship on May 14, Cooper gave him a full report, which was enough to make Cook propose to the Admiralty that the alterations made to accommodate Banks be removed, returning the *Resolution* as close to its original state as possible. Charles Clerke, writing to Banks about these developments, told him that the ship "heel'd within three streaks of her gun ports" even though the wind was light and other ships in the river remained perfectly upright. "By God I'll go to sea in a Grog Tub if desir'd, or in the Resolution as soon as you please," he told Banks; but nevertheless, he thought it was "the most unsafe ship, I ever saw or heard of."

Banks was not convinced, however. When he learned of the changes to be made in the *Resolution,* he protested vehemently and threatened to back out of the voyage.

Banks did not, at first, expect to have to carry out his threat. He merely hoped to convince Cook and the Admiralty to leave the *Resolution* as it was or, better yet, select another ship. But Cook remained firm, and the Admiralty supported him. As it became clear to Banks that he was likely to lose the contest, his protests took on a haughty and overbearing tone. In a letter to Sandwich written at the end of May, he tried to turn the tables on Cook by arguing that the *Resolution,* if returned to its original condition, would be unsafe for a voyage around the world. If allowed to sail, he claimed, it would only be because of the inflexibility of the Navy Board, "who purchas'd her without ever consulting me and now in no degree consider the part which I have taken in the voyage." Banks further contended that if the ship was returned to its original state, the space allotted to the sailors would be reduced, encouraging the spread of scurvy and other illness.

In reality, Banks was concerned about his own quarters, not those of the seamen, but he tried to head off the accusation of self-interest by assuring Sandwich that his personal accommodations were not the issue; rather, he and his companions needed adequate space to carry on their scientific work. He was prepared to go anywhere in the service of science—whether the heart of Africa or the South Pole, "I am equally ready to embark"—and, as he was quick to point out, at his own considerable expense. Banks claimed to have spent more than £5,000 on his preparations already, and "to undertake so extensive a pursuit without any prospect but Distress and disappointment is neither consistent with Prudence or Publick Spirit." When the alterations were completed, Banks visited the ship and was infuriated at the changes. "He *swore & stomp'd* upon the Warfe, like a *Mad Man,* and instantly order'd his servants, and all his things out of the ship," Elliott recalled.

Navy officials saw through Banks's feigned concern for the health of the crew, noting that all he really cared about was his own convenience. They pointed out that the *Resolution* in its original form would provide more space than he had occupied on the *Endeavour;* moreover, he would have the use of a great cabin comparable in size to that of a seventy-four-gun warship—the vessel an admiral would use in commanding a fleet. The Navy

Board took particular offense at the notion that Banks should have been consulted on the choice of ship: "Mr. Banks seems throughout to consider the ships as fitted out wholly for his use, the whole undertaking to depend on him and his People; and himself as the Director and Conductor of the whole; for which he is not qualified, and if granted to him, would have been the greatest disgrace that could be put on His Majesty's Naval officers." Writing about the affair later, Banks repeated his belief that the Navy Board had deliberately altered the *Resolution* in a manner that they knew would make the ship unsafe, as a ploy to get rid of him, "for I now had inadvertently opened to them every idea of discovery which my last voyage had suggested to me, and thence they thought themselves able to follow without my assistance now they had once gotten possession of them."

The friendship between Cook and Banks cooled in the aftermath of Banks's decision, but in the long run they remained on cordial terms. Some months later, when the *Resolution* stopped at the Cape of Good Hope on its way to the Pacific, Cook wrote Banks a conciliatory letter. "Some cross circumstances which happened at the latter part of the equipment of the Resolution created . . . a coolness betwixt you and I," he wrote, "but I can by no means think it was sufficient to me to break off all correspondence with a man I am under many obligations to." Then Cook filled Banks in on the events of the voyage up to that point and wished him success on his own expedition. (Banks, as a sort of consolation prize, was about to embark, with Solander and most of the others he had hired for the South Pacific expedition, on a voyage to Iceland.) Banks never again saw the Pacific, but he remained a patron of scientific exploration and a strong supporter of Cook's work.

For all his tantrums and difficult behavior, Banks had demonstrated the value of including naturalists on voyages of exploration, and his decision to stay home meant a replacement had to be found on short notice. Quiet inquiries within the Admiralty produced the name of a man well versed in botany, zoology, and a variety of other subjects, currently unemployed and only too happy to sign on with Cook. He was Johann Reinhold Forster, a Prussian clergyman whose interests lay more in science, philosophy, and linguistics than in religion. The descendant of an Englishman—his great-great-grandfather had left his native Yorkshire in the 1640s after supporting the losing side in the Civil

War—Forster had returned to the land of his ancestors in 1766 in the hope of supporting himself and his family as a scholar and teacher. Accompanied by his twelve-year-old son George, Forster arrived in London with barely enough money for a month's expenses and no prospects of making a living. But he had built a reputation as a man learned in languages (he reportedly knew seventeen), philology, geography, and natural science; and he carried letters of introduction that helped him move quickly into England's scientific circles.

A few months after arriving in England, Forster took a job as tutor in modern languages and natural history at Warrington Academy, a secure enough position that he felt able to send for his wife and five younger children. At Warrington, Forster found time to pursue his scholarly work, and he established himself as a naturalist of some repute. Teaching adolescent boys, however, was not the ideal profession for one of his temperament—Elliott described him aptly as "a clever, but a litigious quarelsome fellow"—and he was dismissed from the position in 1769, in part for disciplining a student too severely. Another teaching job proved no more satisfactory, and late in 1770 Forster returned to London, where he attempted to eke out a living from writing and translating. One of his major translations was Bougainville's account of his voyage around the world (it remains the only English translation of the work); about the time he finished it, Forster wrote to Banks and Solander telling them that he would be interested in sailing on a Pacific voyage.

In May 1771, a representative of the Admiralty approached Forster informally to ask if he would be interested in going on the voyage in Banks's place. Forster jumped at the chance, but on two conditions: his son George, now eighteen, should accompany him, and the rest of his family should be provided for in his absence. The former condition raised no difficulties; George, despite his youth, was already a well-trained naturalist and a talented draftsman besides. The second condition was solved by transferring Parliament's £4,000 grant intended for James Lind, who had joined Banks in refusing to sail in the *Resolution,* to Forster.

On June 11, the King officially authorized payment to Forster as naturalist on the expedition. In the next two days, Forster consulted both Solander and Banks about equipping himself for the voyage and found both men, not surprisingly, less than eager to help him. Still, he managed to get his baggage on board the

Resolution, anchored in the Thames for last-minute repairs, by June 20. A week later father and son left for Plymouth, where they would join the *Resolution* when the ship docked to take on final provisions. There they met another last-minute addition to the ship's company, William Hodges, a little-known twenty-eight-year-old landscape painter hired to replace Zoffany.

The *Resolution* reached Plymouth on July 3, and, together with the *Adventure,* embarked from Plymouth Sound ten days later, bound for the Cape of Good Hope and then on to the Pacific.

III

CIRCUMNAVIGATING THE ANTARCTIC

10. *To the Antarctic*

Cook's instructions from the Admiralty followed closely the plan for a second voyage that he had outlined toward the end of his first expedition: he was to sail to the Cape of Good Hope and then continue south until he found a continent or concluded that none existed in that part of the ocean. If he found land, he was to explore it in detail, mapping its coastline, collecting specimens of its plant and animal life, and learning as much as possible about its people; if he found nothing, the ships were to continue east around the world and back to the Cape of Good Hope, always staying as far south as possible.

Cook wanted to begin by exploring the area south of the Cape of Good Hope in part because of a discovery made by a French explorer, Lozier Bouvet, on New Year's Day, 1739, Searching for a mythical continent supposedly seen by a Frenchman early in the sixteenth century and subsequently labeled "France Australe," Bouvet found an icy, fog-shrouded bit of land about fifteen hundred miles south and slightly east of the Cape of Good Hope. The fog was too thick for him to determine the extent of his find, but he convinced himself that he was coasting the edge of a continent. Naming his discovery Cape Circumcision after the feast day on which he discovered it, Bouvet returned to France with the intention of mounting a second voyage that would sail east along the coast of this "continent" to Quiros's Espiritu Santo—for he believed that Cape Circumcision and Quiros's discovery were the eastern and western edges of a great continent. Bouvet never made his second voyage, however, and Cape Circumcision joined the list of poorly documented discoveries that might conceivably be part of terra australis. While Cook was skeptical of Bouvet's account, as he was of all tales of southern continents, Cape Circumcision was at least a place to begin his own search.

The *Resolution* and *Adventure* made an uneventful voyage south, stopping briefly at Madeira to purchase wine. The sailors amused themselves during the long days at sea catching birds, fish, and even dolphins by baiting hooks with raw meat and tossing them overboard. The Forsters made a pet of a swallow that followed the ship and eventually roosted on board. In a rainstorm the little bird got so wet it couldn't fly, so George dried its feathers carefully and set it free belowdecks, where it feasted on flies. A day or two later, the bird disappeared; Johann was convinced that one of the sailors caught it and "gave it for a meal to a favourite cat." Forster, who was not adjusting well to life among sailors, railed against the "cruel & illnatured people" he blamed for the death of his pet.

After three months at sea, Cook broke out the sauerkraut as part of his effort to keep his crew free of scurvy. This time the men ate it willingly, and when the ships reached the Cape at the end of October, three and a half months after leaving England, he was pleased to report all hands in good health. The same could not be said for other ships in port. Several had lost dozens of men from illness, with more hospitalized on shore.

Cook and his crew stayed at the Cape of Good Hope nearly a month, replenishing their supplies and getting the ships in order for their Antarctic voyage. Wales and Bayley took their instruments ashore and spent hours making observations to test the chronometers. Because the longitude of the Cape was well established, they could conduct experiments there to test the instruments' accuracy by determining local time through astronomical observations, just as they would at places of unknown longitude, and then comparing their results with the chronometers' readings. If this exercise yielded the correct longitude, the astronomers could be certain that the chronometers were keeping Greenwich time accurately. The watch built according to John

Harrison's specifications worked perfectly, but the *Resolution*'s second chronometer was already losing time.

The Forsters kept busy exploring the terrain, and Johann found to his chagrin that the memory of Joseph Banks remained vivid. Banks had been so generous in paying for specimens, Forster complained, that he had difficulty getting anything for an affordable price. Nevertheless the time spent at the Cape was profitable. He and George met a young Swedish naturalist named Anders Sparrman—like Solander, he had been trained by Linnaeus—and were so impressed with him that they persuaded Cook to take him along. Sparrman, who was then twenty-four, had already been on a voyage to China and was in South Africa on a botanical expedition sponsored by the Swedish government.

Cook got a warm welcome from the Dutch governor of South Africa, who gave him an intriguing bit of news: two French ships sailing across the Indian Ocean about eight months previously had discovered land due south of Mauritius, at about latitude 48°S. The reports were fragmentary and their accuracy open to question, but here was another vague discovery that might be part of a continent; 48°S was only 6° north of the reported latitude of Bouvet's Cape Circumcision. (The explorer was Yves de Kerguelen, who had sailed in search of the same mythical land that eluded Bouvet a generation earlier. He discovered a tiny island southeast of Mauritius, claiming that it was the central part of "France Australe.") Cook also learned that another French expedition, this one commanded by Marion du Fresne, had stopped at the Cape some months earlier enroute to Tahiti to return Ahutoru, the young man who had sailed to France with Bougainville. Luckier than Tupaia, Ahutoru had lived to see Europe, but he died of smallpox at the Cape.

Although it was late in the southern spring when Cook left the Cape, his ships ran into cold, stormy weather within days. The sea was so rough that cooking fires could not be kept burning safely, forcing the men to eat "half-cooked salt meat, with a few tough boiled cabbage leaves for dinner" and cheese for supper. Waves washed over the *Resolution*'s decks and leaked through poorly caulked seams into Johann Forster's cabin. Already annoyed about its small size, Forster began to regret that he had not inspected his quarters and demanded something better before going on board. But he was not the only one suffering; the *Adventure* was leaking too, soaking the crew's sleeping quarters and leaving the men "extreemly fatigued, cold and helpless."

The two ships tossed about for nearly three weeks without a single calm or sunny day to relieve the misery. Then, on December 11, a sailor stationed at the *Adventure*'s masthead cried out that he saw land. As the ships drew closer, what looked like an icy mountain loomed ahead. Pickersgill compared it to a pyramid, with sides about a quarter mile in length, rising at least three hundred feet above the sea. "The sea beat against it with great voilence," he said, "and the spray flew an amazeing hight and had beat hollow caverns in the sides, which had a remarkable blue Icey coulour, from the reflection of the water." Johann Forster was more precise in his description: the mountain of ice was at least twice as high as the *Resolution*'s masthead, or just over four hundred feet—a total of at least 128 million cubic feet of ice.

It was clear to Forster and Cook, once they got close to this icy mass, that it was not land at all, but solid ice—an "ice island," they called it—although others persisted in believing that it was land. As the ships continued their course south, the men saw more ice islands and then huge sheets of ice, lower than the islands but extending much greater distances. Smaller pieces of ice floated in the sea around them, frequently banging the ships' hulls. Furneaux's crew called them "plumpers" from the noise they made.

The air was now at the freezing mark, and the ships were almost constantly battered with rain, sleet, and snow. Ice covered the ropes and sails, and heavy fog frequently reduced visibility to near zero. Under these conditions the ice islands, awe-inspiring at first, became a grave danger. It was worse than navigating among rocks, Pickersgill explained, because men shipwrecked on a rocky islet might have some hope of survival, "but here if we went against an Ice Island there is nothing but immeadate Death." Amazingly, animal life was abundant—seals and several varieties of birds populated the frigid waters—and this encouraged Cook to continue sailing south, since it was generally believed that seals and penguins did not travel far from land. The ice itself, in a perverse way, was a hopeful sign, because prevailing wisdom held that ice must be formed adjacent to land and salt water would not freeze.

Three days later the ships reached a huge wall of ice. In the distance the crew could see what appeared to be land, but there was no way to penetrate the ice to get closer to it. Cook turned east, sailing along the edge of the ice to a point where the ships could get around the mass and sail south again. After about ten miles, they were nearly surrounded by solid ice. It was almost like being on the coast of a continent—narrow streams of water ran through the ice in spots, mountains of ice rose in the distance, and ice islands floated in the sea around them—but it was all nothing but ice, and two more days of sailing convinced Cook and Forster that the hills they had seen in the distance were just more ice too. Afraid of getting trapped, Cook decided it was both useless and dangerous to attempt sailing farther south at this point. Instead he planned to sail north to get away from the ice, then run east some distance, and finally turn south again. In that manner he hoped to get around the ice mass "to satisfy my self whether it joined to any land or no."

Forster, meanwhile, was beginning to form his own theories about ice. From his reading and from conversations with those among the *Resolution*'s crew who had sailed in the North Atlantic, he concluded that ice formations were of three basic types: "ice islands," as the Englishmen called them, which could be hundreds of feet high; even larger masses formed of closely packed chunks of ice called by experienced Arctic sailors "pack ice"; and unbroken sheets three to four feet high and a mile or more in extent, known as "table ice" or "field ice." The first two types, he believed, could move great distances from the point where they were formed and gradually melted, but field ice moved very slowly and remained close to the land where it originated. On this last point Forster followed the conventional wisdom that ice was always formed near land.

How the ice had been created was a more puzzling subject. The usual explanation, Forster wrote, was that large rivers froze, and pieces of the ice eventually broke off and floated north. This might be a reasonable explanation for the formation of pack ice, he believed, but it would not work for massive field ice, which was too extensive to have originated in the mouths of rivers. Instead, he theorized that field ice was formed by the gradual accumulation of snow in bays and inlets. The temperature was so cold in the Antarctic that this snow rarely melted, but kept building up to create huge masses; then, in those rare summers when temperatures were higher than normal, enough of the ice would melt to force huge chunks of it away from land and out to sea. Forster noticed that the ice around these formations was made up of layers about four to six inches thick, supporting his notion that they were created by the progressive buildup of snow.

Chunks of ice taken from the sea and melted produced fresh water, which seemed to prove that ice was formed near land and that seawater would not freeze. The men of the *Resolution* and *Adventure* had ample evidence on this point, because they came to depend on melted ice to replenish their water supply. The machinery designed to distill fresh water from salt worked reasonably well, but slowly, and melting ice was a more efficient method of getting large quantities of water. When the sea was calm enough to permit a boat to row up to the face of an iceberg without being dashed to bits, Cook would send out the boats to chip off chunks of ice. Then, on board ship, some of the ice was packed into casks and left to melt gradually, while the rest melted over fire in huge copper kettles. On one occasion, the crew filled six boatloads, yielding six and a half tons of water—more than they had had on board when they left the Cape of Good Hope. Another watering expedition produced fifteen tons for the *Resolution* and between eight and nine for the *Adventure,* all in less than a day. It was "the most expeditious way of watering I ever met with," Cook thought. Initially, some of the men worried that the water would burst the casks when the ice finally melted; to prove to them that water takes up a smaller volume than ice, Cook filled a small pot with ice and placed it in a warm cabin to melt. The quantity of water was so obviously less than the volume of ice originally placed in the pot that the sailors were convinced.

Forster puzzled over the reasons sea ice produced fresh water. Although he continued to believe that seawater could not freeze, he thought the spray from the ocean must freeze when it washed over icebergs. Why, then, was there not a trace of salt in the water derived from melted ice? The best explanation he could produce was that salt somehow separated itself from water and ran off the icebergs "uncongealed," leaving the fresh water to freeze.

Remarkably, the men managed to retain their sense of humor despite the wretched, wet cold. They compared the icebergs to familiar shapes, like castles, houses, old ruins, towns, even St. Paul's Cathedral, and amused themselves counting penguins (eighty-six on one iceberg) and sometimes shooting them for sport. Even Forster, so quick to accuse the sailors of killing his pet bird weeks earlier, admitted that the crew showed remarkable skill and composure in the face of danger. On Christmas, still sailing south in Cook's effort to get around the pack ice, the men put aside work for their first day of unbuttoned merriment in weeks. Cook made no attempt to cover any distance; "seeing the People were inclinable to celebrate Christmas Day in their own way," he thought it best to let them have their party. Sparrman was amazed that the crew not only got drunk but fought, "in the English fashion, which is called boxing." The sport seemed barbaric to him, but, good scientist that he was, he watched the men closely and described their boxing in his journal in some detail. It would, he said, "throw light on . . . the Christianity and character of the British savages (as well as that of some of our sailors)." Cook told Sparrman that most ship captains would not permit boxing, but on a long and difficult voyage, he thought it best to let the men settle their disputes by fighting rather than nurse grudges for a long time.

By New Year's Day, the ships reached the approximate location of Cape Circumcision, according to Bouvet's reports, but the crew could see nothing except ice. Both Cook and Forster concluded that the French explorer must have seen mountains of ice surrounded by field ice and mistaken it for land—an understandable error, Cook thought, considering that he and his crew had done the same thing three weeks earlier. In fact Bouvet had seen land, but it was a tiny island no more than five miles long; and his longitude was about 7° off, enough to make such a small island extremely difficult to find amid the ice. It was seventy years before anyone saw Cape Circumcision, today known as Bouvet Island, again.

Forster was puzzled that they had sailed amid ice for so long without seeing the land where it originated, but Cook doubted that a continent existed anywhere nearby. Deciding to waste no more time searching, he reversed course and sailed east to look for the land supposedly discovered by French ships south of Mauritius. During the next weeks, Cook and the Forsters continued their study of ice and its relation to land. To compare the temperature at various depths of the ocean, from time to time they lowered a thermometer rigged in a box that would allow water to enter, discovering that the surface of the sea was colder than the air and became progressively colder at lower depths. Often the water temperature was below the freezing point, and sometimes the air temperature was as well, which led Cook and Forster to reconsider whether seawater would freeze. Forster thought perhaps salt water required a colder temperature or that the motion of the sea might prevent its freezing, and he resolved to undertake controlled experiments when he returned to England to test his theories. (Later, when Forster had time to pursue his research on the formation of ice, he found that salt water can in fact freeze and, when melted, produces fresh water. The massive ice in the Antarctic, he decided, was produced simply because the temperature was so cold for so much of the year that the ocean itself froze.)

A rare stretch of good weather allowed the astronomers to resume their work. Wales, Bayley, Cook, and some of the other officers calculated longitude using the lunar-distances method and compared their results with the readings derived from the chronometers, obtaining similar results. The minor discrepancies between the two methods did not concern Cook. It was clearly possible, he concluded, to determine longitude within a degree and a half of accuracy using a chronometer, and usually possible to do even better; and seamen "now have no excuse left for not making themselves acquainted with this useful and necessary part of their Duty." He also took advantage of the good weather and the ample supply of water to have his men clean themselves and their clothes, "a thing that was not a little wanting."

Whenever the ice cleared, Cook directed the ships to sail south for a stretch, still looking for the land behind the ice. On January 17 they crossed the Antarctic Circle, the first men ever to sail that far south. The next day they were surrounded by loose pack ice, and sailors at the masthead counted thirty-eight ice islands in the distance; more serious, they could see an apparently endless mass of solid field ice to the south. Furneaux thought they could get through this "archipelago of ice," but he and Cook saw no useful purpose in attempting it and turned north again. At this point, searching for the rumored French discovery south of Mauritius seemed the more sensible approach to finding a continent.

At the beginning of February they reached the reported location of the French discovery, but land was no more in evidence here than at the spot where Cape Circumcision supposedly lay. "If my friend Monsieur found any Land," Charles Clerke remarked, "he's been confoundedly out in the Latitude & Longitude of it, for we've search'd the spot he represented it in and its Environs too pretty narrowly and the devil an Inch of Land is there." (Cook's expedition missed finding Kerguelen's island by about 5 degrees of longitude.)

Shortly after giving up the search for land, the *Resolution* and *Adventure* lost sight of each other. Cook and Furneaux had a plan worked out in the event of such a separation, and each ship fired guns in an attempt to identify its location for the other. When this failed, Cook sailed to the spot where he had last seen the *Adventure* and cruised around the area for three days. Still there was no sign of the other ship, and finally he continued east, hoping to meet the *Adventure* at Queen Charlotte Sound in New Zealand, which he and Furneaux had fixed as a rendezvous point.

Assuming that the *Adventure* would follow a more northerly course toward New Zealand, Cook headed southeast, intending to make another attempt to cross the Antarctic Circle. Within a few days the *Resolution* was once again surrounded by ice, but, as Cook put it, he and his crew had become so used to the dangers of ice that the moments of fear never lasted long and were "in some measure compencated by the very curious and romantick Views many of the Islands exhibit and which are greatly heightened by the foaming and dashing of the waves against them. . . . in short the whole exhibits a View which . . . at once fills the mind with admiration and horror, the first is occasioned by the beautifullniss of the Picture and the latter by the danger attending it." Only the "pencle of an able Painter" could truly describe the scene, and in fact Hodges was at work drawing it. Even allowing for some degree of romanticization, evident in all of Hodges's work, it was a stunning picture: a frail and tiny *Resolution,* dwarfed by craggy masses of ice. The appalling beauty of the southern sea did not deflect Cook's attention from its dangers, however. Realizing that it would be foolhardy to attempt another crossing of the Antarctic Circle, he shifted course to the northeast, toward New Holland.

Cook always felt compelled to justify himself at some length whenever he chose an easier course over a more difficult or dangerous one, and this occasion was no exception. "After crusing four months in these high Latitudes it must be natural for me to wish to injoy some short repose in a harbour where I can procure some refreshments for my people of which they begin to stand in need," he wrote. Several of the crew were suffering from frostbite on their hands and feet, and symptoms of scurvy were apparent among both men and animals after the long, unbroken stretch at sea; "even the Catts and Doggs were affected so much as to cause their teeth to become loose."

When the mountains of New Zealand's southwestern coast came

into view on March 25, "everybody that was able to crawl on the Masts and yards got up to satisfy their longing senses of a sight allmost forgot," Pickersgill wrote. "Those who were not able, importuned the others as they came down for a discription; without being able to wait untill they could see it off the Deck so much was their attentions ingross'd with it." The need for rest and fresh food was so serious that Cook decided not to sail immediately to Queen Charlotte Sound, his appointed rendezvous with the *Adventure,* but to look first for an anchoring place among the bays he had seen on New Zealand's southwest coast during his voyage in the *Endeavour*. Two days later the *Resolution* sailed into Dusky Bay. Although the entrance was three to four miles broad and small islands were visible within, the Englishmen were hardly prepared for the wild beauty of the setting: not a single bay but an intricate pattern of coves and inlets carved out of sheer cliffs, some connected by narrow passages, some dotted with densely forested islands. The height of the cliffs, the abundance of trees, and the mist that hung almost perpetually over the bay gave the place an otherworldly quality, and also meant that light was dim much of the time; although Cook had named the bay on his first voyage, without entering it, his choice was apt.

Cook chose one of the islands as an anchoring place and was able to get close enough to form a bridge from the ship to land by bending a flexible young tree across the narrow gap, propping up a log parallel to it, and laying a row of planks across both. Immediately the men set about catching fish and gathering plants for food; "the real good taste of the fish, joined to our long abstinence, inclined us to look upon our first meal here, as the most delicious we had ever made in our lives," George Forster recalled, and Cook reduced the normal ration of ship's stores by half, making up the balance with fresh fish and wild fowl. Trees and plants were so thick on the shores and islands of Dusky Bay and the birds so tame that the Englishmen thought the area must never have been inhabited. Birds sat on the ends of their guns, "and perhaps looked at us as new objects, with a curiosity similar to our own." Within twenty-four hours of their arrival, however, they set about leaving their mark on this pristine setting, chopping down trees and hacking away at the dense underbrush to clear enough space to put up tents for the men assigned to shore duty, the armorer's forge, and Wales's portable observatory. For the observatory alone

Wales destroyed "more Trees & curious shrubs & Plants, than would in London have sold for one hundred Pounds."

George Forster waxed eloquent over the power of European "civilization" to impose order on an untamed wilderness: "this Spot, where immense numbers of plants left to themselves lived and decayed by turns, in one confused inanimated heap, . . . we had converted into an active scene, where a hundred and twenty men pursued various branches of employment with unremitted ardor. . . . We felled tall timber-trees, which, but for ourselves, had crumbled to dust with age; our sawyers cut them into planks, or we split them into billets for fuel. By the side of a murmuring rivulet, whose passage into the sea we facilitated, a long range of casks, . . . stood ready to be filled with water. . . . Our caulkers and riggers were stationed on the sides and masts of the vessel, and their occupations gave life to the scene, and struck the ear with various noises, whilst the anvil on the hill resounded with the strokes of the weighty hammer." Even "polite arts" and science flourished, as Hodges painted the scene, Wales pursued his astronomy, and George and his father collected plants and animals. "The superiority of a state of civilization over that of barbarism could not be more clearly stated, than by the alterations and improvements we had made in this place," he concluded.

On their second day at anchor the Englishmen discovered that their bay was not entirely uninhabited when eight people in a small double canoe paddled to within a few hundred yards of the ship. Samuel Gibson, the marine who had attempted to desert at Tahiti and the only man on board who could speak much of the Tahitian language, tried to talk with them, urging them to come on board, but the group would venture no closer and left after about half an hour. Later in the day, Cook and several companions discovered the double canoe beached in a nearby cove. A short stroll inland revealed two small huts—Cooper described them as "miserable habitations being nothing but two small wigwams . . . something resembling Dog Kennels"—and the remnants of a fire, but no people. After leaving gifts at the huts and in the canoes, they returned to the ship; when Cook and the Forsters returned to the cove, now nicknamed Indian Cove, four days later, they found their presents still there, untouched.

Cook, the scientists, and the officers made daily excursions to different parts of Dusky Bay, while the sailors spent their time collecting supplies of fresh food, water, and wood. Among other tasks they devised a way to brew beer from the spruce trees growing near their anchorage. Cook, typically, spent most of his time making a detailed chart of the bay, no simple task because of its size and its many islands and coves, while Johann Forster concentrated on collecting new species of birds, which among all forms of wildlife were his greatest interest. Cook, who did not always take the Forsters with him on his explorations, found time to capture seals, ducks, and a few other kinds of birds to add to their collection, and the sailors on board ship hooked several new types of fish. Some of the men claimed to have seen a yellowish-colored animal about the size of a rabbit, which especially piqued Johann Forster's interest, since Banks had never seen any quadrupeds in New Zealand and Forster was always on the lookout for ways to top Banks's achievement. The mysterious animal never materialized, however, and for good reason; except for rats and domesticated dogs, there are no quadrupeds native to New Zealand.

Plant-gathering, which Forster delegated to his son, was less successful—Forster thought it was because winter was approaching—but still the two naturalists' cabins were crammed with specimens to describe and sketch, a task they took on themselves. (Hodges was a landscape painter, not a scientific draftsman, and had been employed to make a general pictorial record of the voyage.) Johann had complained about his small, cramped cabin from the beginning of the voyage, and now that it had to make do as both work space and living quarters he became even more petulant. The cabin was damp, dirty, and smelly, he claimed; it nestled right under the trees and was therefore always dark, forcing him to burn candles all day; the men piled up lumber in front of his door so he couldn't get in and out easily. The fact that the entire ship was damp and dark much of the time because of the dense trees and almost daily rain was lost on Forster, who preferred to view his problems as a personal affront.

About a week after the Englishmen's first encounter with the local residents, a man and two women unobtrusively appeared on a small island near the *Resolution*'s anchorage. Pickersgill thought he read "Hope, Fear, Dispair, and every other conflict" in the man's face, but George Forster pointed out that if these people were frightened of the English, they could easily have stayed out of sight. That they did not he took as a sign of "openness and honesty," not recognizing that they might well have been motivated by simple curiosity.

The Englishmen tried calling out the Tahitian equivalent of "friend come here," and Cook tossed handkerchiefs as a sign of peace; the man responded with a long speech and a few defiant swings of his club, but refused to move closer or pick up the handkerchiefs. Trying a different tactic, Cook rowed across to the island, alone and unarmed, and handed the man a few sheets of white paper. He "trembled very visibly" but accepted the paper and touched noses with Cook, the traditional New Zealand sign of greeting. Then he called the women to join him and undertook to converse with Cook. "We spent about half an hour in chitchat which was little understood on either side," Cook said, although Johann Forster thought the exchange was "as least as edifying as great many which are usual in the politer circles of civilized nations." Elliott, witnessing Cook's ability to win the confidence of native peoples for the first time, was deeply impressed. "He was *Brave,* uncommonly *cool, Humane,* and *Patient,*" the young man wrote. "He would land alone *unarm'd*—lay *aside* his Arms, and sit down, . . . throwing them *Beads, Knives,* and other little presents then by degrees advancing nearer, til by *Patience,* and *forbearance,* he gained their friendship."

The next morning Cook found the trio waiting for him. At their invitation, he and a few companions followed them home, where they gave Cook a garment made of flax, a belt, beads carved from the bones of birds, and albatross skins. In return, the man asked for an English cloak. Unwilling to part with any of theirs, Cook refused, but when he returned to the ship he had one made from some of the red fabric brought along as part of the expedition's stock of trade goods. At the New Zealanders' home the next day, Cook found the family, which included a third woman and several children, preparing to welcome him with great ceremony. All were dressed in their finest clothes, their hair tied on top of their heads, glistening with oil and decorated with feathers. Cook, who was wearing the red cloak, took it off and presented it to the man, who was so pleased that he immediately gave Cook his club. Then they attempted conversation, Cook having brought along Gibson as interpreter, but with little more success than previously, for Gibson could not understand them despite his reputed command of Tahitian.

A few days later, the New Zealand family paid the English a return visit. This time they came alongside the *Resolution* in their canoe, but declined all invitations to go on board. Thinking they were just bashful, Cook joined them in their canoe and tried to persuade them to come closer, but still they refused. Eventually they landed on the island directly opposite the ship and scrutinized it at length, quite taken with the nails used in its construction. They chatted happily with the sailors on shore, obviously no longer intimidated by the English, and stayed on the island overnight singing and making "strange Gestures, which as far as we could understand," Wales said, "was a Conversation which they held with some Being above the clouds." The next morning they left, not to be seen again for a week.

On their next visit they arrived dressed in new clothes, or what appeared to be new clothes to the English. After some argument among themselves, the man sent two of the women and the children away and approached the ship with the third woman. Standing at the water's edge, he spoke in grave tones for several minutes, "whether to himself, the Ship, Us or some superior Being I dare not even hazard a guess," Wales remarked. Then he and the woman crossed the short bridge to the ship, but instead of walking on the planks spanning the two tree trunks, they took pains to stay on the trunk itself—the one that remained planted in the ground. At the top of the bridge they paused while the man struck the side of the ship with a green branch and intoned another solemn speech. Gingerly, he stepped on board and stamped on the deck as if to test its firmness. Every time he went to a different part of the ship, he repeated the process, waving his branch and stamping the wood under his feet.

The New Zealanders' initial refusal to board the ship and their hesitation when they finally did venture aboard stemmed from a deep distrust of anything so large and elaborate and so obviously man-made. The people in this part of New Zealand had no oral tradition of large ships coming from afar, as the residents of Queen Charlotte Sound did, and it is likely that even the tradition of the great migration from Tahiti had grown dim, since Dusky Bay was so isolated from the main centers of New Zealand settlement. Less than 5 percent of New Zealand's population, or no more than five to seven thousand people, lived south of the area around Queen Charlotte Sound, scattered in small groups, unlike the well-organized tribes of the regions farther north. A ship the size of the *Resolution* was, therefore, totally without precedent; and this solitary family did not have friends and neighbors to help build up their courage and support them in an encounter with the strangers.

No wonder, then, that it took several days for them to work themselves up to boarding the ship, or that they argued about the wisdom of going on board. The man's curiosity was stronger than his fear of the unknown, but before venturing on board he prayed to his god and struck the *Resolution* with a green branch—traditionally a sign of peace—perhaps in an attempt to conciliate the English god. (The people of the North Island had thought the *Endeavour* was a kind of god.) He and the young woman kept their feet on the tree trunk as they crossed over to the *Resolution* because they knew it would support them. How could they be certain that the man-made planks, or the deck of the ship, would not cave in beneath them?

Once on board, the man was fascinated with the ship's decks. The notion of a multiple-storied boat amazed him. He and the young woman also showed great interest in the animals on board, especially the cats. They stroked the animals the wrong way, even after they were shown the right way—struck with the "richness of the furr," George Forster thought. Cook invited the pair to his cabin, which required further thought and discussion. Eventually the man agreed, and once inside he displayed great curiosity about every detail of the captain's quarters. He liked the chairs, especially when he discovered they could be moved from place to place, and asked Cook where he slept. When shown the hammock suspended above the floor, he was "mightily pleased." By now the New Zealanders had gained confidence and ran quite freely around the ship examining every detail, but they would not touch the Englishmen's food, despite repeated invitations.

In the spirit of hospitality, the man offered to anoint Cook's hair with oil in the same manner that he dressed his own. Cook politely declined—the oil, as George Forster explained, "though perhaps held as a delicious perfume, and as the most precious thing the man could bestow, yet seemed to our nostrils not a little offensive"—but the woman was more persistent. She wore a tuft of feathers that had been dipped in oil on a string around her neck, which she insisted on giving to Cook. "He was forced to wear the odoriferous present, in pure civility," Forster wrote. The young woman also bestowed her attentions on Hodges, whose pencils especially intrigued her. She gave him a cape and, over his protests, tried to tie up his hair New Zealand–style.

All shyness vanished, the New Zealanders stayed on board overnight and into the next day, while Cook and the Forsters went off to explore another part of the bay. There they encountered a second group who, like the first, approached the English with a mixture of friendly curiosity and apprehension. Cook's tactic of proffering white paper worked again, and an exchange of native cloth for hatchets cemented good relations. Meanwhile, on board the *Resolution,* the first native guest grew bold enough to request a demonstration of musket fire, and then insisted on firing the gun himself—which he did, three or four times, although his companion pleaded with him to leave it alone. Before leaving the ship, he told some of the crew that he intended to use the hatchets they had given him as weapons. His comment mystified George Forster, who thought the New Zealanders lived in the midst of such abundance that they need never engage in warfare for their survival. Forster could explain the man's bellicosity only as the result of living in a "state of barbarism," and lamented the probability that tools given to the New Zealanders in "the pleasant hope of facilitating the oeconomical operations of these people, and of encouraging some degree of agriculture among them" would be turned to violent purposes. Cook agreed that the people of Dusky Bay must be warlike by nature, even though he thought them much friendlier—more like the Tahitians—than other New Zealanders he had met. Otherwise, he wondered, "why do not they form themselves into some society a thing not only natural to Man, but . . . even observed by the brute creation?"

In the first of his attempts to encourage English-style agriculture in the South Pacific, Cook had his men plant several types of grain and vegetable seeds and release their few remaining geese in a secluded cove where he thought the birds would find adequate food and protection from the local inhabitants. He decided against leaving any sheep or goats, however, because of a shortage of the proper kind of food for them.

At the end of April the Englishmen left Dusky Bay, their health and spirits refreshed after a month on land. Cook was well pleased with his discovery as a potentially useful stopping place for ships, although he admitted that the bay was so remote it would probably not be visited again soon. (Even today, Dusky Bay can be reached only by boat, small aircraft, or foot.) Still, he noted, "we can be no means till what use future ages may make of the discoveries made in the present," and so he charted and described the bay with as much detail as he devoted to other, less obscure places.

About three weeks later Cook sailed into Queen Charlotte Sound, where he found the *Adventure,* exactly according to plan. As Cook suspected, Furneaux had sailed slightly north of the *Resolution*'s track, first searching unsuccessfully for Kerguelen's reported discovery, and then exploring Van Diemen's Land briefly before sailing on to New Zealand. He had anchored in a large bay on the east coast of Van Diemen's Land (known today as Adventure Bay) and arrived at Queen Charlotte Sound about six weeks ahead of the *Resolution.* Furneaux and his men were preparing to settle in for the winter, but Cook did not want to remain idle—he was thoroughly acquainted with that part of New Zealand—and proposed instead a voyage to the tropics for some further exploring before heading back to the Antarctic the following summer.

First he wanted to replenish the expedition's supply of fresh food and continue his program of launching English agriculture. On the *Endeavour*'s voyage, Cook's men had benefited from eating the wild celery that grew abundantly around Queen Charlotte Sound, and now Cook set the *Resolution*'s crew to gather it by the boatload. Furneaux had already planted garden seeds, and Cook planted more, taking care to explain to the inhabitants what he was doing and how it would benefit them. He also put the remaining two sheep ashore at a remote spot where they would be safe from molestation, but was vexed to discover a few days later that both animals were dead, apparently from eating poisonous plants. "Thus all my fine hopes of stocking this Country with a breed of sheep were blasted in a moment," he wrote. Refusing to give up, he left two goats and two pigs, convinced that the New Zealanders would leave them alone out of fear and that, in time, the animals would multiply and stock the region.

Cook did not recognize any of the people who visited the *Resolution,* nor did they remember him. A fort and several dwellings were deserted, which suggested to Cook that the people living at Queen Charlotte Sound in 1770 had moved or been driven elsewhere. He thought it likely that the New Zealanders lived a nomadic existence, moving from place to place as the season or their fancy suited them, a theory strengthened by the appearance of about a hundred men, women, and children who were obviously not residents of Queen Charlotte Sound—the local residents claimed they were enemies—at the English camp. They arrived in several canoes, loaded with possessions, and apparently ready to establish a settlement. "It is very common for them when they even go but a little way to carry their whole property with them," Cook wrote, "every place being equally alike to them if it affords the necessary subsistance so that it can hardly be said that they are ever from home." The scattered families around Dusky Bay must have migrated there in just this gradual, aimless fashion, he believed. This nomadic style of existence was more common on the South Island than the North, Cook theorized, for the people of the North Island appeared to live in organized communities with chiefs, laws, and a well-developed system of mutual defense, while the inhabitants of the South Island were isolated, scattered, and subject to no government except that of their families, forcing them to move about frequently and remain constantly on guard against potential enemies. Yet notwithstanding the unfamiliar faces at Queen Charlotte Sound, everyone asked about Tupaia, including those in the large group, who had clearly traveled some distance. A few people had objects that Cook recognized from the *Endeavour*. He concluded, correctly, that Tupaia's fame had spread through much of New Zealand, and that the items from the *Endeavour* had changed hands many times.

Those who had sailed in the *Endeavour* noticed other differences from their first visit to Queen Charlotte Sound. In 1770, sexual encounters between the sailors and the native women had been infrequent. "The Women of this Country," Cook said, "I always looked upon to be more chaste than the generality of Indian Women." This time, however, the women—with their fathers or brothers—went through the ships by the dozens, offering sex for nails or shirts. Some were reluctant, but the men encouraged them, greedy for the objects the women could earn. Disgusted at this outright prostitution, Cook blamed the situation on the English themselves. "We debauch their Morals already too prone to vice and we interduce among them wants and perhaps diseases which they never before knew and which serves only to disturb that happy tranquillity they and their fore Fathers had injoy'd," he wrote. "If any one denies the truth of this assertion," Cook added, "let him tell me what the Natives of the whole extent of America have gained by the commerce they have had with Europeans."

George Forster, analyzing the situation more closely, observed that married women did not participate in the prostitution and that single women might freely take as many lovers as they liked anyway. So it might be argued that Europeans did not have a negative effect on their "moral characters," but Forster recognized that prostitution developed only because "we created new wants by shewing them iron-tools." On balance, he thought the English had indeed corrupted New Zealanders' morals, an evil he considered even greater than the occasional killing of native peoples. "If these evils were in some measure compensated by the introduction of some real benefit in these countries, or by the abolition of some other immoral customs among their inhabitants, we might at least comfort ourselves, that what they lost on one hand, they gained on the other," Forster thought; "but I fear that hitherto our intercourse has been wholly disadvantageous to the nations of the South Seas." Gifts of iron tools, livestock, and an attempt at English gardening were small compensation for bringing the seeds of greed and dissatisfaction. The only people not harmed by the English, he believed, were those who resisted any close contact with them, like the men of New Holland.

In general, the Englishmen's reactions to the New Zealanders were mixed. The men who sailed on the *Endeavour* had spread tales of cannibalism and ferocious war dances, and the *Adventure*'s crew claimed to have seen some evidence of cannibalism in their two months at Queen Charlotte Sound. But for the most part, the New Zealanders had been friendly and peaceable. Certainly the family at Dusky Bay could not be considered threatening, and the Queen Charlotte Sound residents behaved more like Tahitians than the New Zealanders that the *Endeavour*'s crew recalled.

Whether or not cannibalism existed among the New Zealanders remained a matter of controversy, despite Cook's observations on his previous voyage. Elliott and Burney, among others, accepted the stories without question, labeling the New Zealanders "*desperate, fearless ferocious, Canibals,*" while both Wales and Bayley remained skeptical, having seen no evidence of cannibalism themselves. Wales was downright sarcastic: "Being going to leave this land of Canibals, as it is now generally thought to be, it may be expected that I should record what bloody Massacres I have been a witness of; . . . Truth . . . obliges me to declare, however unpopular it may be, that I have not seen the least signs of any such custom being amongst them." He expressed his doubts about the stories of cannibalism related by the *Endeavour*'s crew, claiming that one could come up with alternative explanations for every incident reported. And Sparrman, although not discounting the possibility that the New Zealanders might be cannibals, noted the similarities between their methods of warfare and those of his Viking forefathers, implying that ancient Europeans could have been cannibals too. After all, they "found delight," he said, in drinking from their enemies' skulls. Rather than condemning the New Zealanders, Europeans should pity their primitive state and hope they become more civilized, for Europeans may well be descended from people "no less barbarous."

George Forster, most remarkably, skirted the whole issue of cannibalism and turned the New Zealanders into independent, democratic warriors. Observing that the chiefs were generally young, strong men, he concluded that they might be elected by their comrades. "The more we consider the warlike disposition of the New Zealanders, and the numerous small parties into which they are divided, this form of government will appear indispensible," he wrote; the New Zealanders understood that the qualities they needed in their chiefs were not necessarily inherited and that "hereditary government has a natural tendency toward despotism." The opposite, of course, was in fact true. Chiefly status in New Zealand, as in Tahiti, was inherited, although the system was not a rigid one and political power and social status were not necessarily synonymous. Just as Banks, the upper-class Englishman, had described Tahitian society as a European-style monarchy, now George Forster—young, idealistic, and highly critical of European political systems—allowed his very different point of view to shape his understanding of New Zealand's government and society.

The Englishmen left Queen Charlotte Sound about three weeks after the *Resolution* arrived, on the next leg of Cook's grand plan to discover a southern continent or prove conclusively that no such continent existed. He intended to sail due east from New Zealand to longitude 135° or 140° west (about 10 to 15 degrees east of Tahiti) and then loop around on a northwesterly course to Tahiti. This region was the last remaining unexplored area within the temperate zones of the South Pacific and therefore the last hope for a southern continent that was not buried under ice. After a short stop at Tahiti, Cook planned to return to New Zealand before crossing the extreme southern Pacific the following summer.

11. *Return to Tahiti*

The *Resolution*'s men were in high spirits as they left New Zealand, with the prospect of a few weeks at Tahiti ahead of them. They even made light of the misery of the previous months; recalling the relative comfort of merchant voyages compared to what they had just been through, some of the sailors joked that the albatrosses following the ship were really the souls of dead merchant ship captains, exiled to the southern Pacific to suffer the cold and storms they had never experienced in life.

The officers couldn't quite bear the idea of going back to salt provisions after the fresh food of New Zealand, so on their second day at sea they had one of the dogs they picked up at Queen Charlotte Sound killed for their dinner—with none of the squeamishness displayed by the men of the *Endeavour* the first time they were served dog meat at Tahiti. The Forsters agreed with Cook that dog tasted just like mutton and couldn't understand why Europeans refused to eat it. Only Wales, who was known for his picky taste in food—he didn't like to eat sea birds because of their oily taste—turned up his nose, and even he was eventually persuaded to try a bit of the roast dog. Some days later, all fresh supplies exhausted, the *Resolution*'s crew dined for the first time on some of the salt beef that had been preserved in a new way. It proved quite palatable, but the sailors, always skeptical of Cook's scientific experiments, labeled the meat "experimental beef," a term they applied to anything out of the ordinary. Beer made from malt was "experimental beer"; water distilled from seawater was "experimental water"; and Wales, Hodges, and the Forsters were "experimental gentlemen."

In two months of sailing through previously unexplored territory, the Englishmen saw no land and no evidence to suggest the presence of a continent in the region. Within a few days after leaving New Zealand, the usual signs of nearby land—clumps of seaweed floating on the ocean's surface, birds flying around the ship—disappeared, and the ocean broke around them in great, rolling swells. From these signs Cook concluded that they would find no large land mass in this part of the ocean. More than ever convinced that the southern continent was a myth, Cook still, in his typically cautious manner (and mindful of the continent partisans at home), refused to abandon his plan to search the extreme southern Pacific the following summer. The possible existence of a continent, he wrote, "is too important a point to be left to conjector, facts must determine it and these can only be had by viseting the remaining unexplored parts of this Sea which will be the work of the remaining part of this Voyage."

But it would not be safe to sail farther south until October, and in the meantime Cook decided to proceed with his plan to spend some weeks at Tahiti, resting his crew and stocking up on fresh provisions. Approaching Tahiti from the northeast, the ships sailed through the Tuamotu Archipelago. Both Cook and Bougainville had been through this area before; Bougainville called it the "Dangerous Archipelago" because it was nearly impossible to see the many small islands and coral reefs until one was almost upon them. Cook studied Bougainville's chart as he navigated among the islands, complaining that the French explorer had been too vague in describing this part of the Pacific, but he agreed that these were indeed dangerous waters and could see no point in stopping at any of the atolls—he called them "half drowned isles"—when they were only a few days' sail from Tahiti. Charles Clerke, who had seen some of the "little paltry islands" when he sailed in the *Dolphin* and remembered Wallis's abortive landing attempts, agreed. Only Johann Forster found any profit in this part of the voyage. An amateur geologist in addition to his other interests, he was curious about the enormous physical difference between these low coral islands and high, mountainous islands like Tahiti. The low islands, he believed, were built up by the progressive accumulation of coral, while the high ones had probably been formed by volcanic action, a reasonably accurate summary of the geological processes involved.

The ships reached Tahiti in mid-August, approaching the island at its southeastern end, the small peninsula called Tahiti-iti. Cook decided to anchor briefly at a small harbor there (called Vaitepiha Bay by the Tahitians) before making the run to Matavai Bay, primarily because the *Adventure*'s men were suffering seriously from scurvy and needed fresh food without delay. The decision nearly proved disastrous, however, as the *Resolution* smashed into a reef while trying to negotiate a narrow opening into the bay. The damage was not serious, but Cook was furious at his crew's carelessness and paced the deck shouting and swearing at his men while they worked to free the ship. Afterward he suffered such severe stomach pains that he could hardly stand, whether because of the stress of the occasion or the intense heat (it was about 95 degrees that day) is unclear.

Meanwhile, even before the two ships were safely anchored, the Tahitians swarmed about them in dozens of canoes. Although Vaitepiha Bay was quite remote from Matavai, many of those who visited the ships remembered Cook and asked about Banks and other men who had been aboard the *Endeavour,* although few, to Cook's surprise, mentioned Tupaia.

The Englishmen made their own inquiries about old friends and got some surprising news. Tuteha and Tepau i Ahurai were dead, killed in a battle with Vehiatua, the chief of Tahiti-iti, and Tuteha's nephew Tu was now chief of the districts around Matavai Bay. (Cook, still thinking of Tahiti-nui as a single "kingdom," mistakenly thought Tu was "king" of that entire part of the island.) Vehiatua had died later, of natural causes, and his teenage son, who took his father's name according to Tahitian custom, now controlled Tahiti-iti.

It was impossible for the English to figure out the reasons for this war, especially since they had an imperfect understanding of Tahitian politics to begin with, and so they simply accepted the changes without giving them much thought. In fact Tuteha had instigated the conflict and persuaded Tu, rather unwillingly, to join him in an attempt to break Vehiatua's power. Tuteha was an ambitious man—he had been responsible for Purea's downfall just before the *Endeavour*'s visit—but this time he overstepped his limits. A skirmish at sea ended in a draw; then Tuteha took his forces into a land battle on the isthmus between Tahiti-nui and Tahiti-iti, where he and Tepau i Ahurai were killed. Tu escaped, and in the end profited from the conflict, as he made peace with Vehiatua, and, released from his uncle's control, built up his own influence as a chief.

The Tahitians also told the Englishmen that another European ship had anchored there a few months earlier. It was a French ship, they said, and had stayed only briefly, but one of the crew had escaped and was still living on the island. Inspired perhaps by the power of suggestion, several of the *Resolution*'s crew claimed to have seen the European over the next few days, but he always ran away when approached. The Forsters doubted the story of the French fugitive, but just in case they were wrong, Johann wrote a letter in French and gave it to one of the Tahitians.

Some days later a man named Tuaha, whom Cook had met on his trip around the island in 1769, came on board, and from him they learned a few more details. Cook showed him a map of Tahiti, and although Tuaha had never seen a map before, he immediately picked out and named every district on the island. He pointed to the spot where the European ship had anchored, just north of Vaitepiha Bay, calling it "pahei no Peppe." *Pahei* is the Tahitian word for ship; *Peppe* Cook interpreted to mean Spain. In fact, a Spanish ship had anchored at Tahiti-iti in late November and early December of the previous year, on a voyage charged with searching for foreign settlements in the South Pacific—Spain was desperately trying to maintain its influence in this part of the world—and with converting the native inhabitants to Christianity. The tale of the deserter, however, was never confirmed.

Political upheavals aside, little had changed at Tahiti. The islanders were eager to trade—the Forsters acquired several new birds and fish from the visiting canoes even before the *Resolution* anchored—and the Englishmen equally eager to acquire souvenirs and women. As on the first voyage, Cook issued strict regulations on trade to prevent his crew from bidding up the price of provisions. No iron was to be given to the Tahitians for anything but food, all trade on board ship or at the appointed trading place on shore was to be supervised by an officer, and no one was to buy "curiosities" until the expedition's food supplies were fully replenished. But despite Cook's efforts, the provisions available for sale were limited to fruits and vegetables; although hogs ran about freely, no one would sell them, claiming they all belonged to Vehiatua. Cook had encountered a similar situation on his first visit, but not on this scale. Vehiatua's control over his people's meat supply appeared absolute.

Sailors looking for sex had no such problems. As soon as Cook and the senior officers went ashore, late on their first day at anchor, women (some, according to George Forster, mere girls of nine or ten) swarmed through the ship. Less rapturous than some of his predecessors in the *Endeavour*, Forster thought the ubiquitous tattoos ruined the women's appearance, and mused on "how little the ideas of ornament of different nations agree, and yet how generally they all have adopted such aids to their personal perfection." The women's real charm, he added, was not physical beauty but their friendly, vivacious manner and "a constant endeavour to please."

From the outset the Tahitians who came on board "took the opportunity of conveying away a number of trifles." Some hit upon an ingenious new tactic, taking coconuts the English had already purchased and throwing them overboard to their friends in canoes below, to be sold again. On the second day at anchor, a Tahitian who styled himself an *arii* was caught stealing a knife and pewter spoon from Cook's cabin—this after Cook had exchanged gifts with him. Cook became so annoyed at the man, and at the Tahitians' thieving in general, that he threw the whole group off the ship and fired a musket over the offending man's head. The accused thief jumped out of his canoe and swam ashore, but Cook would not let the matter rest. Instead he sent a boat to pick up the canoe; when the Tahitians on shore began throwing stones at the unarmed boat, he sent another boat to protect the first and had a cannon fired toward shore; then he seized two canoes. A few hours later "the People were as well reconciled as if nothing had happen'd," according to Cook, but when George Forster went ashore later, he thought the Tahitians seemed "a little more shy or reserved than usual." Over the next two or three days, even Cook had to admit that the islanders had restricted trade, which previous experience taught was the typical Tahitian response to English aggression, and finally he returned the confiscated canoes.

Because Cook planned to stay at Vaitepiha Bay only a few days before moving on to Matavai, he spent most of his time supervising the repair of the ship and acquiring a supply of food. The Forsters, however, and some of the other first-time visitors set out enthusiastically to get acquainted with the Tahitians. The residents of Tahiti-iti had had little direct contact with Europeans and they in turn were eager to study their visitors, never hesitating to ask questions, finger the sailors' clothes, even pull up their shirts to inspect their bodies more closely. Observing the Englishmen's eagerness to learn some of the Tahitian language, as they pointed to objects and asked their names, the Tahitians obliged by becoming language teachers and were, according to George Forster, "much delighted when we could catch the just pronunciation of a word." George proved to be a quick learner. He quickly recognized that the "O" sound at the beginning of most Tahitian words was an article, not part of the word itself, and both he and his father produced reasonably accurate spellings of Tahitian words.

Whenever the Forsters, usually accompanied by Sparrman and Hodges, went exploring, a train of Tahitians followed them, partly out of curiosity, partly to beg gifts, and partly (so the Forsters thought) to keep tabs on the Englishmen's movements. It was a good-natured sort of surveillance, however. On their first expedition it became clear to the Forsters, notwithstanding their limited command of the language, that the Tahitians' conversation focused exclusively on their English visitors. Whenever anyone joined the group, he was introduced and "entertained with a repetition of what we had said and done that morning." Then the newcomer usually asked for a demonstration of musket fire, whereupon the Englishmen would ask the Tahitians to point out a bird as a target. Unfortunately they sometimes picked one far beyond musket range, and to prevent the Tahitians from learning the limited range of their guns, the English pretended not to see the bird until it was close enough to hit.

The group interrupted their walk to visit a Tahitian family at home, where the Forsters' interest in learning how the islanders lived was outdone by their hosts' interest in the English. Like the other Tahitians they had met, this family touched the Englishmen's clothes, examined their arms and hands, and asked their names. They seemed surprised at their visitors' light skin color, the absence of tattoos, and their short fingernails. (Long fingernails were considered a mark of beauty among the Tahitian upper class.) The family patriarch questioned his visitors closely, using a combination of signs and words; he wanted to know their captain's name, how long they planned to stay, and whether they had wives on board. George Forster thought the old man knew the answers already but wanted to hear their replies for himself. Meanwhile, Hodges sketched several of the Tahitians, much to their amusement, for they easily recognized their own portraits.

Continuing their walk, the group visited a *marae* and several more Tahitian homes. Late in the day they met the self-proclaimed chief of the district. An extraordinarily fat man, he was surrounded by a group of servants feeding him. It was not an unusual scene in Tahiti, since the *arii* prized great girth as a sign of beauty and status, and feeding chiefs was part of the elaborate *tabu* system; but George Forster was most disillusioned at this scene of sloth and ostentatious consumption. If Joseph Banks had seen in Tahiti a miniature version of European feudalism, George Forster saw a democratic society of equals, much as he had in New Zealand. Although he had been on the island less than a week, Forster already thought he had found in Tahiti a place where "a whole nation, without being lawless barbarians, aimed at a certain frugal equality in their way of living, and whose hours of enjoyment were justly proportioned to those of labour and rest." But the portly chief being fed by his retainers didn't fit the image; instead, he was "a luxurious individual spending his life in the most sluggish inactivity, and without one benefit to society, like the privileged parasites of more civilized climates, fattening on the superfluous produce of the soil, of which he robbed the labouring multitude." He compared the "indolence" of this Tahitian chief with that observed by travelers to India and other Asian nations, quoting at length the account of the fictional Sir John Mandeville.

Cook and the Forsters made several unsuccessful attempts to meet Vehiatua in their first few days at Tahiti. Finally, after a week, Cook, the Forsters, Sparrman, and several officers encountered the young chief on one of their walks. He seemed fearful at first, but in time, after the usual exchange of presents, he relaxed and asked Cook to stay five months. Cook replied that he could not, as he did not have sufficient provisions for such a long stay; Vehiatua in turn promised a supply of hogs, although none of the Englishmen believed him after the days they had spent trying unsuccessfully to persuade the islanders to trade their animals.

George Forster, alone among the Englishmen, was by now sympathetic to the Tahitians' refusal to give up their hogs, because he already recognized that European visitors were a drain on native resources. The best strategy for Vehiatua to keep "the riches of his subjects . . . and to prevent new wants from prevailing among a happy people," George believed, "was to get rid of us as soon as he could, by denying us the refreshments of which we stood most in need." Indeed, he hoped that eventually Europeans would cease their contact with South Pacific islanders "before the corruption of manners which unhappily characterizes civilized regions, may reach that innocent race of men, who live here fortunate in their ignorance and simplicity." It was a theme that the younger Forster, steeped in the rhetoric of the Enlightenment, would sound many times during the voyage, but one that he recognized as unrealistic; "the dictates of philanthropy do not harmonize with the political systems of Europe!" he thought.

But Vehiatua showed every sign of sincerity in his invitation. He walked about arm in arm with Cook and later returned to the ship, where he greeted all the sailors, inquiring their names and whether they had brought their wives with them. Upon being told there were no English women aboard, he invited the men to choose wives from among his people. One of Vehiatua's companions, described by George Forster as a "fat chief," asked Johann and George if they believed in a god and prayed to him. The Forsters' affirmative answers drew smiles, and the Tahitian said something that suggested they were all in agreement on these important matters. Vehiatua showed particular interest in Cook's and Johann Forster's watches, and wanted to know their purpose. Trying to explain this in signs and a few words took some doing, but the Englishmen finally hit upon an appropriate analogy: the watches measured the days, as the sun did. Then the watches were "little suns," Vehiatua replied, pleased that he understood.

The chief's renewed promises of hogs did nothing to change Cook's mind about moving on, however, and the next morning the *Resolution* and *Adventure* left Vaitepiha for Matavai Bay, where they anchored the next day. The welcoming party all recognized Cook and the other men from the *Endeavour*. One old man greeted Pickersgill, who had sailed in the *Dolphin* as well as in the *Endeavour*, by holding up three fingers to show that it was the lieutenant's third visit to Tahiti. There was none of the initial reticence or slight air of suspicion that the Forsters had detected at Vaitepiha Bay; here the Tahitians crowded the ships, each choosing a friend and exchanging names with him.

Several stayed on board all night, talking with the Englishmen in a combination of words and signs, telling them more stories of the battles in which Tuteha and Tepau i Ahurai were killed and asking questions about Tupaia and the men from the *Endeavour* who had not returned. Their genuine friendliness made a deep impression on George Forster. Obviously these people remembered the horror of the *Dolphin*'s attack and the misunderstandings between themselves and the *Endeavour*'s crew, if not directly then by oral tradition, but they harbored no resentment. Revenge, Forster decided, was not part of the Tahitians' makeup. These favorable first impressions supported his wish to see mankind as basically good; "savage ideas of distrust, malevolence, and revenge, are only the consequences of a gradual depravation of manners." The Tahitians' attack on the *Dolphin,* he believed, was probably caused by some "outrage" committed by the English, or by a desire to repel the English as invaders, and "when they found that Britons were no more savage than themselves," the islanders became friendly.

George's idealism about human nature was seriously shaken the following night when dozens of Tahitian women stayed on board ship. Encounters among sailors and native women at Vaitepiha Bay had taken place on shore or in corners of the ship during the daylight hours, so George was not prepared for the all-night frolics at Matavai Bay. At Vaitepiha he had been disillusioned with the behavior of a chief after observing what he took to be equality among the people; now he felt similarly disturbed at witnessing the women's promiscuity in the wake of his favorable conclusions about the Tahitians' natural goodness of character. George excused the women to some degree by claiming they were all of the lower class and the simplicity of their lives made some of their actions more innocent than they would otherwise be. Still, he was "much hurt" to see "so great a degree of immorality in a nation, otherwise so happy in its simplicity, and in the newness of its wants." It was, he sadly concluded, "a reflection very disgraceful to human nature in general, which, viewed to its greatest advantage here, is nevertheless imperfect."

If the people of Matavai Bay were uninhibited in welcoming the English, the same could not be said of their chief. Tu, or "Otoo," as the English called him, was on the beach when the *Resolution* and *Adventure* sailed into the bay, but by the time the ships anchored he had taken off for his home at Pare. The next day Cook and the Forsters went to Pare to call on him. The meeting was a cordial one; Tu, a tall, attractive youth in his early twenties, remembered Cook and asked about all the officers of the *Endeavour* by name, even though Cook was quite certain that he had never met the young chief on his previous visit.

Tu said he would not visit the ship because he was afraid of its guns—Cook sized him up as a "timerous Prince"—but in fact he ventured to Matavai Bay the next day, bringing gifts of hogs, fish, fruit, and cloth. At first he refused to board the ship, but Cook finally persuaded him, after he had presented the captain with an enormous quantity of tapa cloth and watched while his companions wrapped it around Cook, "encreasing his bulk to a prodigious dimension." Even so, the chief remained timid, and when Cook invited him below decks to his cabin, Tu made his teenage brother go first. Once settled in Cook's quarters, however, he relaxed and surveyed his surroundings. Chairs fascinated him, as they had the Tahitians who visited the ships on other occasions, and he took great interest in the Englishmen's food, although he would not taste it. Tu thought it particularly strange that they put "oil" (butter) on their baked breadfruit and drank what he thought was hot water. He showed great fondness for Johann Forster's dog, prompting Forster to make him a gift of the animal. Afterward the Englishmen never saw Tu without a servant following behind, carrying the dog.

His fears banished, Tu exchanged almost daily visits with Cook during the remaining few days of the *Resolution*'s stay. These friendly encounters did not produce the hoped-for supply of hogs, however, as Tu persistently refused to sell them. Pickersgill, dispatched to Purea's home at Papara, southeast of Matavai Bay, succeeded little better, as the former "queen," apparently much reduced in wealth, claimed she had no hogs to sell. When Pickersgill reported back to Cook, the captain concluded that the two recent wars at Tahiti, one in the interval between the visits of the *Dolphin* and *Endeavour* and the second after the *Endeavour*'s stay, had destroyed most of the island's animal supplies. He saw no reason to stay longer, and the next day, September 2, the ships left for Huahine and Raiatea, where Cook intended to make one last attempt to get a supply of pork.

The Englishmen had stayed at Tahiti just over two weeks, but it was long enough for them to form some judgments about the island and its people, especially as they began with a solid base of previous observations. Cook had supplemented his own observations with a close reading of Bougainville's description of Tahiti (in an English translation of the journal by Johann Forster), and found that the French explorer's conclusions disagreed with his own in several respects. With his usual concern for precision and accuracy, Cook spent some time questioning the Tahitians about those points on which he and Bougainville differed. Most serious was the question of human sacrifice; Bougainville claimed it was a Tahitian practice, but Cook had seen no evidence of it on his previous visit. With Furneaux and Gibson to act as interpreters, he questioned a priest, who confirmed that the islanders did indeed practice human sacrifice and added that the offerings were always "bad men." From this dialogue Cook deduced that the Tahitians deemed human sacrifice necessary on certain rare occasions, and that men who had committed crimes were the victims. Although this was a reasonably accurate interpretation, Cook was not confident that he had learned the truth, because Gibson's and Furneaux's command of the language was limited. Perhaps for this reason the discovery of human sacrifice made little impression on the English, in striking contrast to the horrified discussion evoked by the New Zealanders' cannibalism.

Less scrupulous than Cook about drawing conclusions after short visits hampered by limited language skills, the Forsters did not hesitate to analyze Tahitian society at length. George, eager to see Tahiti as an idyllic society of equals, eventually had to admit that it was in reality organized along class lines. With Banks no doubt in the back of his mind, he even went so far as to say that the island's class sytem "bears some distant relation to those of the feudal systems of Europe." But unlike Banks, Forster detected a society in which the essential simplicity of life and the ease with which people could acquire the necessities of existence reduced class distinctions to the merely ceremonial. Admittedly, the upper class had some material goods that the lower classes lacked, but on the whole, Forster thought, there was less distance between the highest and lowest classes in Tahiti than between a "reputable tradesman and a labourer" in England. Moreover, the affection that the Tahitians displayed toward their chiefs led him to believe that they perceived themselves as one large family, with the "king" considered the "father of his people." James Burney, on the other hand, like Banks a member of the English upper class, compared Tahitian government to that of Poland—a symbol of oppression in his mind—arguing that the *arii* and the *teuteu,* the upper and lower classes, corresponded to lords and peasants.

Even Forster had difficulty reconciling the behavior of the Tahitian chiefs with his view of the island as essentially an equal society. Yet for him evidence of chiefly luxury did not shake his fundamental belief in Tahitian equality, but only made him question how long it would last. The chiefs' "indolence," he believed, was already driving Tahitian society toward its "destruction." He speculated that labor would gradually fall more heavily on the people of the lower classes, who would become "ill-shaped" and darker-skinned because of their long hours of work in the sun. Prostitution of lower-class women would make their children smaller in size, while the chiefly class continued to multiply, becoming larger and "purer" in color because of their better diet and greater leisure. (The Englishmen were preoccupied with skin color in all their contact with South Pacific Islanders. Burney, for example, was disappointed in the Tahitians' appearance, because he expected them to be nearly as white as Europeans; and he, like most of the English, associated lighter color with higher status.) Eventually, Forster argued, "the common people will perceive these grievances, and the causes which produced them; and a proper sense of the general rights of mankind awaking in them, will bring on a revolution. This is the natural circle of human affairs." The introduction of "foreign luxuries" would only hasten this evolutionary process.

If Forster's vision of the impending social conflict in Tahiti smacked too much of the incipient revolutionary (he later became a partisan of the French Revolution), his belief that European "discovery" of the island would hasten social changes was a perceptive one. Unlike most of his contemporaries, Forster deplored this prospect, because the changes he foresaw as a result of European contact were all negative. Indeed, he went so far as to question the morality of scientific discovery: "If the knowledge of a few individuals can only be acquired at such a price as the happiness of nations," he wrote, "it were better for the discoverers, and the discovered, that the South Sea had still remained unknown to Europe and its restless inhabitants."

Few of the *Resolution*'s men were as thoughtful as George Forster. The exuberant young Elliott spoke for most of his compatriots when he said that Tahiti was a "*Paradise* . . . the women *beautiful,*" with "the finest form'd *Hands,* fingers, and Arms, that *I ever saw.*" Wales, on the other hand, debunked the image of Tahiti as paradise. Contradicting prevailing shipboard opinion was a role he obviously enjoyed; he had scorned his colleagues' belief that the New Zealanders practiced cannibalism, and now he argued that Tahiti was really no more beautiful than England (whether out of conviction or just to be difficult is hard to tell). The woman, far from being Venuses, were "masculine," their hair "cut short in the bowl-dish fashion of the country People in England." Their eyes were too prominent, their noses flat, their lips thick; the older women's breasts "hang down to their Navals." Wales did, however, defend the Tahitian woman against charges of loose morals. Tahiti might have proportionately more prostitutes than England, but anyone who described the island's women solely on the basis of those who visited the *Resolution* and *Adventure* might just as well characterize English women from observation of those "which he might meet with on board the Ships in Plymouth Sound." Cook agreed with Wales on the last point, but he also took an unusually pragmatic approach to the whole matter of Tahitian sexuality. Far from viewing the young women's promiscuity as an unfortunate blot on an otherwise happy society, as the Forsters did, he observed that "Incontency in unmarried people can hardly be call'd a Vice sence neither the state or Individuals are the least injured by it." (It is worth noting, however, that he made this comment after learning that the Tahitians apparently had a way of curing venereal disease.)

John Marra, the gunner's mate, had probably the most realistic and sympathetic understanding of the Tahitians. More sophisticated than Elliott and unhampered by George Forster's command of political theory, Wales's jaded sarcasm, or Cook's obsession with theft, he tried to view the Tahitians in their own terms. Marra could defend their propensity for theft, writing "Is it not very natural, when a people see a company of strangers come among them, and without ceremony cut down their trees, gather their fruits, seize their animals, and, in short, take whatever they want, that such a people should use as little ceremony with the strangers, as the strangers do with them; if so, against whom is the criminality to be charged, the christian or the savage?" Cook would have argued that he asked permission to cut down trees and paid for the fruit and animals, but Marra realized that the Tahitians' notions of payment and trade were different from the Englishmen's and that Cook's legalistic way of dealing with the islanders, although a sincere attempt at fairness, was nevertheless incomprehensible to them.

Cook was fond of the Tahitians, and he came away from his second visit to the island feeling even more warmly toward them than he had after his first voyage. "The more one is acquainted with these people," he wrote, "the better one likes them, to give them their due I must say they are the most obligeing and benevolent people I ever met with." Theft was their only vice, he believed (having concluded that sexual promiscuity among the young was not a vice), and even on that score he adopted a more tolerant attitude this time around. When the Tahitians stole items of some value—weapons or hardware from the ship, for example—he went to considerable lengths to get them returned, but otherwise he thought it was not worth bothering, since punitive measures inevitably upset the islanders and put a stop to trade. This approach marked a significant change in attitude on Cook's part, one that persisted throughout the voyage. On an initial visit (as at Vaitepiha Bay, for example) or an extended stay he felt it important to take a strong stand on theft as a means of establishing and maintaining his authority, but if the visit was to be short, as it was this time at Matavai, the practical need to keep peace with the inhabitants to ensure a steady flow of trade took precedence over principle.

When the two ships sailed, each had a Tahitian on board. That their countrymen Tupaia and Ahutoru had gone off in European ships and never returned did not seem to deter more of the islanders from begging to go to England, and Furneaux agreed to take a young man named Mai, or Omai, as the English called him, in the *Adventure*. (The sailors called him Jack.) Cook discouraged the idea, since he doubted Mai's intelligence and didn't think he would be as useful to the English as Tupaia had been, but he did not veto Furneaux's decision. At the last minute, Cook was persuaded to take a man named Porio on the *Resolution*.

Mai's motive for joining the English expedition was a combination of a desire for personal glory—going to England would set him above the common mass of Tahitians—and an interest in revenge. His native island, Raiatea, was under the control of the chief of neighboring Bora Bora, and Mai often talked about bringing back guns from England to arm his fellow Raiateans to drive out their conquerors. George Forster found this desire for revenge unsettling. Since all the Society Islanders had as much as they needed for subsistence and seemed to live happy lives, he could see no possible motive for conquest except "a spirit of ambition." Such desires ill accorded with the "simplicity and generous character" of the Tahitians, Forster thought; "it gave us pain to be convinced, that great imperfections cannot be excluded from the best of human societies."

Less is known of Porio, but he too apparently wanted to sail with the English as a way of setting himself above his fellow Tahitians. He insisted upon having an English name—the sailors dubbed him Tom—and would dress only in English clothes. At Huahine, Porio refused to speak Tahitian with the islanders; but he didn't know any English, so he spoke nonsense syllables in an attempt to impress the islanders with his command of their visitors' language.

At Huahine Cook found that little had changed from his previous visit. Ori, the chief who had exchanged names with Cook and treated him with such hospitality before, remained in his position of influence. He and Cook greeted each other emotionally, and Ori again adopted the name "Cookee" for the duration of the Englishmen's stay. Even more gratifying to Cook was the plentiful supply of hogs. Each morning Ori sent a generous quantity of provisions as a gift for Cook's table, and Pickersgill managed a brisk trade with the island's residents. After five days Cook moved on to Raiatea, where he found food supplies even more abundant than at Huahine; in nine days the Englishmen bought five hundred hogs. They consumed about one hundred and kept the remainder alive, penned up on the ships' decks, for future consumption.

While the chiefs of these islands were well disposed toward the English, not all of their subjects were so welcoming. Theft was rife, as it had been at Tahiti, and more serious incidents occurred as well. Sparrman went out alone to "botanize" at Huahine, armed only with his hunting knife, and was attacked by two men who offered to serve as guides. They tore off his clothes, wrested his knife away, and chased him along the beach; he finally got away from them and made it back to the ship, frightened and nearly naked. Cook didn't take the incident too seriously—he was more upset with Sparrman for going off alone than with the islanders for attacking him—but Sparrman thought his life had been in danger, and the Forsters took his side against Cook.

A few days later, at Raiatea, George Forster was the victim of a similar attack. On an expedition with his father and several others, George decided to hire a canoe to take them back to the ship. He went off to negotiate the arrangement, only to have his gun seized from him. Johann, watching in horror, was convinced that his son would be injured or killed, and raced to George's defense. Although the attacker quickly handed the gun back and ran off, Johann gave chase and fired at him. The affair ended without further incident, but when the group returned to the ship, some of the Forsters' companions criticized Johann's conduct to Cook, who reprimanded him for attacking an island resident unnecessarily.

Already annoyed with Cook over his handling of the attack on Sparrman, Johann proceeded to make a major issue out of the threat to George. (George himself didn't even mention the incident in his journal.) He accused Cook of treating his scientists unfavorably, refusing to defend them against native hostilities even though he went to great lengths to get back the most minor items stolen from the naval crew. Cook, on the other hand, thought Forster exaggerated the seriousness of the attacks on Sparrman and George, who in his opinion were at least partly to blame for allowing themselves to get into situations where they were vulnerable to attack. In fact, Cook was not playing favorites, as Johann would have it; he often blamed theft on sailors' carelessness, just as he blamed the recent attacks partly on Sparrman and George Forster, and he did not always insist on getting stolen goods back, despite what Forster claimed. But Forster interpreted all untoward circumstances, whether it be a cramped cabin or the most recent incidents, as personal injuries, and he berated Cook for his lack of consideration. Cook, who by now was exceedingly tired of Forster's petulance, ordered Johann to get out of his cabin and stay out.

The two men eventually patched up their disagreement, but it took considerable doing. Shortly before the ships sailed from Raiatea, Cook asked Furneaux to tell George Forster that he wanted to resolve the quarrel, but Johann would not budge until Cook personally invited him back to his cabin. The next day Cook visited Forster's cabin to issue the requested invitation, and after a lengthy conversation, according to Forster, "we both yielded without giving any thing up of honour."

In mid-September the *Resolution* and *Adventure* sailed from

Raiatea. Porio left the ship abruptly, without telling anyone that he had changed his mind about going on with them, while George Forster was approached by a Raiatean named Hitihiti (the English spelled his name Odiddy) who also wanted to join the voyage, and Cook agreed to take him.

From Raiatea, Cook decided to follow a slightly different course back to New Zealand, one that he thought would take them to a group of islands discovered by Abel Tasman in the 1640s and not seen by Europeans since. In the interests of completing the map of the Pacific, Cook wanted to locate these islands and then make a brief stop at New Zealand to fortify his crew for another Antarctic cruise.

12. The Second Antarctic Summer

On his great voyage of 1642, after discovering Van Diemen's Land and New Zealand, Abel Tasman made a wide sweep to the northeast, discovering three islands which he named after cities in his Dutch homeland: Amsterdam, Middelburg, and Rotterdam. Cook had read Tasman's account of the voyage carefully, and he knew that the islands, if charted reasonably accurately, lay some sixteen hundred miles west and slightly south of the Society Islands.

Two weeks after leaving Raiatea, the Englishmen approached islands fitting Tasman's description of Middelburg and Amsterdam. Anchoring in a small bay at the former, they were quickly surrounded by dozens of outrigger canoes much like those of the Tahitians, only smaller. The islanders rushed to board the ships, rubbing noses in good-natured greeting—to the astonishment of the English, who assumed these men had never seen Europeans and any tradition of Tasman's visit would have vanished after more than a century. The Forsters learned that the island was called "Ea-Oowhe" (Eua) and the larger island in the distance, Tasman's Amsterdam, was known as "Tonga-Tabboo" (Tongatapu).

Eua was small and flat, with none of the grandeur of Tahiti's mountains and luxuriant rain forests, yet every inch of it was cultivated with exquisite care in broad green lawns and gardens. From the ship the island looked like "a very large Park, laid out by design," Wales thought. The first house Cook visited was a small but neatly constructed dwelling set on a broad expanse of lawn and surrounded by flowers and shrubbery. Fences made of woven reeds divided the land into separate gardens; wooden doors allowed access from one parcel of land to the next. This evidence of private property and good agricultural habits drew enthusiastic praise from the English. "The regularity of their plantations, and excellency of their Fences here I think is truly admirable," Charles Clerke wrote; and George Forster thought them evidence of a "higher degree of civilization" than he had yet seen in the Pacific.

The houses showed similar evidence of care in their construction. Floors were neatly covered with woven mats, and movable partitions could set off separate areas for sleeping, allowing a higher degree of privacy than in Tahiti. The islanders appeared to respect privacy in other ways, too, allowing the Englishmen to wander about at will without trailing after them in crowds as the Tahitians always did.

Walking about the island, the Forsters could see that Eua was a much less wealthy island than Tahiti and its neighbors. Canoes and houses were smaller, although they were more carefully constructed, with greater detail of ornament; clothing was simpler in design. They noticed few hogs, fowls, and breadfruit trees, and deduced that yams and bananas were the staples of the local diet. The islanders were little inclined to trade food—the Forsters' excursion suggested it was because they didn't have much to spare—so Cook decided to leave after a day and sail on to Tongatapu (where Tasman had obtained abundant provisions), but not before giving the local chief a supply of garden seeds and explaining, as best he could, what to do with them.

At Tongatapu the Englishmen met an equally friendly reception. The islanders came to trade as soon as the ships dropped anchor, although to Cook's disappointment they offered only cloth and "curiosities." In an attempt to get them to exchange food instead, he prohibited the purchase of anything else, and Johann Forster, who had picked up the words for breadfruit, hogs, coconuts, and other foodstuffs at Eua, asked for them by name. This strategy succeeded, as the next day the islanders brought canoes full of provisions.

The Englishmen spent three productive days at Tongatapu restocking their food supplies and exploring the island. Like Eua, the entire island was laid out in neatly fenced, well-tended gardens and lawns. Impressed that every inch of land was used to best advantage, with roads and fences taking up the least possible amount of space, Cook compared Tongatapu to "one of the most fertile plains in Europe." Craftsmanship and attention to detail were apparent everywhere, in the carefully built canoes and houses, the quality of tools, and the beautifully designed baskets so tightly woven that they would hold water.

During their first day exploring the island, Cook, the Forsters, and several officers all went their separate ways. Cook visited a religious shrine—a simple rectangular building set on an artificially constructed hill in the midst of an elegantly manicured lawn—while the Forsters walked all over the island collecting new species of plants and birds, including several parakeets. Hitihiti was delighted at the opportunity to purchase a quantity of bright red feathers, which, as he explained to the Englishmen, were highly prized in Tahiti. Some of the sailors, thinking perhaps of future trips to Tahiti, had the foresight to buy feathers themselves.

The Tongans, like the residents of Eua, charmed the English with their warm hospitality, but they also displayed a propensity for theft and considerable ingenuity in carrying it out. Wales had his shoes snatched when, after wading ashore through the surf, he dropped them on the sand momentarily before putting them on his feet. (A chief immediately retrieved the shoes and handed them back.) An ambitious group tried to steal one of the *Resolution*'s boats, while another man got into the master's cabin and took the logbook, a naval almanac, and several other volumes, escaping in a canoe. When a group of sailors followed him in one of the boats he tossed the books overboard. A second boat raced to the spot and retrieved the books from the shallow water, but the first group continued to chase the thief, who had by now jumped out of his canoe. He played with his pursuers, diving underwater whenever they got close to him; once he unhinged the rudder and pulled off the tiller. A sailor grabbed the boat hook, snared the thief in the ribs, and hauled him into the boat, but the Tongan had the last laugh as he jumped overboard and swam to shore despite his injury.

Cook thought theft was less serious at Tongatapu than at other islands he had visited, but he adopted harsh tactics against thieves nevertheless—a further example of his belief in the importance of establishing his authority from the outset over groups of people unaccustomed to dealing with Europeans. He ordered muskets fired at a man who took an officer's jacket, and had other offenders tied to the shrouds and flogged. None of the islanders seemed concerned, however; they offered no sympathy to the men who were beaten, apparently believing that thieves deserved their punishment, and did not panic at the sound of gunfire. There was no mass flight and no halt to trade. Amazingly, George Forster thought, nothing affected the friendliness of these people.

It was quickly obvious to the English that the people of Eua and Tongatapu bore a striking resemblance to those of Tahiti. They were physically similar, if somewhat shorter and thinner; Cook thought their skin color darker than that of the highest-ranking Tahitians but lighter than the lowest-ranking. Both men and women tattooed their bodies extensively (the men even tatooed their genitals), although the patterns were different from either the Tahitians' or the New Zealanders'. Canoes and cloth were similar in design and manufacture, and as far as the Englishmen could determine, forms of religion and government were much like the Tahitians' as well. Young, unmarried Tongans enjoyed the same sexual freedom as their Tahitian counterparts, although George Forster thought prostitution a little less prevalent than at Tahiti.

At first the Tongan language sounded quite different from Tahitian to most of the Englishmen and even to Hitihiti, but the Forsters, with their talent for linguistics, noticed similarities immediately. Once they demonstrated the parallels among several words, Hitihiti picked up the language quite easily. Clearly the Tongans shared the same origins as the people of the Society Islands and New Zealand, yet the differences among the three island groups were as striking as the similarities, leading the Forsters (and Cook, to a lesser extent) to theorize about the reasons cultures develop as they do.

The islanders' extensive and painstaking cultivation of land and the absence of sharp class distinctions most strikingly distinguished Eua and Tongatapu from the Society Islands, in the eyes of the English. On the latter point, the Englishmen observed that the islanders showed great respect to their chiefs, but the outward manifestations of class differences at Tahiti—lighter skin color, larger size, more elaborate clothing—were lacking. George

Forster, quite logically, explained these variations by examining differences in the environments of the two sets of islands. Tahiti and its neighbors were large, lush islands with abundant trees, vegetation, and rainfall. Wood for houses and boats was plentiful, and food grew everywhere with a minimum of effort. Eua and Tongatapu, in contrast, were smaller, had fewer trees, and much less rainfall; consequently, Forster argued, their residents built smaller houses and canoes and had to work harder at growing their food. Forced to be more frugal with their natural resources, they used every square foot of land for some productive purpose.

Forster pushed his analysis further to explain social and political differences as well. Because the Tongans had to work at growing their food, their society was not only a more industrious but also a more equal one. There was no surplus food, and even chiefs did some physical labor, so the sloth and "luxury" of the Tahitian chiefs—to say nothing of their enormous girth—had never been permitted to develop. Moreover, Forster believed, the habits instilled by continuous agricultural labor carried over to other areas of life. Instead of devoting their free time to mere pleasure, the Tongans worked on their boats and baskets and tools, which displayed much more care and skill than those of the Tahitians.

This society of industrious equals, of course, drew high praise from George Forster. But he had some difficulty explaining why this apparently ideal state did not have a democratic form of government, maintaining instead a hereditary system similar to that of the Society Islands, and why the islanders treated their chiefs with such great deference. The only explanation he could devise stemmed from his observation that the Tongan chiefs, unlike those of Tahiti, seemed to expect nothing from their people except the honor due to their station—or "servile submission," as Forster viewed it. They demanded no tribute, no hours of labor, no personal service; and so the burden of chiefly support weighed lightly on the islands' people. Still, Forster thought that the very existence of hereditary chiefs and the respect shown to them "seems likely to facilitate the introduction of luxury"; and "luxury," in Forster's scheme of things, was the entering wedge for despotism.

Cook and Johann Forster both thought the Tongans displayed a "higher state of civilization" than the Tahitians or any other Pacific islanders they had met, a judgment they based primarily on the islanders' highly developed agriculture and crafts and on their generally polite behavior toward the English. Forster, like his son, took a particular interest in such questions, and enumerated the characteristics of Eua and Tongatapu that made their people qualify, in his mind, as a "civilized" society. They had a form of religion with places for worship, priests, and formal prayers. "They exercise all the Social virtues to one another, which are usual among the civilized nations," especially charity, for the islanders shared whatever they had and cheerfully helped each other with work. Their crafts were highly developed, if allowance was made for the simplicity of their tools; their clothing was varied in style, and their land fenced and cultivated. They had music, drama, and dance—simple in form, perhaps, but not unpleasing. In short, the Tongans had all the elements of a highly developed civilization, in embryo, so to speak. Pointing out that Europe, hundreds of years earlier, had been no more sophisticated than present-day Tonga, Forster suggested that lack of iron tools was the major impediment to a more complex culture in the islands.

What Forster saw at Tongatapu was, in effect, the history of humanity compressed: here were men and women in an early stage of development, preserved by isolation from acquiring the technology that would push them into the next stage of human history. Such ideas had been brewing in Forster's mind throughout the voyage. The brief stop at Eua and Tongatapu helped crystallize his thinking, because the islands were similar enough to what he had seen before to be clearly a part of the same broad culture and yet different enough to suggest the possibility of change in that culture—change that he was inclined to explain as part of a fundamental process of human evolution.

The *Resolution* and *Adventure* left Tongatapu the first week in October, bound for New Zealand. One midshipman grumbled about Cook's decision to make another stop there, since they had gotten plenty of food at Tongatapu, but Cook was loath to give up any opportunity to stock up on provisions, and he also wanted to leave livestock and garden seeds on the North Island, as he had earlier at Queen Charlotte Sound. Moreover, the stop at New Zealand was not really a detour in Cook's mind, since he intended to sail across the entire southern Pacific at the highest possible latitude, from the point where he had turned north toward New Zealand the previous summer, east to Cape Horn. Such thoroughness seemed excessive to his crew, but Cook was determined not to leave unexplored any corner of the ocean which armchair geographers, as he liked to call them, might later claim as the location of a continent.

Strong winds made sailing hazardous when the ships reached New Zealand, forcing Cook to give up his plan to land on the North Island. The Englishmen did get close enough to shore to see crowds eyeing the ships, however, and a few men ventured out in their canoes. Many remembered the *Endeavour,* not all favorably. One group, visiting the ships off Cape Kidnappers, repeatedly said *matte* (death) and *puppuhe,* their imitation of the sound of gunfire. Cook gave animals to one chief with instructions about their care; the man was more interested in nails, but promised Cook not to kill the animals.

The stormy weather persisted, and the two ships became separated as they approached Cook Strait at the end of October. Having previously agreed with Furneaux to stop at Queen Charlotte Sound, Cook proceeded on to his familiar anchoring place, hoping the *Adventure* would catch up. Once settled at Ship Cove, Cook learned that his plans to supply New Zealand with domestic animals were off to a poor start. He found one of the pigs that Furneaux had left the previous May, but the New Zealanders said the other animals had been killed and eaten. "All our endeavours for stocking this Country with usefull Animals are likely to be frusterated by the very people whom we meant to serve," Cook complained, but he refused to give up and gave more animals to the men who had cared for the last remaining pig. The garden was doing a bit better. Rats had eaten the peas and beans, and the New Zealanders had dug up the potatoes (probably because potatoes were the only vegetable that resembled anything they were accustomed to eat), but the cabbages, carrots, onions, and parsley were thriving.

The Englishmen spent the better part of November at Queen Charlotte Sound, overhauling the *Resolution* for its next venture into icy waters, which gave the Forsters plenty of time for botanizing. But despite energetic hikes through most of the hills around the sound, they collected fewer than thirty new species of plants and very few new animals—a disappointing haul for Johann, who felt keenly that Banks had already beaten him to most of the flora and fauna of the South Pacific.

Wales had his own problems. His calculations of longitude differed considerably from those Bayley had made on their previous visit, and although he recalculated his observations several times, he was unable to explain the discrepancy. It was an especially vexing problem, since Queen Charlotte Sound was one of the few places in the South Pacific visited often enough by Europeans to have its longitude fixed precisely.

There were the usual incidents of petty theft, on both sides this time. When a group of New Zealanders accused one of the sailors of stealing from them, Cook had him flogged, explaining that "it has ever been a maxim with me to punish the least crimes any of my people have commited against these uncivilized Nations. Their robing us with impunity is by no means a sufficient reason why we should treat them in the same manner." He believed the best way to preserve good relations with the local inhabitants was to demonstrate the power of guns "to convince them of the Superiority they give you over them"; then "a regard for their own safety" would deter them from mounting an attack, "and strict honisty and gentle treatment on your part will make it their intrest not to do it." Cook's strategy probably did help prevent deliberate attacks, but nothing stopped theft. Cook himself was a victim on this visit to Queen Charlotte Sound, when a chief took a handkerchief from his pocket even while haranguing bystanders about the evils of stealing from the English.

Toward the end of their stay at Queen Charlotte Sound, several officers walking along shore encountered a group of New Zealanders feasting. Nearby they noticed the severed head of a young man and a human heart impaled on a stick attached to the end of a canoe. Pickersgill bought the head and brought it back to the ship, where, at his invitation, the New Zealanders on board sliced off a piece, broiled it, and ate it with obvious great pleasure. This incontrovertible proof of cannibalism—even Wales, who had gone to such pains earlier to deny that the New Zealanders ate human flesh, was finally convinced—evoked a variety of responses from the men watching. Some, whom Johann Forster described as those "whose hardened Souls had made them unfeeling against all humanity," laughed. Others were indignant, railing against the cruelty of human nature. And some became sick. Hitihiti had the strongest reaction: "he became perfectly motionless, and seemed as if Metamorphosed into the Statue of Horror." Then he burst into tears, lectured the New Zealanders on their cruelty, and went to his cabin, where nothing would console him.

Cook, Sparrman, and the Forsters took a more dispassionate view, considering cannibalism as yet another cultural trait to be analyzed. Cook attempted to explain the practice as part of the evolution of human societies. Man is a "savage" in his "original state," he wrote, and some vestiges of savagery survive even as man becomes civilized; the custom of eating enemies killed in battle was a prime example. The absence of religion and a "settled form of government" (which Cook falsely believed characteristic of New Zealanders) made it especially difficult to end the custom of cannibalism, he believed. Sparrman thought cannibalism was characteristic of many primitive cultures, citing examples from the early histories of Germany, Mexico, Brazil, and North America to support his argument, and George Forster agreed, theorizing that "increasing civilization" would eventually put a stop to the practice among New Zealanders. At the same time he questioned whether Europeans, who were "too much polished" to eat their fellow men, were any more righteous when they killed men in war, sometimes on the flimsiest of reasons. He and his father also criticized European explorers, themselves included, for encouraging cannibalism among the New Zealanders, after several women told them that their husbands had recently gone off to battle. The Forsters thought these wars stemmed from the New Zealanders' desire to get more cloth, tools, and other objects to offer the English in trade, which they could do most easily by stealing from their neighbors. The English, by offering new and exotic goods, generated ever-greater materialistic desires among the New Zealanders, with unfortunate results—in this case, an increased incidence of cannibalism.

Just before leaving Queen Charlotte Sound late in November, Cook buried a bottle at the base of a tree with a note for Furneaux and nailed a sign on the trunk saying "look underneath." But he and his crew were not to see the *Adventure* again for the rest of the voyage.

Sailing through Cook Strait to the Pacific, the *Resolution* passed a wide, deep bay on the southern tip of the North Island, one that Cook had not seen on his voyage in the *Endeavour*. He named it Palliser Bay, after his friend and patron Sir Hugh Palliser. Johann Forster, not usually of an imperialist turn of mind, thought it would make an excellent location for a European settlement, particularly if the residents were taught to make canvas and rope from their native flax, which could then be marketed in the East Indies. George thought that Europeans might well settle here at some time in the future after the nations of Europe lost their American colonies. "If it were ever possible for Europeans to have humanity enough to acknowledge the indigenous tribes of the South Sea as their brethren," he hoped, "we might have settlements which would not be defiled with the blood of innocent nations."

Two weeks' sail southeast from New Zealand did nothing to change Cook's opinion that terra australis was a myth. At about 55°S, great rolling swells broke over the ship from the southwest, indicating that nothing but open ocean lay in that direction. If a continent existed south of New Zealand, Cook concluded, it must be much farther south, at latitude 60° or more. (The coast of Antarctica at that point lies at about 70°.) Johann Forster agreed. Even before leaving New Zealand, he had announced his conviction that they would see no land en route to Cape Horn.

Like Cook, Forster believed that what was essentially a negative voyage of discovery could still have a major scientific impact. Proving that no southern continent existed would be significant in itself, and there were other accomplishments too: exploring Antarctic waters, demonstrating the possibility of obtaining fresh water from sea ice, locating good places to refresh ships' crews in the Pacific, and stocking Pacific islands with livestock. Some weeks later, however, sailing through unending fog and sleet and surrounded by icebergs, he was less sanguine. The voyage as a whole might be significant, but he personally had little to show for his year and a half at sea. Forster always felt that he traveled in the shadow of Banks, and now, depressed by the cold and the endless days at sea, he began to feel even sorrier for himself than usual. He had probably found nothing that was not previously discovered by Banks and Solander (this was, of course, not true); he would earn little from the voyage, because over half of his stipend had been paid to outfit himself and George (also not true); he would gain no recognition for his efforts, because Banks and Solander had a long head start on publishing the results of their voyage. (He could not know at this point that Banks, with his short attention span, had moved on to other things and would never publish a full account of his travels in the *Endeavour*.)

This was Forster at his most gloomy, but one could hardly blame him for being discouraged. After almost two months at sea and two weeks of sailing constantly among icebergs, hovering around the Antarctic Circle halfway across the Pacific with the prospect of more weeks of the same, everyone on board was depressed. The only provisions left were two-year-old salt meat and moldy biscuit. Waves washed over the ship constantly, through cracks, below decks, and into the men's sleeping quarters; the ship was always damp and cold, looking "more like a subterraneous mansion for the dead than a habitation for the living." Snow and sleet froze on the rigging as it fell, and icicles dangled from the masts. Ropes and sails were stiff, like wires and sheets of metal; it took enormous effort from the freezing, weakened men just to get the sails up and down. Despite careful navigation, the *Resolution* often banged into icebergs—"those confounded Ice Isles," Clerke called them—sending jolts through the entire ship. Once, in mid-December, they came perilously close to shipwreck. "According to the old proverb a miss is as good as a mile," Cook remarked, "but our situation requires more misses than we can expect." Prudence dictated a shift to the north.

Only Hitihiti was enjoying himself. Snow and ice, entirely new to him, appeared nothing short of miraculous. The snow he called white rain; hailstones were white stones; icebergs were white land. No one could convince him that the icebergs were not, in fact, land. Equally amazing was the constant daylight around the Antarctic Circle. He would never make his friends believe "the wonder of petrified rain, and of perpetual day," he told George Forster. To keep track of what he saw, Hitihiti kept his own sort of journal: a bundle of twigs of different lengths. At every island they visited he selected a twig to represent the discovery, and although the twigs all looked alike to the Englishmen, he could recount every bit of land sighted, in the proper order, by referring to his bundle.

On Christmas day the *Resolution* was just north of the Antarctic Circle. The men had been saving their daily allowance of grog and gave themselves up to a day of revelry despite the miserable cold—or perhaps because of it. George Forster looked askance at the proceedings and took the opportunity to pontificate on the uncivilized behavior of common seamen, but Cook, as was his habit, let his men indulge themselves in their one day of merriment. Those who stayed sober enough had some sport counting icebergs; Cook counted one hundred large ones and "innumerable" smaller ones from deck, while Elliott claimed to see three hundred from the masthead. Whales splashing around the ship and penguins lined up in rows on the larger icebergs offered some diversion too. A few of the officers amused themselves shooting at the penguins, "after which they whirld off three deep and March down to ye water in a rank; they seemd to perform their Evolutions so well that they only wanted the use of arms to cut a figure on Whimbleton Common."

But these were momentary distractions, while the reality of cold and ice continued unabated. Although he was now convinced that no land lay to the south of them in this part of the Pacific, Cook would not give up the search, and in mid-January he steered south again to make another attempt to get below the Antarctic Circle. At this point morale reached its nadir. The order to turn south drew complaints from nearly everyone on board, and Johann Forster, whose mood only worsened as the days went by, grumbled about Cook's insistence on pursuing the search for a continent that, in all likelihood, did not exist. "There are people," he wrote, with Cook obviously in mind, "who are hardened to all feelings, & will give no ear to the dictates of humanity & reason; false ideas of *virtue & good conduct* are to them, to leave nothing to *chance,* & future discoverers, by their *perseverance;* which costs of lives of the poor Sailors or at least their healths. These people should be constantly employed by Government upon such Schemes: as for instance the N.W. or N.E. Passage; there they will find a career to give to their genius full Scope; but wo! the poor Crew under them."

Forster could be expected to look at events in the worst possible light, but he was not alone now. According to the usually cheerful George, everyone was depressed—especially since Cook's insistence on another swing south made it obvious that they would not get to Cape Horn and across the South Atlantic this summer, but would have to spend another winter in the tropics before returning to England. At least that was what everyone assumed would happen. In reality Cook was close-mouthed about his plans, making matters worse by perpetuating uncertainty. This was typical of Cook, whose self-assurance when it came to matters of exploration and navigation led him, much of the time, to follow his own dictates without consulting his officers. He enjoyed the confidence and respect of his crew to such a degree that his aloofness usually didn't matter, but at this point morale was so poor that his unwillingness to discuss the future of the voyage was deeply resented. No one wanted to eat, so sick were they of salt meat; no one made sport of counting icebergs or penguins anymore. "We rather vegetated than lived," George said; "we withered, and became indifferent to all that animates the soul at other times."

The *Resolution* crossed the Antarctic Circle for the third time on January 26 and pushed steadily south for the next three days. "God knows how far we shall still go on," Johann Forster complained, "if Ice or Land does not stop us, we are in a fair way to go to the pole & take a trip round the world in five minutes." But on January 30, early in the morning, the men on watch saw a wall of ice extending across the horizon. Mountains of ice rose from the sea to the low-lying clouds, making it almost impossible to distinguish the ice from the clouds hanging over it. Cook counted ninety-seven of these ice mountains. As the ship drew closer, the crew could see that the edge of the ice field was made up of loose chunks so closely packed that it would have been impossible to steer the ship through them; beyond was an unbroken sheet of ice, so massive that Cook speculated it might extend all the way to the South Pole. It was likely, he thought, that the ice islands they had already passed were once part of this great mass before breaking off and drifting north. Clearly the ship could go no farther, and Cook ordered the crew to reverse course. At 71°S, they had reached the southernmost point of the voyage. "It is so far South, as ever any man in future times shall choose to go," Johann Forster thought; and for once he was right. No ship has ever sailed farther south in that part of the Pacific.

Even Cook was ready to turn around. "I will not say it was impossible anywhere to get in among this Ice," he wrote, "but I will assert that the bare attempting of it would be a very dangerous enterprise and what I believe no man in my situation would have thought of. I whose ambition leads me not only farther than any other man has been before me, but as far as I think it possible for man to go, was not sorry at meeting with this interruption, as it in some measure relieved us from the dangers and hardships, inseparable with the Navigation of the Southern Polar regions." Since it was most unlikely that any land existed between this point and Cape Horn, he decided to sail directly north to the tropics. Regardless of what Forster might think, Cook was ambitious and stubborn but not to the point of foolishness. And he was as tired of the Antarctic as any of his men.

But he was not quite ready to go home. Cook's men were correct in suspecting that he intended to continue the voyage for another year. Although he might easily have run east to Cape Horn and been home by spring or early summer, "for me at this time to have quited this Southern Pacifick Ocean, with a good Ship, expressly sent out on discoveries, a healthy crew and not in want of either Stores or Provisions, would have been betraying not only a want of perseverance, but judgement, in supposeing the South Pacific Ocean to have been so well explored that nothing remained to be done in it." He might already have proved beyond a doubt that terra australis was a fiction, but "there is however room for very large Islands, and many of those formerly discover'd within the Southern Tropick are very imperfectly explored and there situations as imperfectly known."

Specifically, he decided to look first for Easter Island, discovered by the Dutch explorer Jacob Roggeveen in 1722 and charted differently on every map Cook had, and then for the islands in the western Pacific discovered by Mendaña and Quiros and rediscovered by Bougainville. Then he would proceed to Cape Horn in time to cross the southern Atlantic the following summer. In short, what Cook proposed was a great loop west across the Pacific and then back again, all for the purpose of confirming old discoveries and scouting the few remaining unexplored parts of the South Pacific.

It was a wise decision to turn north, since several of the crew were ill, some with early symptoms of scurvy. Johann Forster thought it would take two months to reach the tropical islands and that many men would die first, a typically pessimistic Forster prediction. In fact no one died and the illness among the crew remained mild, but toward the end of February Cook himself fell ill, much more seriously than any of his crew. He suffered from severe stomach pains and an obstructed bowel, and for several days could barely eat or walk. Attempts to diagnose Cook's ailment from the distance of two hundred years have produced conflicting opinions. An intestinal obstruction, perhaps caused by a roundworm infestation picked up somewhere in the tropics, is most likely. For several days Cook's shipmates feared for his life, and he remained in a weakened condition for the better part of a month. When he had recovered enough to be able to eat, Johann Forster gave up his Tahitian dog to provide fresh meat and broth for the captain.

In mid-March, three months after the Englishmen's last sight of land, a small island came into view. Judging from its location and Roggeveen's description, Cook thought it must be Easter Island, and a closer look confirmed his judgment. Even from a distance the massive stone statues that were the island's most famous characteristic stood out in bold relief against the horizon. For the men of the *Resolution,* this discovery came none too soon. Easter Island was a tiny island, but it was land, with promise of fresh food, water, and a few days' rest after their ordeal in the Antarctic.

13. *Island Hopping Across the Pacific*

From the ship Easter Island did not look promising. Small, with low hills sparsely covered by scrubby plants, it had no physical features of interest apart from its remarkable stone statues. But the inhabitants were friendly—one man swam to the ship soon after the *Resolution* anchored, climbed aboard, and stayed the night—and willing to trade food, although they did not have a great deal to offer.

Cook was still too sick to leave his cabin, so he sent Pickersgill to explore the island. With the scientists and a handful of officers, Pickersgill tramped for miles, searching for water and food and talking with the inhabitants. It was immediately obvious that the Easter Islanders bore a striking resemblance, both physically and linguistically, to the people of the Society Islands, Tongatapu, and New Zealand. Their language was immediately recognizable, their features and coloring similar, their bodies tattooed in the familiar patterns, and their clothing made from *tapa* cloth like that of Tahiti, although "infinitely worse manufactur'd," according to Clerke. Further investigation revealed that they cooked their food just as the Tahitians did, and that their weapons closely resembled the New Zealanders'.

Pickersgill's report astonished Cook. "It is extraordinary," he thought, "that the same Nation should have spread themselves over all the isles in this Vast Ocean from New Zealand to this Island which is almost a fourth part of the circumference of the Globe." At some point, obviously, Polynesians had traveled enormous distances by canoe, but now the residents of each island group knew nothing of their fellows on more distant islands, except "what is recorded in antiquated tradition." Time, Cook observed, had turned the Polynesians into "different Nations each having adopted some pecular custom or habit . . . never the less a carefull observer will soon see the Affinity each has to the other."

Barren, devoid of both the lush vegetation of Tahiti and the carefully cultivated gardens of Tongatapu and Eua, Easter Island was no South Pacific paradise, however. It had no trees of any size, only about twenty species of plants by the Forsters' count, no streams, and, judging from the dryness of the land, little rainfall. The only water Pickersgill and his party could find was stagnant and unpalatable. The population was small, not more than seven hundred or eight hundred according to Johann Forster's estimate, and the inhabitants obviously had to struggle to make a living from their island's meager resources; their cloth, houses, and canoes were crudely made compared to those of other Pacific islands.

The obvious poverty of the Easter Islanders made their stone statues, which ranged from fifteen to thirty-five feet high, even more incredible, and the Englishmen concluded that they must be the work of a much earlier generation, since the present inhabitants seemed incapable of anything so ambitious. Pickersgill suggested two theories: either men from another area had visited the island and built the statues at some ancient time, or else the islanders themselves had once been more prosperous and more highly skilled in craftsmanship. War or some other great misfortune, he believed, must have impoverished the island and left its people struggling for survival, to the point that they lost the skills that had allowed them to create such sculptures. In fact, archaeological evidence indicates that the statues were built in the latter half of the seventeenth century, as memorials to the islanders' ancestors, and that a battle between two factions on the island stopped the work suddenly—so suddenly that, as the Englishmen observed, work on some of the statues remained unfinished.

The sheer physical feat of erecting the massive figures was an even greater mystery. "How they cou'd form these Images . . . without any kind of Metal Tool . . . or the least knowledge of Mechanic Powers . . . is to me the most wonderful matter my Travels have ever yet brought me acquainted with," Clerke wrote. It remains a mystery to modern scholars as well. The statues were carved out of volcanic rock in a prone or slightly sloping position, face up, with the craftsmen chipping away at the sides and back until the completed sculpture was finally separated from its bed of rock. How the figures were moved is not clear. The islanders have always told anthropologists that it was by force of *mana;* certain men on the island commanded the statues to move. Modern scholars have debated about more tangible methods, including rolling the statues on logs (but as Cook noted, there were no trees) and dragging them with ropes. There is, however, little agreement on how the feat was accomplished.

Scarcity of fresh food and water induced Cook to leave Easter Island after five days. Even that little time had helped restore the crew's health and spirits, but it was important to get to a place with more abundant provisions soon. Cook had seen enough to satisfy himself that this was indeed Roggeveen's Easter Island, and probably also the island discovered by the English explorer John Davis in 1686. Davis's discovery, usually just referred to as "Davis Land," had often been cited by proponents of a southern continent as evidence of the continent's eastern extremity. Cook was, as usual, pleased to clear up a little geographical mystery, but otherwise did not attach much significance to his latest find. "No Nation will ever contend for the honour of the discovery of Easter Island," he thought, "as there is hardly an Island in this sea which affords less refreshments and conveniences for Shiping than it does." Hitihiti had the last word: adding another stick to his bundle, he said that "the people were good, but the island very bad."

From Easter Island the *Resolution* continued northwest toward Tahiti, but instead of taking the most direct route to the Society Islands, Cook decided to sail farther north first, to look for a group of small, mountainous islands discovered by Mendaña in 1595 and named by him the Marquesas. Barely three weeks later, at a point about eight hundred miles northeast of Tahiti, the explorers spotted a small island several miles ahead, then another, then three more. Johann Forster pored over the accounts of Mendaña and other early explorers, trying to figure out exactly which island was which, and grumbling about the sloppy practices of earlier explorers; "had all the former Navigators taken the prudent Step to inquire the Natives, for the Names of the Islands they saw, we might be able to ascertain with certainty, what are new discoveries & what not," he wrote. Cook, less concerned about identifying each island precisely, was convinced from his own reading that the cluster of islands ahead was indeed Mendaña's Marquesas.

The *Resolution* anchored the following day at one of the larger islands, called by Mendaña Santa Christina and by its residents Tahu Ata. A dozen canoes surrounded the ships within minutes, and although it took some coaxing to get them to come alongside, in time the islanders were persuaded to bring fish and breadfruit in exchange for nails. Several men returned the next morning with more fruit and a small pig, which they sold for a broken knife. "How much I wished to see another in its Throat! may be guessed by those who have not tasted fresh meat for four or 5 Months," Wales exclaimed.

The islanders quickly lost all remaining reserve and visited the ship freely, bringing plentiful supplies for trade, although some took whatever the Englishmen offered in payment without surrendering the goods they had brought to sell. Cook made little fuss until one of the men grabbed an iron stanchion and jumped overboard into his canoe. He ordered shots fired over the canoe, but his officers "took better aim than I ever intend" and killed the man. Immediately the rest of the islanders paddled back to shore, and when a group of sailors landed on the island later to get water, everyone on the beach fled. The next morning Cook went ashore and eventually persuaded the Marquesans to resume trade; George Forster, indignant at the murder, thought it was because he had somehow convinced them that death was a just punishment for the man's theft.

When trade halted abruptly once again, four days later, Cook learned that some of the younger officers had been trading for themselves the previous day, offering the red feathers acquired at Tongatapu for "curiosities." The Marquesans valued red feathers above almost anything else, and once they discovered that the English had a supply they would accept nothing else in payment. Unfortunately the Englishmen didn't have enough feathers to purchase a sufficient quantity of provisions, and Cook noted angrily that "the fine prospect we had of geting a plentifull supply of refreshments of these people [was] frustrated, . . . which will ever be the case so long as every one is allowed to make exchanges for what he pleaseth and in what manner he please's." He saw no alternative but to sail immediately for the Society Islands to get the food they needed.

Although four days was not enough time to learn much about the Marquesas, especially since the ship visited only one of several islands in the group, it was obvious to the English that they had reached yet another outpost of Polynesia. They thought the Marquesans were the most physically attractive Pacific islanders they had yet encountered, notwithstanding their practice of tattooing their bodies from head to foot. Cook called them "as fine a race of people as any in this Sea or perhaps any whatever," and others echoed his sentiments.

George Forster considered the Marquesans less civilized than the Tahitians, which had both positive and negative connotations in his mind. They had few visible distinctions of social rank, always a favorable characteristic from Forster's point of view; none was fat and lazy, like many of the Tahitian *arii,* nor did any seem stunted from hard labor, like the Tahitian *teuteu.* All were "active, very healthy, and beautifully made." (This physical well-being was the major reason that Cook and others thought the Marquesans were such attractive people, but only Forster made it into a statement on social class and civilization.) The Tahitians had more "comforts and conveniences" and greater skill in arts, Forster thought; but such evidence of superior civilization had to be weighed against its concomitant disadvantages, notably a higher degree of social stratification and political inequality. The young Forster remained confused about which was really superior—the more complex civilization of the Tahitians, or the simpler but apparently more just and equal society of the Marquesans. Although Forster was generalizing on the basis of minimal observation, modern anthropologists have confirmed his impression that Marquesan society was less highly stratified than that of Tahiti or Tonga. The Marquesans had only two social classes, in contrast to the three-tiered system of the other islands; and the chiefs had much less control over their people and were more likely to engage in productive work themselves.

The *Resolution* anchored in Matavai Bay four days after leaving Tahu Ata. Recalling the difficulty they had getting food during their last visit to Tahiti, Cook intended to stay just long enough to check the accuracy of his chronometers (by comparing their readings with the known longitude of Matavai Bay) before going on to Huahine and Raiatea for provisions. But it was quickly obvious that conditions had changed for the better. New houses had been constructed, new canoes dotted the bay, hogs were everywhere, and when the Tahitians discovered the Englishmen's new treasure—red feathers—they eagerly offered food in abundance. Johann Forster credited English tools with the improvements in the island, while his son, more realistically, attributed them to recovery from war. Cook changed his plans and decided to stay several days to stock up on food supplies.

No one on the *Resolution* was happier to see Tahiti than Hitihiti, who had never been to the island even though many of his friends and relatives had lived there at one time or another. (His sister greeted him when they landed at Matavai Bay.) Eager to show off his treasures and tell his stories of distant and exotic places, he thought his experiences as a traveler and his newly acquired possessions would give him new status among his countrymen, and he was not disappointed. The Tahitians "courted and looked upon [him] as a prodigy," heaping presents on him and asking for gifts in return from Tonga, the Marquesas, and Easter Island, even when the items in question were just like those made at Tahiti. (Unscrupulous sailors caught on to this fact and began selling objects they had acquired from one Tahitian to others.) Just as Hitihiti had predicted, no one believed his stories about snow and ice, but they did accept his account of cannibalism in New Zealand. A few of his new friends came on board to see the head Pickersgill had bought; George Forster was surprised when they all used the same Tahitian word, which he translated as "man eater," to identify it. Asking other Tahitians about this incident, he learned that according to island tradition, at some ancient time the Tahitians had also been cannibals.

Tu visited soon after the *Resolution* anchored, and one of Purea's friends, a chief named Potatau, came all the way from Papara, drawn by reports of red feathers. So eager was he to obtain some of this precious commodity that he offered his wife to Cook, much to the disgust of George Forster, who thought this fascination with such a useless item proved "the existence of a great degree of luxury" among the Tahitians. George was equally perturbed at the behavior of the sailors and the Tahitian women. "The excesses of the night were incredible," he wrote after the first day at an-

Tahitian war canoes assembled in the harbor at Pare. Engraving based on a drawing by William Hodges. *Bancroft Library, University of California, Berkeley*

Matavai Bay, Tahiti. Oil by William Hodges. *Yale University Center for British Art, Paul Mellon Collection*

Tu, the Tahitian chief. Chalk drawing by William Hodges. *National Library of Australia*

Hitihiti, the Tahitian who sailed on the *Resolution* for part of Cook's second voyage. Chalk drawing by William Hodges. *National Library of Australia*

Scene at Tongatapu. Watercolor by William Hodges. *National Library of Australia, Rex Nan Kivell Collection*

Cartoon of Joseph Banks, spoofing his passion for collecting insects (among other natural phenomena) and for the South Pacific. *University Library, Kenneth Webster Collection, University of California, Los Angeles*

Engraving of Mai at his formal presentation to King George III and Queen Charlotte. *National Library of Australia, Rex Nan Kivell Collection*

Mai. Drawing by Nathaniel Dance. *Public Archives of Canada*

Mai. Oil by Joshua Reynolds. *Castle Howard, York, England*

OPPOSITE PAGE:

Top. The inasi ceremony witnessed by Cook at Tongatapu. Engraving based on a drawing by John Webber. *Bancroft Library, University of California, Berkeley*

Center. Kealakekua Bay, Hawaii. Watercolor by John Webber. *Dixson Library, State Library of New South Wales*

Bottom. The *Resolution*'s crew shooting walruses in the Arctic. Watercolor by John Webber. *Dixson Library, State Library of New South Wales*

The scene just before Cook's death. Watercolor by James Cleveley. *The Mariners' Museum, Newport News, Virginia*

Cook's death. Oil by George Carter. *National Library of Australia, Rex Nan Kivell Collection*

Cook's death. Oil by Johann Zoffany. *National Maritime Museum, Greenwich, London*

"The Apotheosis of Captain Cook." Engraving based on a drawing by Philippe Jacques de Loutherbourg. *Department of Prints and Drawings, British Museum*

The Hawaiian chief Kianna. Engraving from John Meares's account of his voyage. *Bancroft Library, University of California, Berkeley*

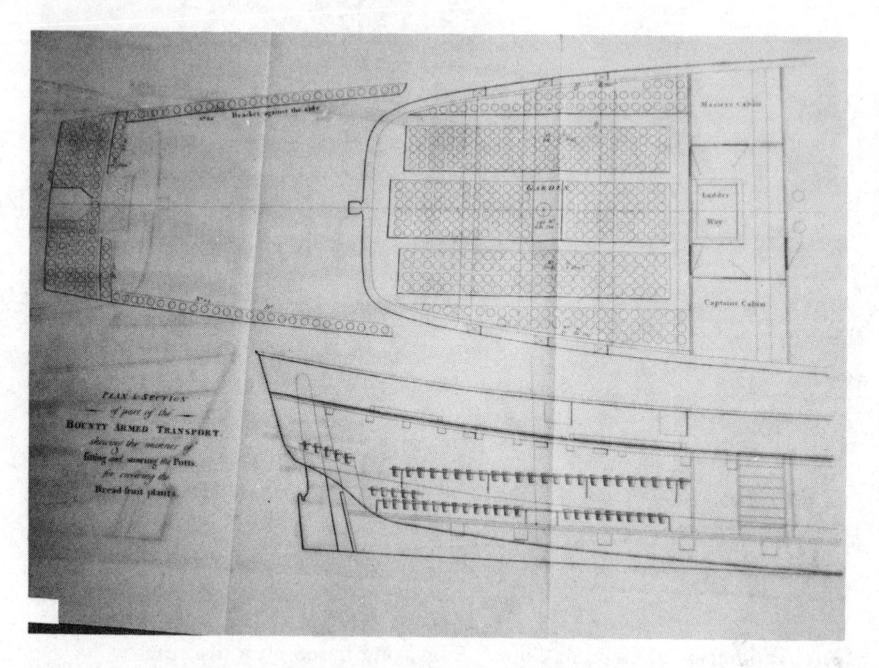

Joseph Banks's sketch of the hull of the *Bounty* showing plans for stowing breadfruit plants. Engraving from Bligh's account of his voyage. *Sutro Library, San Francisco*

chor. Some women had their favorite men from the last visit, but others became adept at playing one against another to collect more presents.

After settling in at Matavai Bay, Cook, the Forsters, and several officers visited Tu at Pare. Approaching the harbor there, they were astonished to find it crammed with huge, doubled-hulled war canoes, each crowded with men dressed in long, richly colored garments, short capes made of feathers, and red or yellow feathered helmets. Hundreds of people lined the shore. Cook counted 159 war canoes, fifty to ninety feet long, with elaborately carved prows rising several feet in the air and draped with flags and streamers. There were, in addition, at least an equal number of smaller canoes loaded with food, many with thatched houses built on platforms between the hulls.

It was a formidable array of Tahitian military force, and Cook and his companions were momentarily taken aback by it, since they had no inkling that the islanders were capable of mounting such an impressive display. But the Tahitians on shore cried out *"Tiyo no Tu!"* ("friend of Tu"), and Cook decided to land. Tu was nowhere in evidence, but his uncle Tii, Potatau, and an older chief named Towha, who turned out to be the leader of the warriors, greeted the English warmly. (The English henceforth called Towha "the admiral of the fleet.") Potatau told Cook that the fleet had been summoned to attack the neighboring island of Eimeo (today called Moorea), whose chief was normally under Tu's influence but had lately been showing signs of independence. The practical Cook wondered how these men could fight wearing such heavy clothes, while the more intellectual George Forster compared the Tahitian fleet to those of the ancient Greeks.

The warriors did not plan to fight immediately, however, but were reviewing their ranks in preparation for an expedition some months hence—the Englishmen never succeeded in finding out exactly when. They did learn, to their amazement, that the fleet was drawn only from the districts around Pare, which forced them to reconsider their estimates of Tahiti's population. Cook calculated that 7,760 men, at a minimum, were required to man the fleet in the harbor; some of his officers thought his figure was low. The Forsters extrapolated from that figure a population for the district, based on assumptions about the size of families, and then multiplied their estimates by the number of districts on the island. George concluded that 120,000 was a conservative estimate of the island's population. Cook's estimate was much higher, upward of 200,000; modern experts believe the population was considerably less than either man's estimate, more likely around 35,000.

Tu and Towha visited the *Resolution* the next day. It was the first time Towha had ever seen a European ship, and he toured every inch of it, showing particular interest in its construction. Tu took pride in showing the older chief around and demonstrating English ways to him; he explained to Towha how to eat his food with a knife and fork, salt his meat, and drink wine from a goblet.

Towha endeared himself to the English a few days later when a Tahitian who attempted to steal a water cask was caught and confined on board ship. Tu begged for the man's release, but Cook considered the matter sufficiently serious to justify making an example of the thief in the hope of discouraging further incidents. In Towha's presence he lectured Tu on theft, reminding him that the English never took anything from the Tahitians without paying for it. Any Englishman who broke this rule was punished; therefore it was only fair that Tahitians who broke the rules should also be punished. In the long run, Cook continued, such a policy would be in the Tahitians' best interests, since continued theft might eventually provoke the English to bloodshed. Then Cook had the accused man tied to a post on the beach and flogged with two dozen lashes. The Tahitians who watched were terrified and ran away, but Towha called them back and made a long speech. The gist of it, as far as the English could tell, was a repetition of what Cook had told Tu: the English were their friends, they never stole from the Tahitians despite their superior military power, and stealing from friends was an unjust act that called for punishment. Such reasoning from a Tahitian, especially one who obviously had great influence over the islanders, guaranteed him the Englishmen's respect.

Later that day, George Forster and Hodges went back to Pare with Towha, who took the opportunity to satisfy his curiosity about the English. He wanted to know about their government and the ranks of their chiefs; he had heard about Banks and thought, from the influence he had wielded during the *Endeavour*'s visit, that he must be the English king's brother. Cook, Towha believed, was the "high admiral"—his own counterpart among the English. He asked about their food, and about plants and animals, concluding that England must be a poor country if it had no coconut or breadfruit trees. At a banquet for his guests, as a matter of courtesy he decided to eat in English fashion, and, after ordering his attendants to bring a large knife and several bamboo sticks, he carved the meat and gave each of his guests a stick to use as a fork. Then he cut his breadfruit into small pieces and ate a bit of it with each piece of fish, instead of scooping it up by handfuls, the normal Tahitian practice.

The visits back and forth between Pare and Matavai Bay continued for several days, while the Tahitians provided an uninterrupted supply of pigs and breadfruit for trade. Johann Forster, determined to outdo Banks, spent his days hiking to the tops of the mountains around the bay collecting plants, and despite his grumbling that Banks had already identified every new species at Tahiti, he managed to find a few that Banks had missed—although at the expense of injuring his hip on the narrow, slippery mountain paths.

In the midst of this pleasant routine, one evening a sentry put his musket down, walked away from the post briefly, and returned to find his gun stolen. Tu learned about the theft even before Cook did and retreated to Pare, while most of the residents of Matavai Bay scattered in all directions. Cook blamed the theft on the sentry's negligence and later had the man flogged, but the theft of a gun was a serious offense even given the circumstances, and Cook was determined to get it back. He dispatched Hitihiti to calm Tu's fears and sent two boats to take several canoes hostage. When Hitihiti returned to the ship with the news that someone from Tahiti-iti had taken the gun and Tu had no way of getting it back, Cook released the canoes and sent Tii, who had accompanied Hitihiti, to inform Tu that all was forgiven. Still the Tahitians stayed away from the Englishmen, and the next morning Tii asked Cook to visit Tu himself. Understanding by now that the Tahitians had cut off trade on Tu's orders, Cook realized that the flow of provisions would not be restored until the chief was placated. It was less clear to him that Tu expected Cook to bring his overtures of peace in person—as king to king, in effect—but Cook went, and trade was restored.

This incident was a classic example of English-Tahitian conflict over property. The Tahitians typically reacted to a serious theft by disappearing and cutting off trade, while the English took hostages, seized Tahitian property, or both. Cook tried to use the threat of punishment as well, but it was usually impossible to identify the person responsible for any specific theft, and his sense of fair play ruled out random punishment. The Tahitians won these struggles of will most of the time, because the Englishmen needed their food and the islanders knew it. After two previous visits to Tahiti, Cook finally recognized this fact, but still he was surprised that the Tahitians did not understand that the English ultimately had the upper hand because of their superior weapons—"that all their War Canoes, on which the strength of their Nation depends, thier houses and even the very fruit they refuse to supply us with are intirely in our power." This was a power he was unwilling to use, however, at least for the moment, because he believed the Tahitians' friendship for the English was the product of "their own good Natured and benevolent disposition, gentle treatment on our part, and the dread of our fire Arms." Harsh treatment would soon change their good-natured behavior, and an excessive use of firearms would only stimulate revenge.

On one of his last visits to the *Resolution,* Tu asked Johann Forster and Hodges to stay in Tahiti, promising to make them chiefs of Matavai and Pare. The Tahitians also tried to persuade Hitihiti to stay, with more success. He had married and made friends, and was disposed to settle at Tahiti; still, several Englishmen tried to convince him to continue the voyage, tempting him with promises of new experiences and riches in England. Cook felt obliged to be more realistic and told him that he would probably never return to the Pacific if he went to England. Torn between curiosity and the entreaties of his friends, Hitihiti finally compromised by agreeing to go with the English as far as Raiatea, his home island. He would dispose of his possessions there and return to Tahiti when the opportunity offered.

There was no shortage of Tahitian youths who wanted to take Hitihiti's place and sail to "Bretannee." Johann Forster, still trying to outdo Banks, wanted to take a boy home with him as a servant; Cook at first agreed, but then the idea caught on, and so many officers wanted to acquire their own Tahitian servants that Cook, who worried about what would happen to them in England, denied all such proposals. He also refused Tu's request to send a young man as far as Tongatapu to collect red feathers, explaining that there was no way to ensure that the man would ever return to Tahiti. Cook did, however, agree to give passage to several men from Bora Bora, who were visiting Tahiti and wanted to sail with the English to Raiatea on their way home.

Going to England may have become the fashion among Tahitian youths, but at least one English sailor preferred Tahiti. Just as the *Resolution* was about to sail, members of the crew spotted John Marra, the gunner's mate who later wrote so sympathetically about the Tahitian people, swimming toward shore. Cook had him seized and confined in irons until the ship was clear of Matavai Bay, but then released him without further punishment. Cook liked Marra; he had recruited him in Batavia on his previous voyage, and had found him a reliable sailor. He also knew that Marra had no family or permanent residence, and understood the appeal of Tahiti to a man in those circumstances. "Where then can Such a Man spend his days better than at one of these isles where he can injoy all the necessaries and some of the luxuries of life in ease and Plenty?" he wrote. George Forster, who agreed that life at Tahiti would be more pleasant than the lot of an English seaman, was nevertheless convinced that any Englishman, "being born and bred up in an active sphere of life, acquainted with numberless subjects, utterly unknown to the Tahitians . . . would shortly have been tired of an uninterrupted tranquility and continual sameness, suited only to a people whose notions are simple and confined."

The *Resolution* left Tahiti in mid-May, headed for Huahine and Raiatea, where they found that little had changed in a year. The elderly Ori still held sway at Huahine and was as friendly as ever to the English, but his subjects again displayed their hostility in a series of apparently unprovoked attacks. First the Forsters' servant was attacked, leaving Johann vowing revenge. A few days later two men guiding Anderson (the surgeon's mate and an amateur naturalist), Vancouver, and another midshipman on a tour of the island seized the nails and hatchets that the three had brought along as trade goods. And when Clerke, Cooper, and a third officer went off on a hunting party, they were set upon by a group of men who stole their possessions, ripped off their clothes, and beat them up. Cook protested to Ori, but was not inclined to press the matter, especially since he thought the young officers had brought their troubles on themselves by carelessness and a "Vain opinion that fire Arms rendered them invincible." This attitude did not sit well with Johann Forster, who complained once again about Cook's leniency toward the islanders and proposed shooting one as an example. Then, he claimed, "the rest would be so alarmed, that no thefts would be committed, & it would cause more honesty among them & greater security for Europeans."

Shortly before the expedition left for Raiatea, Hitihiti brought a message from Ori urging Cook to assemble an armed party to attack the men who had victimized the English. When Cook visited Ori to learn more, the chief told him that the men had formed "a sort of Banditti" to rob the English. Cook resisted Ori's proposal, since he thought the men would simply flee into the island's mountainous interior, but Ori insisted they were preparing to fight. After discussing the matter with his officers, Cook decided to do as Ori asked, since word of a refusal might get to Raiatea and encourage the people there to commit even greater outrages; so he took forty-eight men and marched with Ori and a group of supposedly loyal islanders into the hills in search of the rumored "Banditti." As Cook had suspected, however, they found no armed force waiting for them.

When the Englishmen sailed for Raiatea, they were accompanied by a fleet of about sixty canoes carrying members of the *arioi,* an elite society of *arii* dedicated to conceiving and performing theatrical entertainments, on a visit to the neighboring island. Soon after they arrived, the group staged a performance, called a *heiva,* in this case a pantomime done in such exaggerated style that it was impossible to mistake the actors' meaning. The Englishmen were especially amused at a scene depicting childbirth in which the parts of both laboring woman and newborn child were played by men, although they were surprised that the women in the audience could watch the performance without embarrassment. Adept at improvisation, the actors added skits about the English, including one bit about a Tahitian girl who ran off with the English only to receive a thorough scolding when she returned home.

The *arioi* lived according to a code requiring them to remain unmarried and kill any children born of their sexual liaisons, a situation that both fascinated and shocked the Forsters (as it had Banks, who had also become acquainted with several members of the group). When a pregnant woman told Johann that her child would be killed at birth because its father was a member of the *arioi,* Forster argued with her, trying to convince her that this was an evil thing to do. Asked why, he replied that "Eatooa"—god in the language of the Society Islands—would be angry if she did such a thing, to which she replied that "the Bretannia Eatooa might be so, but that theirs was not."

George tried to explain the group's existence in rational terms. Already persuaded that the chiefs of the Society Islands, better fed and healthier than people of lower status, would produce more and stronger children, he believed the chiefly class over time would become too large for the islands' population to support. Therefore, he theorized, the *arioi,* with its prohibition on marriage, must have been formed originally as a way of reducing the birthrate by keeping some young chiefs celibate. George thought it was probably also intended to foster a group of warriors, who were required to remain celibate since sex would "enervate" them. (He was very likely correct about the connection between infanticide and need for population control, but the European notion of sex as enervating would have been laughable to the Society Islanders.) In time, George guessed, the *arioi* found it too restrictive to abstain from all sexual relations, but retained the ban on marriage and killed any children that resulted from their casual liaisons. No longer the elite class of warriors, "they have almost wholly lost the original chaste and sober spirit of the order," having become the "most luxurious" people in the islands. George consoled himself at this evidence of failing among people who were otherwise so attractive by pointing out that infanticide and the other excesses of the *arioi* were confined to a very small group—and added that those who advertised abortion services in English newspapers were just as criminal.

Cook was ready to leave Raiatea by the first of June, but delayed his departure when some of the islanders brought word that two ships had arrived at Huahine three days before, one commanded by Furneaux and the other by Banks. Cook didn't believe the story—Furneaux would certainly have sailed immediately for Raiatea when he learned that Cook had recently left Huahine, and the notion of Banks and Furneaux somehow hooking up together was preposterous—but in case there was some grain of truth in it he left a note directing Furneaux to meet him at the island Tasman had called Rotterdam, where Cook intended to go next. (Cook suspected the Raiateans had invented the tale to keep the Englishmen at their island a little longer.) The *Resolution* finally sailed on June 4, with Hitihiti staying on board until they were nearly out of the harbor. Before he left the ship, he asked Cook to "Tattaow some Parou"—write some words—for him, and Cook obliged with a testimonial for the benefit of any Europeans who might someday land at Raiatea.

Before beginning his search for the islands Mendaña and Quiros had discovered, Cook planned to stop at Rotterdam Island, the third of the group discovered by Tasman and the only one the *Resolution*'s crew had not visited on their swing across the Pacific the previous year. (Its native name was Nomuka.)

When the Englishmen reached Nomuka about three weeks later, they found the inhabitants decidedly less welcoming than their neighbors at Eua and Tongatapu. The first officers to land on the island had hardly climbed out of their boat when a man stole one of their muskets along with several other objects. Clerke, faring no better here than at Huahine, had his gun stolen later in the day, and another islander made off with some of the cooper's tools. Following his usual belief in the necessity of impressing new acquaintances with the Englishmen's superior strength, Cook ordered all the marines ashore, had two large canoes seized, and fired at one man who attempted to resist, even though Clerke's gun had been returned by the time the marines landed. The islanders fled at the sound of gunfire, but some returned later with the other musket, at which point Cook restored their canoes and attempted to explain that he had seized the boats only to get the guns back.

Cook's strategy in this instance had the desired effect, as the people of Nomuka proved as friendly as the other Tongans, bringing steady supplies of food to the ship for trade—all controlled by a single man, who paddled around the bay taking fruit and vegetables from other canoes and selling them to the English. The sailors dubbed him the "Custom house officer." One hospitable couple greeted Cook on one of his visits ashore and offered him a young woman. Embarrassed, Cook made signs that he had nothing to give her, "and thought by that means to have come of with flying Colours but I was misstaken, for I was made to understand I might retire with her on credit." He declined again, only to be rewarded with abuse from the older woman. "I understood very little of what she said," he reported, "but her actions were expressive enough and shew'd that her words were to this effect, Sneering in my face and saying, what sort of a man are you thus to refuse the embraces of so fine a young Woman." The young woman he could resist, but the old woman's tirade so unsettled him that he got into his boat and went back to the ship.

Cook's biggest problem during the brief stay at Nomuka turned out not to be native theft or a shortage of provisions, but a bitter quarrel between his two chief scientists. On the second day at the island, the chronometer stopped when Wales forgot to wind it. Cook brushed the incident aside as unimportant, because they had made enough lunar observations to reset the watch accurately; but Johann Forster, who didn't like Wales, needled him about it. Wales, already upset at his own forgetfulness, claimed in his defense that he had been unable to wind the instrument because Cook and Cooper, who had the keys to the cabin where it was kept, were both on the island at the time he normally wound it. Forster pointed out that Cook always left the keys with someone else when he was away from the ship, and threatened to include an account of the incident when he wrote his book about the voyage. Wales didn't like Forster any more than Forster liked him, and from this point on the two men spoke to each other only when they couldn't avoid it.

Shortly before leaving Nomuka at the end of June, Cook named it along with Eua and Tongatapu the "Friendly Archipelago" because "a lasting friendship seems to subsist among the Inhabitants and their Courtesy to Strangers intitles them to that Name."

As he sailed west, Cook left the familiar territory he had traversed many times before and headed into a little-known part of the Pacific. The Spanish discoveries in that region had eluded subsequent explorers for two centuries, providing grist for the "continent mongers," as Joseph Banks had called them. Although Bougainville had visited the western Pacific, discovering several islands he thought were the same as those seen by Mendaña and Quiros, he stayed only briefly and his findings were therefore inconclusive. Cook intended, as his last task before turning south for the final leg of his Antarctic voyage, to put this confusion to rest.

14. *The Islands of Quiros*

Mendaña and Quiros, exploring the western Pacific in the late sixteenth and early seventeenth centuries, visited a series of islands in the western Pacific that they took to be the fringe of a great southern continent. The first of these finds, discovered by Mendaña in 1568, appeared so massive that he thought it must be the fabled Ophir of King Solomon, until further investigation revealed it to be just an island—and a rather small one at that. Convinced nonetheless that a continent must be nearby, Mendaña returned to Peru determined to mount another expedition to establish a settlement in the Solomon Islands, as he called them, and find the southern continent.

It took twenty-six years, but Mendaña finally persuaded the Spanish government to sponsor another voyage, and in 1595 he sailed to the Pacific again, with Quiros as pilot and more than three hundred men, women, and children as potential colonists. Mendaña set his course too far south and missed the Solomon Islands, happening instead upon another island he named Santa Cruz. He attempted to establish his colony there, but sickness, near-mutiny, and hostile inhabitants contributed to turn the expedition into a disaster. Mendaña and most of the would-be colonists died. Quiros survived to return to Peru and then to Spain, where he talked government officials into trying again; in 1605 he sailed from Peru with two ships.

Instead of risking another landing at Santa Cruz, Quiros sailed farther south in search of a more hospitable site. When he discovered what appeared to be a vast, fertile land with a huge bay—large enough for a fleet of a thousand ships, he claimed—Quiros thought surely he had found the southern continent at last. Naming it Austrialia del Espiritu Santo, he claimed the land for Spain and appointed officials for the new colony; but after three weeks, he decided the islanders were too hostile to permit peaceful cohabitation and ordered his ships to sail on. The two ships were separated shortly afterward, and Quiros turned back to Mexico while Luis Vaez de Torres, in command of the second ship, continued east, discovering the strait between New Guinea and New Holland and returning to Spain by way of the Cape of Good Hope.

Bougainville rediscovered Austrialia del Espiritu Santo in 1768, proving it to be merely an island. He did not pursue his discoveries in this area, however, but continued west to the edge of the Great Barrier Reef, where he turned north and rediscovered the Solomon Islands. Although Bougainville was reasonably certain of the identity of Espiritu Santo, he did not connect his second landfall with Mendaña's Solomons.

When the Englishmen approached a cluster of small islands in mid-July, Johann Forster set to work trying to determine which of them had been seen previously by Mendaña, Quiros, and Bougainville. As they drew closer to the first—a small, mountainous bit of land, well cultivated and heavily forested—Forster recognized it as one Quiros had called Manicolo (Malekula). At night hundreds of fires flickered in the darkness, a sure sign of a large population. Closer inspection revealed a good harbor, and Cook decided to anchor.

Spanish relations with the inhabitants of these islands had been violent, and Bougainville's men had also been attacked, so the English were cautious as they prepared to land; but their reception at Malekula was friendly. As soon as the *Resolution* anchored, men came out to the ship in canoes, waving green branches and pouring water over their heads. Most were armed with spears or bows and arrows, but they made no threatening gestures and eagerly exchanged their arrows for Tahitian cloth, talking loudly and enthusiastically in a language utterly unintelligible to the English.

The Malekulans were quite unlike any South Pacific islanders the Englishmen had encountered. Short—few stood taller than five feet four—and slender, they were much darker than the Tahitians or Tongans with "frizzled and woolly" black hair, flat noses, and prominent cheekbones. Most painted their faces and chests and sported ornaments of bone or wood in their earlobes and noses. They wore no clothing except for a bit of cloth wrapped around the genitals and fastened to a string tied tightly around their waists—so tightly, Cook remarked, "that it was a wonder to us how they could endure it." The Englishmen, whose standards of beauty favored light skin, tall men, and European facial features, thought the Malekulans were the ugliest people they had ever seen. Indeed, remarked George Forster, he and others made "an ill-natured comparison between them and monkies."

More canoes visited the *Resolution* that evening, until Cook finally had to send them away, since the islanders seemed quite prepared to spend the night on board. Failing that, scores of them gathered on shore, beating drums, singing, and dancing almost until dawn; after an hour or two of quiet, they were back at the ship before breakfast. Several men went on board and made themselves at home, some climbing the shrouds as nimbly as if they did it all the time. They wanted everything they saw, but were not disappointed when they were refused; remarkably, no one stole anything. Closer acquaintance forced the English to take a more positive view of these men. They were not only extraordinarily good-natured, but obviously intelligent, catching the meaning of signs and gestures without difficulty.

The Forsters were even more impressed when they went ashore later in the day and sat down with a group to learn more of the language, which was altogether different from the Polynesian dialects. At first the Malekulans were surprised at the Forsters' eagerness to study the language and puzzled at their attempts to capture sounds by putting them down on paper; but they were delighted to cooperate, and taught the two men about eighty words of their language. They in turn wanted to learn English, and unlike the Tahitians could pronounce the words they were taught perfectly. Surprised at such linguistic facility, the Forsters tried out a few words of other European languages, including what they considered the most difficult sounds to pronounce, and were amazed that the Malekulans got them all right. "What they wanted in personal attraction was amply made up in acuteness of understanding," George had to admit. The islanders were not eager to have the English trespass too far on their territory, however. When George and Sparrman walked a little distance inland in search of plants, several men made emphatic signs that the pair should return to the beach.

George thought the Malekulans were "a race totally distinct" from the Society Islanders, judging them lower than Polynesians in terms of "civilization" despite their intelligence. His views were shaped in part by a deeply ingrained belief that the physical characteristics of a group of people were related to the quality of its culture; by this standard, the short, dark, flat-nosed Malekulans were presumed inferior to the tall, lighter-skinned Polynesians. George was also influenced in his opinion by the simplicity of Malekulan society. (That he had only two days to observe it did not stop him from making generalizations.) The islanders lived in small groups, with no apparent government or distinctions of class; they devoted nearly all their time to producing the bare necessities of life. Moreover, George thought the Malekulans treated their women badly—they carried their children on their backs, and he did not approve of men who expected women to carry burdens—and the treatment of women was, in his mind, a barometer of civilization.

The Malekulans were quite willing to let their visitors have wood and water, but they offered no food other than small quantities of fruit. George thought it was because they had no food to spare, and he was probably right; others observed that the English had nothing that the Malekulans wanted in trade. Whatever the reason, there was no point in staying longer, and Cook decided to leave after two days. As the *Resolution* headed out of the harbor, however, the islanders suddenly became more eager to trade, to the point of offering their bows and spears, which they had refused to give up before. Marbled paper was their favorite form of payment. One man, Wales noted with amusement, "converted it, before our Eyes into a covering for the only part which is covered about them." The Malekulans were scrupulously honest in trade. It was difficult for the canoes to keep up with a ship under sail, and the *Resolution* got well ahead of the islanders before some of the men had paid for their purchases; "it is almost incredible to conceive what efforts they made to come up with us again & deliver it," Wales said.

Cook proceeded south from Malekula, charting the outlines of the several small islands they passed. He did not see any reason to anchor, as the crew had collected plenty of wood and water at Malekula, and he did not want to jeopardize the completion of his plan by spending any unnecessary time in port. Johann Forster did not view the situation in the same light, however. His discoveries could occur only on land, and every island passed without stopping was an opportunity lost as far as he was concerned. "We are to float on the water for ever, to have very few relaxations a shore," he complained, blaming Cook for caring only about his own achievements, with no concern for the work of others. Forster accused Cook of deliberately denying him the chance to explore these islands out of a desire for his own self-aggrandizement; "lands should be discovered," he wrote, "but none of its productions, because people think that if any man but they themselves makes discoveries, their reputation & fame would decrease in the same proportion, as that of others gets a little addition."

This was an altogether unfair statement, telling in its illumination of Forster's own anxieties and self-doubt, but it did make clear the differences of opinion about the purpose of this segment of the voyage. Cook had, often enough, spent extended periods of time on land and devoted considerable attention himself to collecting information about the natural history of his discoveries; but his aim on this winter's cruise was to cover as much territory as possible in order to produce an accurate map of the Pacific, stopping only when necessary to rest the crew, gather provisions, and make lunar observations to test the accuracy of the chronometers. Forster simply had to make do with what he could collect in a few days here and there.

In fact, shortly after Forster's outburst, Cook was ready to stop again, and he directed the *Resolution* toward another small, mountainous, well-wooded island about 175 miles south of Malekula. (Cook later learned that it was called Eromanga.) The inhabitants appeared friendly at first, offering yams, coconuts, and a bamboo container filled with water in response to sign-language requests, but when Cook returned to his boat, the islanders tried to seize it. Cook raised his musket; the islanders let loose a volley of darts and arrows; the marines accompanying Cook opened fire. Four islanders and two Englishmen were injured in the ensuing melee. Given this inauspicious beginning and the likelihood of finding other islands nearby, Cook decided to move on. Even this brief stop, however, had permitted some hasty observations, and several of the Englishmen were surprised at how much these men differed from those of Malekula despite the two islands' proximity. Taller and lighter-skinned, the Eromangans spoke a language that sounded completely unlike the Malekulan tongue.

From Eromanga Cook headed for another island visible to the south, guided through the night by a bright fire near its highest point. The next morning, as molten lava and ash poured from the summit of the island accompanied by a low, rumbling noise almost like the sound of thunder, the Englishmen discovered that what they had taken for fire was actually the eruption of a volcano. After the violence at Eromanga, the Englishmen were nervous about landing, and at first their fears seemed justified. Dozens of armed men gathered on shore and more paddled out in canoes, some with coconuts to trade and others eager "for carrying off every thing they could lay their hands upon." One man tried to take the rings off the rudder; another grabbed the anchor buoys. Cook ordered muskets fired over their heads, with no effect whatsoever. A cannon blast dispersed them temporarily, but they soon returned, as bold as before, until more gunfire finally drove them off. When Cook went ashore later in the day, he met a crowd of armed men lined up in two rows on the beach; but they made no move to attack as he distributed gifts and directed the men accompanying him to fill two water casks from a nearby pond.

Cook remained wary, and the next day he had the *Resolution* anchored closer to shore and turned broadside so its cannon could be fired easily toward the beach in the event of an attack. Meanwhile hundreds of men were gathering on shore, lining up in two groups as they had the day before; Cook estimated the crowd at about a thousand, while Clerke put it at three to four thousand. When a few came out to the ship to trade and refused to pay for what they were given, Cook was actually pleased, "as I wanted a pretence to shew the Multitude on shore the effect of our fire arms without materially hurting any of them." He ordered muskets fired, and then larger guns, which got rid of the thieves but had no effect on the crowd on the beach. "They seem'd to think it sport," Cook wrote.

With his marines and sailors in three boats, Cook approached the beach again, still worried that he might be walking into a trap. He signaled the crowd to move back, but they ignored him, and he decided to try another display of firepower. Gunshots unsettled the islanders only momentarily, however; it took a volley from the *Resolution*'s cannon to get them to leave the beach. Then Cook had his men mark two lines on each side of the beach with stakes and rope, making a space about fifty yards wide that would be off-limits to the islanders. Gradually small groups returned, still armed—when Cook made signs for them to put down the weapons, they replied with signs asking the English to drop theirs first—but otherwise unthreatening. They even climbed coconut trees and picked the fruit for their visitors, asking for nothing in return.

George Forster thought the islanders viewed the English as "invaders" and were determined to defend their property. In fact, it is more likely that they thought the Englishmen were their ancestors returning from the dead; the strangers' white color was confirming evidence. The two groups lined up on either side of the beach represented two different clans or tribes living on the island. They were not eager for the "ancestors" to stay, but were not about to attack them.

By the next afternoon only a handful of men remained on the beach. Johann Forster thought they had retreated inland out of fear of the English, but George more accurately surmised that they lived on another part of the island, and, having assured themselves that the strangers were not a serious threat, felt free to go home. Enough remained close to the beach, however, to keep tabs on the English and to allow the Forsters to pursue their inquiries about the islands' inhabitants. When Johann asked the name of the island, his informants replied "Tanna." Cook adopted it, in keeping with his practice of using native names for islands when he could discover them, although in fact *tanna* means earth or ground; the inhabitants had no single name for their island.

The few men who remained close to the ship offered sugar cane, coconuts, yams, and bananas for trade, although in small quantities. Like the Malekulans, they showed no interest in iron as payment, but preferred Tahitian cloth, New Zealand greenstone, mother-of-pearl shells, and, most of all, bits of tortoiseshell. Some of the Englishmen tried to buy their weapons as souvenirs, but the islanders indicated that they would trade their arms only in exchange for the English guns, "which I thought but reasonable," Wales remarked.

Over the next few days, some of the islanders became bold enough to visit the ship. They were "a stout well limbed race of people, better & stronger than those at Mallicolo," the English thought, dark-skinned with "woolly" black hair plaited in innumerable short, narrow braids. Their clothing, like that of all the other men the English had seen in these islands, was limited to a minimal covering for the genitals, and they painted their faces with red, black, and white designs. Cook invited one particularly lively and curious young man to dine. Perfectly comfortable with his hosts, he tried all their food, although one bite of salt pork and one glass of wine were enough for him. (He preferred yams and apple pie.) "His manners at table were extremely becoming and decent," George Forster thought, notwithstanding his insistence on using a hair ornament as a fork. The dogs on board delighted the young man—he had never seen quadrupeds before, and called them all hogs—so Cook gave him two. And Johann Forster found him useful as a language teacher.

At first Forster thought the language completely different from any he had yet heard, including those of Eromanga and Malekula, although a few words struck him as similar to the equivalent terms in the language of Tongatapu. (This was a perceptive observation, as there is a strong Polynesian influence on that part of Tanna where the English had anchored.) Further study persuaded him that there were two, perhaps three, different languages spoken on the island: one similar to the Tongan language and at least one other that was something like what he had heard at Eromanga. Forster drew his conclusions from listening to people pronounce the same words, including numbers, in two completely different languages; he surmised that the island had been populated at various times by people from different parts of the Pacific, with more recent arrivals either driving out the earlier inhabitants or intermarrying with them. Later he learned that one of the neighboring islands was called "Eetonga," and he watched another man using a stone hatchet very much like those used in the Society and Tongan Islands—further evidence, Forster thought, that Polynesians had settled on Tanna at some point.

Cook learned that the Polynesian-sounding language was spoken at a small nearby island, leading him to suggest that it had been settled by Polynesians who, over time, had mingled with the residents of Tanna. Forster and Cook were quite accurate in their reasoning. The island Cook learned about was Futuna, a tiny Polynesian outpost about fifty miles east of Tanna. Occasional exchanges between the two islands encouraged the persistence of Polynesian influence on Tanna, begun when migrants from Tonga and Samoa settled there and intermarried with the indigenous population. In fact the Tannese spoke six different languages, plus several dialects, but oddly enough, despite the ancient long-distance migrations, they had no knowledge of islands as close as Malekula and Eromanga. "I took no little pains to know how far their geographical knowlidge extended and did not find that it exceeded the limmits of their horizon," Cook said.

Given the islanders' cordial behavior—after a few days they called many of the English by name, and the ropes and stakes on the beach became unnecessary—the Forsters were surprised one day when, on a walk toward the eastern point of the harbor, they encountered a group of armed men blocking their path and urging them to turn back. To avoid trouble, they complied, trying again two days later only to be confronted at the same spot. This time they ignored the men and pushed on, but the islanders pleaded with them to go back, making signs that they would be killed and eaten. When the Forsters pretended not to understand, the men became more graphic, biting their arms to make their point absolutely clear.

These incidents only made the Forsters more curious about that part of the island and the reasons for the inhabitants' reluctance to allow visitors there. Unwilling to accept defeat, they made yet another attempt to walk out to the point, with the same results. By now George was convinced that the Tannese were cannibals, and he theorized that they feuded with each other or with men from other islands. Their initial hostile behavior toward the English, he thought, supported his point. More evidence came when some of the islanders asked if the Englishmen were cannibals. Cook, on the other hand, was skeptical; the Tannese had plenty of meat and no one had actually seen them eat human flesh. Their efforts to keep the English away from the eastern side of the harbor were "only owing to a desire they on every occasion shew'd of fixing bounds to our excursions," he thought, and understandably so; "Its impossible for them to know our real design . . . in what other light can they than at first look upon us but as invaders of their Country; time and some acquaintance with us can only convince them of their mistake."

In fact the Forsters were close to the mark in their interpretation of these events. The eastern point of the harbor was a sacred place, one the islanders did not want them to visit. The Tannese were indeed cannibals, and were divided into several warring tribes; by allying themselves with the first men they met, the Englishmen became enemies of the other tribes. The men who urged the Forsters to stay close to the beach were trying not only to steer them away from a sacred place but also to protect them.

The Englishmen encountered similar problems when they attempted to get a closer look at the volcano. Although many of the islands they had visited were of volcanic origin, Tanna was the first with a still-active volcano—it erupted almost every day—and their stay on the island offered a unique opportunity to undertake some geological investigation. On one of his walks, Johann Forster came upon an area where the earth was covered with a thin, brittle crust hot to the touch. Steam escaped from its fissures, and little piles of sulfur gave off a nauseating smell. Later, with George, Cook, and Wales, he went back to measure the temperature of the ground, discovering that a thermometer buried in the earth for one minute registered 210°F. These steam vents were at least six or seven miles from the point where the actual volcanic eruptions occurred. Hoping to get a better look at their source, the group proceeded in the direction of the volcano itself, but soon encountered men who firmly pointed them back toward the beach.

The Englishmen spent a little over two weeks at Tanna, their longest stay on any of the western Pacific Islands and their only significant opportunity to study the people of the region. Brief as the visits to Malekula and Eromanga had been, however, they were useful in demonstrating the differences among tribes living relatively close to one another. After traveling over thousands of miles of ocean and discovering people living on widely scattered islands who were culturally similar and spoke variants of the same language, in one small corner of the Pacific the Englishmen found men and women who resembled neither the Polynesians nor their own neighbors. Moreover, as Cook had discovered in questioning the Tannese, they knew little of islands beyond those they could see from their own, itself a striking contrast to the extensive geographical knowledge of the seafaring Polynesians.

On the whole, the Englishmen viewed the people of Tanna, Eromanga, and Malekula as far more primitive than the Polynesians. Their houses were "mere sheds" and their canoes small, rudimentary craft dug out of a single tree trunk. They knew nothing of iron and had no interest in it—at least until the end of the Englishmen's stay at Tanna, when a few men began to ask for hatchets and large nails. Their only cloth was a simple, coarse fabric, and their clothes were minimal. Even their weapons, in which the men seemed to take great pride, were simply made. In short, Cook believed, "These people seem to have as few Arts as most I have seen."

The Forsters concluded that the Tannese had a rudimentary government at most; they lived in small villages, each apparently independent of the others, and influence within villages appeared to depend only on age and strength, with no clear distinctions of political or social rank. They found no evidence of religion, apart from a suspicion that the eastern end of the island from which they had been barred was a sacred place. These observations, together with his view that the women were treated badly, led George to accord the Tannese a low rating on his scale of civilized peoples. And yet they presented him with a dilemma, because in other ways they did not fit his stereotype of a primitive culture. Their intelligence and curiosity could not be denied; they lived in "fixed habitations" and cultivated their land, whereas primitive men were usually nomadic hunters; they ate a variety of food and devoted considerable care to preparing it. And their music, to his ears, was much more highly developed than that of any other island he had yet visited. "It cannot be disputed," he wrote, "that a predilection for harmonious sounds implies great sensibility, and must prepare the way for civilization." At Tanna George was confronted with a society that defied his ideas about the progress of civilization by combining some elements of the simplest sort of culture with other elements of one much more complex. His only solution was to describe Tanna as more primitive than the islands of Polynesia but ripe for progress toward greater civilization.

From Tanna Cook turned north, charting the west coast of the island as well as those of Eromanga and Malekula. Just north of Malekula he discovered a new and much larger island, which, judging by its appearance and location, was clearly Espiritu Santo. On the northern side he sailed into a large bay, recognizing it as the grand harbor Quiros had described as fit for a fleet of a thousand ships. Not all of Cook's officers agreed, however, because it was perfectly obvious to them that this bay, large as it was, did not match Quiros's enthusiastic description. Nor was the land all that he had claimed, although Cook noted charitably that the extent of the island and its mountainous appearance might easily allow it to be mistaken for a continent. Charles Clerke, on the other hand, criticized Quiros's "pompous" description. "I firmly believe Mr. Quiros Zeal and warmth for his own favourite projects has carried him too far in the qualities he had attributed to this Country," Clerke wrote. "For fine and fertile as it certainly is, I'm afraid he's given it to prolific a Soil and luxuriant a Clime."

A strong wind blowing directly down the bay made it difficult for the *Resolution* to get close enough to shore to anchor, and Cook decided to continue his circuit of the island without attempting to land, much to Johann Forster's annoyance. Espiritu Santo was the largest single island the Englishmen had seen since New Zealand, and one previously unexplored by naturalists. To Forster, Cook's refusal to land there was one more example of the captain's total disregard for the scientific work of the voyage. Even George complained that "the study of nature was only made the secondary object in this voyage, which, contrary to its original intent, was so contrived in the execution, as to produce little more than a new track on the chart of the southern hemisphere. We were therefore obliged to look upon those moments, as particularly fortunate, when the urgent wants of the crew, and the interest of the sciences, happened to coincide." Unfortunately for the Forsters, this was not one of those times. "We were in want of little we could expect to find here and had no time to spend in amusements" was the way Cook saw it.

After naming the cluster of islands from Espiritu Santo to Tanna the New Hebrides, Cook headed for New Zealand, where he intended to make a final stop for provisions and repairs before returning to the Antarctic. He expected to find nothing in his path—Bougainville had sailed these waters and found nothing—but four days later the crew discovered land about twenty miles distant, stretching across the horizon as far as they could see. A range of steep hills ran nearly its full extent. At that distance, it was impossible to tell whether the land was an island, a series of islands, or part of a continent; by the next morning, the *Resolution* drew close enough for the crew to make out a single, long island with an unbroken line of breakers just off shore. Closer inspection revealed a barrier reef running its full length, but Cook spotted a narrow opening and sent two boats to find a passage through it. Four hundred miles southwest of the New Hebrides and over a thousand miles north of New Zealand, Cook had made an entirely new discovery, one of the few in a career devoted primarily to rediscovering, confirming, extending, and charting the finds of earlier explorers.

While the *Resolution* coasted outside the reef, waiting for the boat's crew to find a safe passage, several of the island's inhabitants paddled out in their canoes. They were unarmed and showed no surprise or fear, even though they had presumably never seen Europeans. Later they lined up their canoes on each side of the passage through the reef, marking a safe channel for the ship. When Cook and several others went ashore the next day, dozens of men crowded on the beach gave them a friendly welcome.

Given its location, George Forster thought this island had perhaps been a link in the chain of Polynesian migration from the Society Islands to New Zealand, and so he expected to find men who resembled the Tahitians and spoke a similar language. But he was soon proved wrong. Not only were these islanders clearly not Polynesians, they were quite different from the people of the New Hebrides as well. Although they resembled the Tannese in color and physical features, they were much taller—some over six feet—and more robust, the finest physical specimens of "any woolly Headed Nation we ever met with," in Clerke's opinion. Their canoes were similar to those of the Tongan Islands, and to complete the cultural confusion, they spoke a language unlike any the English had heard before, "a circumstance sufficient to discourage the greatest and most indefatigable genealogist," George Forster wrote.

A totally new discovery was not something Cook could pass over in haste, particularly when it was an island as large as this one, so the Englishmen stayed nine days. Some, under Pickersgill's direction, charted the island's coastline. Wales observed an eclipse of the sun. And the Forsters finally had their chance to botanize in virgin territory, although not without experiencing some frustrations. Trouble brewed on their second day at anchor, when Cook and Wales went to a small offshore island to observe the eclipse. Forster wanted to visit the main island to look for plants; Cooper, in charge during Cook's absence, announced that the boat going ashore was ready to leave just as Forster was about to sit down to dinner. He asked Cooper to let him take another boat with a small crew later in the afternoon, but Cooper refused on the grounds that he could not divert men from their work. This struck Forster as specious reasoning, since he thought most of the crew stood around doing nothing much of the time, and he accused Cooper of deliberately sabotaging his work by seeing that the boat left at an inconvenient time and refusing to provide another. The incident only fed Forster's growing paranoia; not just Cook and Wales but everyone in the ship was jealous of him, he thought, because they all realized that he was earning a great deal more money on the voyage than they were.

The next day Forster was more successful, collecting several new species of plants. One was from a genus that up to now had been discovered only in Africa; another was known to grow only in America; several were linked to Asian species. These were exciting discoveries for Forster, since they suggested that "this Isle therefore connects the 3 Continents Asia, Africa & America." Over the next several days he expanded his collections, most notably with a number of specimens similar to varieties Banks had found in New Holland, but his good results did nothing to temper his bitterness at what he perceived as a concerted effort among others to thwart his scientific work. And as Forster grew increasingly resentful of his shipmates, they gave up all pretense of cooperation. When Anderson, the surgeon's mate who was also an amateur naturalist, returned from an excursion along the island's coast with several plants, reputedly of new species, he refused to show them to either Johann or George. On top of Cook's stubbornness about stopping at islands, Cooper's refusal to provide a boat when Johann wanted it, and the continuing tension with Wales, this impertinence from a petty officer was too much to bear. Even the usually unflappable George was inclined to agree with his father's conviction that the men of the *Resolution* were collectively campaigning against them.

In all their excursions ashore, the Englishmen were unable to learn the native name for the island—as in the New Hebrides, the inhabitants had names for particular districts, but not for the island as a whole—and Cook decided to call it New Caledonia. Notwithstanding Johann Forster's botanical finds, it was a poor island, devoid of the coconut and breadfruit trees that provided the staple foods of many Pacific islands, and the islanders offered no food for trade, although they were quite willing to barter anything else, including weapons. "Nature has been less bountifull to it than any other Tropical island we know in this Sea," Cook wrote; "the sterility of the Country will apologize for the Natives not contributing to the wants of the Navigator." Obviously they had no food to spare; most of their land was either too steep or too swampy to be cultivated, and the few patches that could be farmed had to be irrigated with water from rivers.

The people of New Caledonia impressed the Englishmen a great deal, more so than any of the other western Pacific islanders, with their striking physical appearance and their friendly, trusting manner. They allowed their visitors to wander around unescorted and showed no fear of firearms, although they watched in amazement when the sailors shot birds. Women and children, rarely seen in the New Hebrides, talked to the Englishmen freely. Most remarkably, the islanders never stole anything, and as Wales pointed out, it was not because they had no interest in English possessions. In those few other places where theft had not been a problem, it was because the inhabitants either were afraid of the English or had no interest in their belongings; but the New Caledonians respected English property in spite of their curiosity.

Still, it took the English some time to shed the caution they had adopted in dealing with the men of the New Hebrides, and the New Caledonians were at times puzzled by their visitors' lack of trust. Pickersgill once led an exploring party on an overnight excursion, stopping at a small offshore island where he and his companions spread out their wet clothes to dry. Several curious men crowded around them, so the sailors, following their experiences at Malekula and Tanna, drew a line around the clothes and indicated that the islanders were not to cross it. The men agreed, but, amused at such behavior, one drew a circle around himself and made signs to everyone to stay away. Later that evening the islanders were even more astonished to watch the Englishmen gnawing meat from bones around their campfire. All conversation stopped, as they "looked with great surprize, and some marks of disgust, at our people." Eventually the problem became clear: they assumed the sailors were eating the remains of a man. Pickersgill tried to convince them otherwise, but sign language failed him; New Caledonia had no quadrupeds, and these men could not believe that bones of that size could be anything other than human.

The same qualities that made the New Caledonians attractive to the English also made them puzzling. Why were they so friendly, when they had never seen Europeans before and all other Pacific islanders, with the exception of those at Tongatapu and Eua, had displayed some level of mistrust or hostility toward their first European visitors? Why were they so tall and strong, when their island was poor and their diet limited? And why were they so different from the people of any other land near them? The English tried to explain cultural differences in environmental terms and relate the people of one island or group of islands to others nearby, but the New Caledonians defied explanation on these grounds.

These difficulties were only part of a series of questions raised by the *Resolution*'s most recent sweep across the Pacific. In their travels the Englishmen had encountered two quite different groups of people, each with many variations. The first, and the ones with whom they had the greater experience, were Polynesians, who were physically attractive to European eyes and extraordinarily friendly. They were excellent navigators, as the Europeans could tell from their boat construction, their familiarity with other islands, and the fact that they had populated a vast area of the Pacific over the centuries. The English had spent enough time with Polynesians, especially Tahitians, to gain a rudimentary understanding of their culture and to form friendships that persisted over the intervals between visits to their islands. In short, the Polynesians had begun to take shape in the English mind as a complex culture—as people with families, government, religion, art, ambition, even intellectual curiosity.

Having begun to form some conclusions about Pacific islanders based on their experiences with Polynesians, the *Resolution*'s crew then encountered a wholly different set of people. Grouped today under the general term "Melanesian" (Greek for "black islands"), these men and women appeared ugly to the English because of their small stature, dark skin, and frizzy hair; and they were, with the exception of the New Caledonians, suspicious or downright hostile. Although some of them displayed considerable curiosity about the English, they appeared to live in a much more restricted world, linguistically and geographically isolated, and in a much more primitive state of culture. As a result of these qualities, and the brevity of their acquaintance with Melanesian groups, the Englishmen never felt the kind of personal attachment that developed between themselves and the Polynesians. The difference was apparent even in Hodges's drawings; his Melanesian portraits do not convey the personalities of their subjects in the way that his pictures of Tahitians and other Polynesians do.

In trying to understand these various groups and explain differences among them, some of the Englishmen employed methods that would eventually be accepted as the basic tools of anthropologists: direct observation of all aspects of material culture and use of native informants to elicit more abstract information. Banks was a master at these techniques, but he had confined his attention primarily to the Tahitians and, to a lesser extent, the New Zealanders. On this second voyage, Cook and the Forsters—the only men on board with more than a superficial interest in understanding native peoples—worked with a much broader range of cultures. As on the first voyage, Cook confined his attention mostly to the obvious physical characteristics of the places he visited. He did not trust native informants, as a rule, and was in any case more interested in the tangible aspects of culture. His attempts at comparing the various places he visited were limited, for the most part, to cataloguing differences in houses, boats, dress, food, and so on; and his explanations of the differences, when he attempted them at all, were superficial.

The Forsters, on the other hand, were more concerned with discoveries on land than on sea, and as the voyage progressed they became most interested in studying people, just as Banks had on the first voyage. Their method, however, was quite different. Unlike Banks, who learned by partaking of native life as much as an outsider could, the Forsters were more detached and analytical in their work. In part these differences reflected the three men's temperaments. Neither Forster could enter into other people's lives in the way that Banks had; George was open and gregarious enough, but too quick to explain rather than listen, and Johann, although an excellent observer, was simply incapable of feeling empathy with other human beings. Even more significant, however, were fundamental differences in method. Banks adopted an essentially empirical approach, explaining his observations in terms familiar to him—hence the analogy between Tahiti and feudal Europe, for example—as anyone would in trying to make sense of new information. (Admittedly some people, Cook included, were better than Banks at setting aside their own preconceptions when looking at unfamiliar cultures.) The Forsters worked in a more deliberately theoretical way, attempting above all else to compare and categorize the cultures they observed in terms of levels of civilization.

The differences they discovered, therefore, had to be explained systematically. At the most basic level, they attributed cultural differences to environmental conditions, and in many respects they were correct. The Tongans were accomplished farmers because their land did not provide enough food naturally, as Tahiti did. The Tahitians had highly developed arts and an elaborate class structure in part because they had a surplus of the necessities of life. But environment alone could not always explain differences among the islands—as George Forster discovered, for example, at New Caledonia, where limited food supplies logically would suggest that the people would be small when they were in fact among the largest he had seen in the Pacific. This observation, on top of other hard-to-explain variations among the islands, led him to the reasonable if rather obvious conclusion that "the different characters of nations seem therefore to depend upon a multitude of different causes."

When the voyage was over and Johann Forster had time to sort out and reflect on the material he had collected, he concluded that fundamental racial distinctions among the South Pacific islanders, along with environmental characteristics, explained the differences in their cultures. He divided the Pacific islanders into two racial groups: the first included the Society and Tongan Islands, the Marquesas, New Zealand, and Easter Island; the second, the New Hebrides and New Caledonia. He based this division, which corresponds roughly to modern anthropologists' distinction between Polynesians and Melanesians, primarily on the obvious physical differences among the islanders, noting further the variations within each broad group.

Forster theorized that the two groups had their origins in different parts of Asia: the darker-skinned people in New Guinea and the Spice Islands, and the lighter-skinned groups on the Malay peninsula. As the latter group moved across the ocean, they encountered the darker-skinned people, who had migrated eastward earlier, and either forced them far into the interior of their islands or conquered them and made them servants. The *teuteu* of Tahiti, Forster believed, were the remnants of the original dark-skinned settlers. On other islands, as in the New Hebrides, the Malay people had made few inroads.

Debates have raged over the origins and migration patterns of Pacific peoples ever since Forster's time, but modern scholarship supports his general theory, although not the imagined drama between light- and dark-skinned people on Tahiti and other islands. The Polynesians are descended from men and women who migrated from coastal southeast Asia sometime in the second century B.C. Known as "Lapita" people after their distinctive pottery, they worked their way gradually eastward, through the previously settled Melanesian islands to Tonga and Samoa, where they settled around 1200–1000 B.C. From there, about two thousand years ago, some moved west—hence the Polynesian influences Cook and the Forsters noticed in the New Hebrides—while others sailed east to the Society Islands, the Marquesas, and other islands in central Polynesia. (Still others made their way southwest to New Caledonia, where they mixed with earlier settlers, creating the unique culture that confused the Forsters.) Not surprisingly, the central Polynesians developed a distinct dialect and different customs over the centuries, which they spread to New Zealand, Easter Island, and the Hawaiian Islands in subsequent migrations. It was no accident that Hitihiti had more trouble understanding the Tongans than the New Zealanders; because of the Polynesians' unusual migration patterns, the New Zealanders were culturally closer to the Tahitians than to the Tongans, although geographically much more distant.

Not satisfied with merely classifying, Forster went on to create a hierarchy of civilizations within each group. Among the lighter-skinned people, he ranked the Society Islanders highest, followed by the Marquesans, Tongans, Easter Islanders, and finally the New Zealanders. (This ranking represented a shift from his original thinking, when he accorded the Tongans a higher place.) Among the darker-skinned people, who as a group he placed below the Polynesians, he ranked the New Caledonians first, then the Tannese, and finally the Malekulans, basing his distinctions on gradations of skin color, quality of physique, and cultural traits such as the development of agriculture, religion, and the arts.

To explain differences in placement along this continuum of civilization, Forster resorted to environmental factors again, especially climate; men and women living in warmer climates were more "civilized," he argued, than those living in colder zones—witness the people of Easter Island and New Zealand. Though obsessed with racial characteristics, he never suggested that they were an impediment to cultural progress. Rather, he viewed the Pacific—and indeed, the world as a whole—marching along gradually toward higher levels of civilization. In fact, he liked to compare certain features of the Pacific islands to the early history of Europe: Tahitian war canoes and fighting methods resembled those of the ancient Greeks, for example, and he suspected that cannibalism had once been universal. Forster summed up his point of view by comparing the advancement of nations with the growth of the individual: while individuals passed through the stages of infancy, childhood, adolescence, and manhood, so nations went through the four stages of animalism, savagery, barbarism, and civilization.

In advancing these explanations, Forster was hardly original. The use of environmental conditions, especially climate, to explain differences in culture had been common among Europeans from ancient times, and a belief in the inexorable progress of civilization was especially characteristic of eighteenth-century intellectuals. Both sets of ideas were based upon a conviction that all societies evolved according to a common set of principles, although at different historical periods, depending on specific local conditions; the great value of studying primitive societies, for intellectuals like Forster, was the opportunity they provided to observe the various stages of development through which, presumably, European societies had already passed.

In his fascination with the development of civilization, however, Forster came down squarely against those European intellectuals who romanticized the "noble savage" and found virtue in simple societies. The most primitive peoples of the Pacific were happy, he admitted, but theirs was a happiness born of ignorance, "founded on mere sensuality . . . transitory and delusive." It was significant, he thought, that the more "savage" people he encountered showed no curiosity about England and no desire to go there, while the Tahitians did. (Forster failed to recognize that the intellectuals who idealized "primitive" man based their raptures primarily on what they knew of Tahiti, which he considered the most civilized of Pacific cultures.) He agreed that complex societies had serious failings, but denied that primitive cultures were therefore better. On the contrary, it was the "wish of humanity, and of real goodness" to hope they would progress toward civilization without falling into "those evils, which abuses, luxury and vice have introduced among our societies."

George Forster also rejected the noble savage myth, notwithstanding his frequent praise for the simplicity of Polynesian societies, although for somewhat different reasons. Blasting "fanciful writers of novels, who cannot shed their ideals and who are used to dreaming idly of children of nature, the golden age, original goodness and simplicity and an innate feeling that everything belongs to everybody," he pointed out that Pacific islanders shared more of the flaws of modern society than many Europeans realized. Both Forsters were in some sense attacking a straw man from the English point of view, however, certainly as far as the explorers themselves were concerned; although Cook and others remarked on the happiness and simplicity of the lives of Pacific islanders, even the most primitive, none suggested that this way of life was preferable to their own. Quite the contrary: they wanted to bring civilization to the Pacific, although most did not employ such explicit terms as Johann Forster. They planted gardens, installed livestock, and congratulated themselves on the progress Tahitians had made in rebuilding their island with the aid of English iron. Acknowledging these gifts, Johann Forster lamented that the English had not been able to do more to "communicate intellectual, moral, or social improvements."

Nevertheless, the Englishmen sometimes had doubts about the value of what they were doing. George especially lamented the negative influence of Europeans, observing that they introduced new desires with their gifts and new diseases with their friendship. In a remarkably prescient observation, he once remarked that Europeans should be especially careful about introducing any contagious disease; smallpox, in particular, "would undoubtedly make dreadful havock, and go near to destroy the whole race of Taheitians." Cook and others made similar comments, although less often. But the Forsters clung to their belief that the benefits of European ways could be spread without the vices, and George in any case recognized the naiveté of thinking that Pacific peoples could have remained forever undiscovered by others.

The *Resolution* left New Caledonia in mid-September. Once again expecting an easy passage with no interruptions, Cook was surprised a week later when his crew spotted an island in the distance, its shores lined with tall pillars reminiscent of ships' masts. The officers and scientists debated the nature of the pillars, with Cook and most of the others assuming they must be trees (although no foliage was visible at this distance) and Johann Forster insisting they were made of stone. The officers ridiculed Forster—they had never seen or heard of such stone formations—and got in response a lecture on basalt pillars in Ireland and Germany, which so antagonized them that nothing would have persuaded them to admit Forster might be correct. The argument continued for the five days required to reach the island, with Forster becoming so certain of his position that he bet twelve bottles of wine on the outcome.

The officers felt themselves vindicated when Cook and Forster landed on the island to find that the pillars were unusually large spruce trees. Cook's analogy to ships' masts had been apt; the timber was ideal for masts and yards, which the *Resolution* by now badly needed. After taking what they could use, the Englishmen moved on, only to discover a much larger island with similar trees a few days later. Both were uninhabited. Cook named the first the Isle of Pines and the second Norfolk Island.

A week after leaving Norfolk Island the Englishmen anchored in Queen Charlotte Sound, where they immediately found evidence that other Europeans had visited recently: trees had been chopped down and onions, which the New Zealanders refused to eat, had been harvested from Cook's garden. Cook concluded that the *Adventure* had preceded them. Oddly enough, however, none of the New Zealanders were anywhere to be seen. Five days after the *Resolution* anchored, two canoes finally approached Ship Cove but paddled hastily away as soon as the men in them saw the *Resolution*. Later in the day a shore party encountered another group of New Zealanders, who also seemed eager to avoid the English. These incidents puzzled Cook, because the people of Queen Charlotte Sound had always been friendly before, and certainly Europeans were no longer strangers to them.

With a certain amount of coaxing, the New Zealand men were finally drawn into conversation and told their visitors an extraordinary story. Some weeks earlier a ship had visited the Sound. Its crew fought with the New Zealanders and killed several of them, but in the end the native men had killed and eaten the Europeans. Although the details were fuzzy, Cook and his companions worried that the tale was true and the ship in question the *Adventure*. Over the next few days, they heard several variations; some said the ship had been wrecked, others that a battle had taken place. The time of the incident varied from three weeks to two months previously.

In an attempt to settle the matter, Cook invited two men he considered reliable into his cabin and questioned them in some detail. He and George Forster drew a picture of Queen Charlotte Sound on a large piece of paper and cut two smaller pieces into shapes representing ships—the *Resolution* and *Adventure*. They moved the paper "ships" into and out of the map of the Sound, to indicate their previous visits. Then, after a short pause, they moved the *Resolution* into the Sound again; but the two New Zealanders stopped them, took the *Adventure,* moved it into and out of the Sound, and then counted on their fingers to show the number of months the ship had been gone. This exchange seemed to discount the tales of shipwreck and massacre, but Cook and the others remained concerned about the stories that some Europeans had been killed, even after other native informants assured them that the accounts were false and that no Europeans had been murdered.

The *Resolution*'s crew stayed at Queen Charlotte Sound three weeks, just long enough for Cook to complete his program of stocking the area with animals and for Wales to finish the observations needed to establish the Sound's longitude precisely. Johann Forster, who no longer even attempted to conceal his dislike of Wales, took the opportunity to call the astronomer's competence into question again, complaining that "it must seem very strange, that a Man who has been 3 times here, & each time during 3 weeks at least, could not settle the Longitude in that time." Cook, however, was pleased that Wales's diligence in recording numerous separate observations had allowed him to correct a small error in the longitude calculated by Green during the *Endeavour*'s visit. The error was not large enough to affect navigation, Cook admitted, but his passion for accuracy was such that he thought the correction worth the time spent on it.

From Queen Charlotte Sound Cook directed his course across the Pacific to Cape Horn, aiming to reach the coast near the west entrance to the Strait of Magellan and then sail south to chart the coast between the Strait and the southern tip of the Cape. It was an uneventful, even boring, voyage. "I never was makeing a passage any where of such length, or even much shorter, where so few intresting circumstance[s] occured," Cook remarked after they reached the Strait; but he found compensation for the last long, monotonous haul across the Pacific in the certainty that he had completed the task he had set for himself in as thorough a fashion as anyone could possibly expect. "I have now done with the SOUTHERN PACIFIC OCEAN," he wrote, "and flatter my self that no one will think that I have left it unexplor'd, or that more could have been done in one voyage towards obtaining that end than has been done in this."

After charting the west coast of Tierra del Fuego and celebrating Christmas in a desolate harbor several miles north of Cape Horn, on New Year's Day the Englishmen reached the tip of the Cape and sailed into the Atlantic on the last leg of their search for a southern continent.

Five days later they woke up to a familiar sight: a huge mass of ice on the horizon. So accustomed were the crew to the massive icebergs of Antarctic seas that they assumed this was another "ice island," but a closer look showed it to be land. The coastline, which extended as far as the eye could see, was a series of sheer cliffs broken by narrow bays, with high mountains visible inland. The extent of the land, the height of the mountains, even the sheer quantity of ice suggested this was no mere island, but part of a continent. After three futile winters, it appeared that the explorers had finally found their southern continent, but one so cold and barren that it was of no use to anyone. Cook called it a "savage and horrible" land; "the Wild rocks raised their lofty summits till they were lost in the Clouds and the Vallies laid buried in everlasting Snow. Not a tree or shrub was to be seen, no not even big enough to make a tooth-pick."

Further exploration revealed that this was not the edge of a continent after all, but just another island, albeit a large one. Cook named it the Isle of Georgia after the English king. Ten days later the group encountered more islands, smaller than Georgia but equally mountainous and ice-clad, which Cook thought might be part of a peninsula extending north from a continent. Because he believed the icebergs found throughout the Antarctic were formed near land, Cook deduced that a continent must lie farther south, surrounding the South Pole. He found additional proof for his theory in the fact that his expedition had encountered ice at more northerly latitudes in the Indian and South Atlantic Oceans than in the South Pacific; if ice was formed in the open ocean, one would expect to find it at a consistent latitude all around the globe. Cook was remarkably accurate in his deductions. The Antarctic continent does indeed extend farther north in the areas he indicated, and the point where the *Resolution* reached the southernmost latitude of its journey was in that part of the Pacific where the Antarctic coast lies farthest south.

The risks of navigating in Antarctic seas and the slim possibility of finding any land of significance in the South Atlantic persuaded Cook that he could not justify further exploration at these latitudes. He thought briefly about going back to the Indian Ocean to search again for the land supposedly discovered by Kerguelen, but decided to sail directly to the Cape of Good Hope instead. "My people were yet healthy and would have gone wherever I had thought proper to lead them," he wrote, but he worried about their contracting scurvy when supplies of preventives were running low. "Besides," he went on, "it would have been cruel in me to have continued the Fatigues and hardships they were continually exposed to longer than absolutely necessary, their behavour throughout the whole voyage merited every indulgence which was in my power to give them."

Moreover, Cook himself was heartily sick of the Antarctic. Normally a man of restrained emotion, he never allowed himself any open expression of depression or discouragement. Toward the end of the terrible winter of 1773-74, when the endless frigid days pushed morale on board to its lowest point and Cook himself fell ill, he voiced no complaints and went through elaborate justifications for leaving a difficult task when he finally decided to turn north. But now his language betrayed his repugnance for the unrelenting icy landscape—the islands discovered in recent days were "savage," "wild," "doomed by Nature never once to feel the warmth of the Suns rays, but to lie for ever buried under everlasting snow and ice"—and this time Cook did not belabor his decision to leave the Antarctic.

En route to the Cape, Cook reflected on the accomplishments of his voyage. Having nearly circumnavigated the globe in latitudes farther south than anyone had ever attempted before, and in addition crisscrossed the Pacific at every conceivable point where a continent might lie, he believed "the intention of the Voyage has in every respect been fully Answered, and a final end put to the searching after a Southern Continent, which has at times ingrossed the attention of some of the Maritime Powers for near two Centuries past and the Geographers of all ages." It was not that he denied the existence of a continent; indeed, as he had remarked earlier, he thought a continent probably did exist near the South Pole and that he and his crew might even have seen part of it under all the layers of ice. But such a continent was of no use to anyone, Cook thought. It was incomprehensible to him that a land completely covered and surrounded by snow and ice could offer any benefit to mankind, or that future generations would waste any time trying to find it. "The risk one runs in exploreing a coast in these unknown and Icy Seas, is so very great, that I can be bold to say, that no man will ever venture farther than I have done and that the lands which may lie to the South will never be explored," he concluded.

Shortly before arriving at the Cape of Good Hope, the *Resolution* encountered a Dutch ship with several English sailors on board, who told a gruesome tale about a boat's crew from the *Adventure* being killed and eaten in New Zealand. After the stories recounted by the people at Queen Charlotte Sound this was disturbing news indeed, but Cook refused to believe it until he had heard more. At the Cape he found waiting for him a letter from Furneaux confirming the unhappy story of the murder of his men.

As Cook surmised, the *Adventure* had anchored in Queen Charlotte Sound shortly after the *Resolution* left at the end of November 1773. There were the usual incidents of native theft, but nothing out of the ordinary, when, several days into their stay, a boat sent to a nearby cove to cut grass for the animals did not return. The next morning, Furneaux sent another boat with a crew headed by James Burney to search for the missing sailors. Entering a small cove, Burney saw a double canoe hauled up on the beach and two New Zealanders nearby; they ran into the woods as soon as they spotted the English boat. In the canoe Burney and his men found shoes that obviously belonged to their fellow sailors; nearby were a dozen or more baskets filled with human flesh. Further searching revealed more shoes and a hand with the initials TH tattooed on it. (One of the missing men was Thomas Hill.) The fate of the boat's crew was undeniable. Despite the fact that the Englishmen had never yet experienced violence at the hands of the New Zealanders, these ten men had been killed and devoured.

Coming after several friendly encounters between Englishmen and New Zealanders, the "massacre at Grass Cove," as it came to be called, puzzled the English. Burney thought the New Zealanders had attacked in spontaneous reaction to a quarrel, or perhaps because the Englishmen were outnumbered and isolated from their fellows. Bayley blamed John Rowe, who had been in charge of the boat's crew, speculating that he placed too much confidence in the New Zealanders because he had lived in North America for many years and had become accustomed to dealing with native tribes there; Rowe might have allowed his men to fall unwittingly into a trap. No one suggested that the Englishmen could have instigated the quarrel, although Cook implied that the New Zealanders would not have attacked without provocation, commenting that they were "Brave, Noble, Open and benevolent . . . but they are a people that will never put up with an insult if they have an oppertunity to resent it."

From the Cape of Good Hope the journey home was a familiar and uneventful one. The *Resolution* touched briefly at St. Helena, where Cook got his first taste of what Hawkesworth had done to the account of his first voyage. Many of the people of St. Helena were offended at statements attributed to Cook about their ill-treatment of slaves; Cook had in fact never made such statements—Hawkesworth had taken them from Banks's journal and written them as if they were Cook's words—and he was angry that the book had been published without his seeing it first.

After two more brief stops, at Ascension and Fernando de Noronha Islands in the mid-Atlantic, the *Resolution* reached Plymouth on July 29, 1775. The following day the Englishmen anchored at Spithead, near Portsmouth, where Cook left his ship for London in company with the Forsters, Wales, Hodges, and several officers. Remarkably, as Cook noted with considerable pride, only four men had died on the journey, one from tuberculosis, two by drowning, and one from injuries sustained when he fell down a hatchway. Elliott, putting aside the weeks of misery in the Antarctic, wrote that "their never was a ship, where for *so long a period,* under such circumstances, more *happiness, order,* and *obedience* was enjoy'd."

15. Mai

The *Adventure* beat the *Resolution* back to England by a year, bringing not only new tales of the South Pacific but also a live specimen: Mai, the young man Furneaux had taken on board at Raiatea. By the time Cook returned to England, the first part of his voyage was familiar news, and Mai had become the talk of London.

When Furneaux arrived in London, he went straight to the Admiralty to report on his voyage, taking Mai with him. Admiralty officials immediately summoned Joseph Banks, partly to serve as interpreter—Mai could speak practically no English, and Furneaux's Tahitian was hardly any better—but also because he remained England's ranking expert on the South Pacific. Banks's command of Tahitian had faded with disuse, but he quickly took charge of Mai, with the result that Banks once again upstaged the sailors and, together with Mai, became the focus of curiosity about the voyage.

Within days of the *Adventure*'s arrival, Banks and Mai had an audience with King George, an occasion that proved a fertile subject for London newspapers. According to one report, upon being told to greet the King by kneeling and kissing his hand, Mai replied, "what, won't he *eat me* when he has got me down?" Another claimed the young man was so flustered when he met the

King that he forgot everything he had been told and simply held out his hand with the words "How do ye do?"—a shockingly informal way to greet the British monarch. A more fanciful writer had Mai saying to King George, "Sir, You are King of England, Otaheite, Ulhietea [Raiatea], and Bola Bola: I am your Subject, & am come here for Gunpowder to destroy the Inhabitants of Bola Bola, who are our Enemies." In short, Mai was portrayed to the English public as an exotic buffoon. Journalists loved to exaggerate and poke fun at his fractured English, his untutored behavior, and his confusion at European customs. To the English, Mai was initially viewed as one of a series of exotics trotted out to be discussed and gawked at, much as one might talk about a trained animal or some faraway natural wonder.

Englishmen in the late eighteenth century had, in fact, developed quite a taste for unusual humans. A ship captain named George Cartwright brought four men from Labrador to London late in 1772; like Mai, they met the King, dined with members of the Royal Society, received the attentions of Banks and Solander, and went to the theater and opera. For a brief time they were so popular that Cartwright displayed them at public exhibitions twice a week. Four princes from India visiting London shortly before Mai's arrival also drew huge crowds. And in some quarters, at least, word had filtered across the Channel about Ahutoru, the Tahitian brought to France by Bougainville, who received similarly exuberant attentions from Parisians in 1769.

But the reactions to Mai changed as the weeks went by. Once past his first awkward introductions to English society, he adapted remarkably well to his new home. Unlike the other visitors paraded before the English public, Mai took an interest in his surroundings, asked questions, responded with enthusiasm to new acquaintances and new situations, and learned quickly (or perhaps knew instinctively) how to ingratiate himself with the people he met. Within two or three weeks of his arrival, the newspaper stories changed their tone. The young man was not so simple and ignorant as the earlier accounts had suggested; "his deportment is genteel, and resembles so much that of well bred people here" that it was difficult to believe he had so recently come from an "uncivilized" society in the South Pacific. This writer noted in particular that Mai's behavior was polite, his appearance clean and "decent," and his table manners excellent.

Thoughtful observers who actually met and talked with Mai

confirmed the reports of his affability and good behavior. Solander was enthusiastic about the young man's virtues, and Banks continued to introduce him to his fashionable friends. Lord Sandwich frequently entertained him at his country estate. Fanny Burney, whose brother James had been a lieutenant on the *Adventure*, met Mai on several social occasions. An avid diarist (and later a highly successful novelist), she was one of the few to commit her personal reflections to paper. Like the other socialites who met Mai, Burney remarked on his mastery of the proper social graces, but she was even more impressed with his natural courtesy and concern for others. "Indeed he seems to shame Education," she wrote, "for his manners are so extremely graceful, and he is so polite, attentive, and easy, that you would have thought he came from some foreign Court." Mai managed to foster these favorable impressions without speaking much, since his minimal command of English improved only slightly during his stay in England. His attempts at conversation, consequently, usually amused his listeners but did not detract from their generally favorable impression of him.

Mai's skill at adapting to English society was the key to his success, although contemporary Englishmen did not see the situation in quite those terms. He embraced English culture with enthusiasm, professing loyalty to King George, adopting English dress, eating English food, mimicking the social customs, attempting to learn the language. He managed to reinforce Englishmen's belief in the superiority of their own culture while piquing their curiosity with his foreignness. The fact that he did not look or act *too* foreign, of course, helped him win acceptance too. Almost all commentators, from the hack journalists to Fanny Burney and Daniel Solander, remarked on Mai's skin color and appearance; most agreed that he was dark and not especially handsome, but "well made" and attractive in a rugged sort of way. One writer said "his complexion much resembles that of a European accustomed to hot climates." Burney found him darker than she expected, but with a "pleasing countenance." A fiction gradually developed, no doubt encouraged by Mai's proficiency at the social graces, that he held high rank in his native society. The opposite was true, as Cook and his officers knew perfectly well, but the fabricated version of Mai's background probably also eased his acceptance in English society.

Mai came to be perceived as the embodiment of natural virtue. Fanny Burney made the point when she said "he seems to shame

Education." Here was a supposedly uncivilized man who, upon minimal exposure to polite society, behaved with such courtesy and good breeding that he compared favorably with men who had lived all their lives among the English upper class. It seemed impossible to well-bred Englishmen that Mai could have learned "civilized" behavior in such a short time; the obvious explanation was that, in his simple, good-natured way, he knew instinctively how to behave and needed only the appropriate setting for his natural good manners to flourish. Burney thought he had "an understanding far superior to the common race of *us cultivated gentry*. He could not else have borne so well the way of Life into which he is thrown, without some practice." A gentleman of her acquaintance was, she claimed, "a meer *pedantic booky*" by comparison, despite his upper-class background. Mai, on the other hand, "with no tutor but Nature," behaved "like a man who had all his life studied *the Graces,* and atended with unremitting application and deligence to form his manners. . . . I think this shows how much more *nature* can do without *art,* than *art* with all her refinement unassisted by nature."

Others used the image of a South Pacific native captivating English society to make an argument about the virtue of simplicity and the corruptions of civilization. Mai was a perfect subject for critics of English society. His familiar manner of addressing King George might have caused laughter, but some used it as an example of the young man's inherent sense of equality with his fellow man, and praised him for it. One writer thought that the kings and queens of Europe would find it difficult to answer a simple "How do ye do?" because of all the tyrannical deeds for which they were responsible. "It might be very well that all petitions and remonstrances for the future should conclude with the famous question," he remarked.

Another writer, masquerading as a Tahitian who had lived some time in England, used the public fascination with Mai as an opportunity to criticize supposedly civilized European behavior. Tahitians, he argued, were considered "barbarous" because they "do not live in a regular manner or method of policy or religion," but such reasoning was specious. "We practise those virtues you only teach," the hypothetical Tahitian wrote, "are enemies to luxury, strangers to adultery, . . . never go to war but from a principle of self preservation or self defence . . . and whilst we entertain the most sublime ideas of an Almighty Being, do not cut the throats of each other for differing in the manner of worshipping him." Moreover, Tahitian men did not "preach chastity" to women while attempting to violate it. The English, in sum, had no right to label Tahitians uncivilized. Indeed, the writer claimed, "you deserve the appellation yourself."

A writer pretending to be Mai took on English society at length in a pamphlet called "An Epistle from Omiah." In the guise of a Tahitian observing English society, the author attacked everything from English class structure to women's behavior. He went through the standard critique of western European society: men cared only for money, which could never buy happiness but only increased their desire for more; religion was hypocritical and formulaic, with no concern for true moral virtue; the upper classes prospered at the expense of keeping the majority of the population in misery; wars periodically killed thousands for no good purpose. He criticized the English legal system for making justice dependent on the skill of lawyers (and therefore on money) and satirized scientists for creating "systems of philosophy" that bore little relation to truth. Women, this "Omiah" claimed, embodied society's decadence in their excessive makeup—he likened it to tattooing—and frivolous dress. He accused English women of displaying a false sense of sexual morality, enticing men with their actions but withholding sexual relations out of an exalted sense of virtue.

English artists also contributed to this image of Mai's virtuous simplicity by painting him in a classical manner. The Tahitian style of dress, which involved draping yards of *tapa* in a variety of ways around the body, encouraged such a portrayal by its similarity to the drapery of ancient statues. Nathaniel Dance (commissioned by Banks to paint portraits of both Mai and Cook) produced a realistic image of Mai's face, judging from a comparison with drawings made by Hodges during the voyage, and included ethnographic details like tattoos on his hands, a wooden headrest, and a feather ornament; but his pose and his ankle-length, carefully draped *tapa* robe show the influence of a classical ideal in painting. A group painting of Mai, Banks, and Solander by William Parry also shows the Tahitian in a Romanized version of native dress. Joshua Reynolds, one of the most distinguished English artists of the eighteenth century, completed the transformation of Mai by minimizing the ethnographic details employed by Dance, slimming his chubby face, lightening his skin color slightly, and adding a turban. With a background painted to suggest a tropical scene, the overall effect is to portray an eastern exotic of great dignity.

Mai was a convenient vehicle for critics of English society, but he himself did not utter a word against the customs of a nation that lavished luxuries on him and turned him into an overnight celebrity. On the contrary, he loved the attention, boasted about women's interest in him—he told Fanny Burney gleefully on one occasion that he was on his way to visit twelve women in an evening—and did his best to master English dress and manners. Mai was even capable of imitating English snobbery, as Banks discovered when they went to the popular Sadlers Wells Theatre. Mai professed to enjoy the show, but then asked if Lord Sandwich or other members of the nobility patronized the theater. When Banks replied that they did not, Mai refused to go again.

Those who knew Mai best, Cook included, thought he was stupid and selfish, if good-natured; he had neither the intellect nor the inclination to understand, much less criticize, English society. Instead he tried merely to imitate it, and at that he was extraordinarily skillful, as indeed a number of his fellow Society Islanders had shown themselves to be during the course of the English visits there. Mai's motives for going to England were inspired mostly by vanity. He wanted to acquire English guns and go home to defeat the "Bolabola men"; close questioning from some of his English hosts revealed too that he had suffered ridicule from some of his compatriots, mostly because of his broad, flat nose, and he thought a triumphal return to the islands, laden with English finery, would earn their respect.

Once the initial flurry of excitement over Mai wore off, Banks and Solander took him to the country retreat of the Baron Dimsdale, who had made a reputation as a practitioner of smallpox inoculation. Mindful of the deaths of the Labrador men and Ahutoru from smallpox, they wanted to protect Mai from a similar fate. Afterward Sandwich entertained him for a week at his country estate outside London, where Mai captivated his host's houseguests by cooking a meal Polynesian-style, baking pork and vegetables in an underground pit. Banks took him on a month's tour of northern England, concluding with a visit to Cambridge, where Mai met the people who impressed him most deeply: university professors, who seemed to him nearly the equivalent of the gods in his own society.

Upon returning from their travels, Banks arranged for Mai to board at a house in his neighborhood. Thomas Andrews, who had been the surgeon aboard the *Adventure,* was to live with him as companion and interpreter—a necessity, since Mai was far from fluent enough in English to get along entirely on his own. By the end of 1774, he settled down to a relatively quiet life, no longer courted by fashionable society.

By the time the *Resolution* returned to England in July 1775, a year after the *Adventure,* public enthusiasm for news of the South Pacific had diminished considerably, in large part because Mai upstaged anything Cook might produce, but also because none of the accomplishments of the second voyage quite compared in the public mind with the weeks at Tahiti and New Zealand during the first. (The relatively recent publication of Hawkesworth's volumes, late in 1773, helped keep the *Endeavour*'s triumphs fresh in people's minds.) In addition, more significant news took precedence over Cook's return; in April 1775, hostilities had broken out between British troops and colonial militia in Massachusetts, and by midsummer the British government was in the midst of preparing for war against the rebellious Americans. Furneaux, in fact, was about to sail to North America as captain of a frigate, and Burney was already there.

Only a few small notices in the newspapers announced Cook's arrival at Spithead on July 30, and some of those reported that he had arrived in the *Endeavour*. Another item invented a discovery—an island 160 miles long and 146 wide, with a "delightful" climate and fertile soil—but on the whole the journalists found little to say about the *Resolution*. But if the public response was muted, in official circles Cook was greeted with enthusiasm. The Lords of the Admiralty were prepared for his return, because he had sent them letters and some of his officers' journals with three ship captains leaving the Cape of Good Hope ahead of the *Resolution*. Banks and Solander were kept informed too, first by Sandwich and later by Clerke, who dashed off a note to Banks from Spithead announcing his return from "our continent hunting expedition."

Banks and Sandwich were on a yacht trip down the Channel when Cook arrived in London. Sandwich rushed back to the capital, but Banks continued his trip, perhaps reluctant to face Cook after he had so abruptly withdrawn from the voyage. But if Banks worried that Cook harbored ill feelings against him, he misjudged his old friend. Solander, waiting at the Admiralty when Cook made his appearance, reported that the captain asked warmly after Banks and said his presence was the only thing that could have improved the voyage. Banks did not return to London for another month, but his name was so closely associated with the natural history of the South Pacific that several men from the *Resolution* called at his house in his absence, offering to sell the "curiosities" they had collected. And although Cook sent all the official collections from the voyage to the British Museum, in care of Solander, he had four casks marked especially for Banks.

After checking in at the Admiralty, Cook went home to his family to learn that his baby son George, born shortly before he had left England, had survived only four months. His oldest son James, now twelve, had left home to enroll in the Naval Academy at Portsmouth, and eleven-year-old Nathaniel was about to follow his brother there. A week later, Cook had his official presentation at court, and shortly thereafter he was promoted to post-captain. More honors followed in the fall, when Cook was nominated to become a Fellow of the Royal Society, in a statement signed by twenty-five Fellows (instead of the usual three to six), including Banks, Solander, Nevil Maskelyne, Johann Forster, and Admiralty Secretary Philip Stephens. Wales was nominated soon after, and both men were officially admitted early in 1776. In July, the Society awarded Cook its Copley Medal, given annually to an individual who had made outstanding contributions to science.

Cook's official reports on the voyage skimmed over his remarkable circumnavigation in high latitudes and focused instead on issues concerning the health of seamen. He and his men had endured much longer periods of time at sea between ports than on the previous voyage, providing an ideal opportunity for testing scurvy preventives. Although Cook continued to advocate regular consumption of sauerkraut, he now correctly believed that eating fresh provisions as much as possible, by putting the crew on a diet of local produce at every stop and stocking the ship with whatever would keep for a few days at sea, was the most effective means of combating the disease. It was also important to take on fresh water whenever possible, even if the ship already had an adequate supply, because water stored in casks quickly became fetid. Cook summarized his findings on the health of seamen in a report to the Royal Society, which was published in the *Philosophical Transactions* early in 1776.

In a letter to his old friend John Walker, Cook wrote more generally about the events of his voyage and allowed himself a bit of boasting about his accomplishments. The *Resolution*'s voyage, he thought, should be the last such expedition to the Pacific, "as we are now sure that no Southern Continent exists there, unless so near the Pole that the Coast cannot be Navigated for Ice and therefore not worth the discovery." Already there was talk, however, of another voyage—this time to take Mai home, in a refurbished *Resolution* with Charles Clerke in command. Cook supported the idea, since he did not believe that it was fair to keep Mai or anyone like him in England indefinitely, but could not imagine that such a voyage might have useful purposes beyond returning Mai to the Society Islands.

Whatever its mission, Cook did not expect to command a return voyage to the South Pacific. At his request, he had received a prestigious appointment to the naval hospital at Greenwich, a position that would ensure him a comfortable living until his retirement. Nevertheless, he asked the Admiralty Secretary to appoint him with the understanding that he could give up the post whenever he might be recalled to active service. Cook was quite candid about his misgivings: "Months ago the whole Southern hemisphere was hardly big enough for me and now I am going to be confined within the limits of Greenwich Hospital, which are far too small for an active mind like mine."

It is not clear why Cook asked for the Greenwich Hospital position; he admitted to Walker that the job would be "a fine retreat and pretty income," but was not at all sure that he would enjoy "ease and retirement." Since Cook believed there was little more to be accomplished in the Pacific, it is most likely that he considered the Greenwich post preferable to a voyage intended primarily to take Mai home or another, even less interesting, command. Weariness after two long, dangerous voyages and concern about his family may also have influenced his decision. Whatever his reasons, he intended to stay home for some time and was content to let Clerke head the next Pacific expedition.

Cook spent much of his time over the next several months getting his journal ready for publication. His experience with Hawkesworth had so disturbed him that he decided to prepare the account of the *Resolution*'s voyage himself. Unfortunately, he had a different set of problems to contend with this time. First there was an unauthorized account of the voyage advertised in the press within six weeks after the *Resolution* docked. Rumor had it that the author was Robert Anderson, the ship's gunner, and Cook visited Anderson to find out if the rumors were true. Anderson denied authorship, but after a bit of investigation learned that John Marra, the gunner's mate who had attempted to desert at Tahiti, was responsible. Marra told Anderson that two other seamen had offered their journals to booksellers, but the manuscripts were so badly written that no one could read them. All three journals had escaped Cook when he collected his crew's logs and journals before docking. Persuaded that Marra's book was an isolated incident and not significant enough to affect his own publication, Cook decided not to attempt to suppress its appearance. In fact Marra's account, published in 1776, proved popular; a pirate edition appeared in Dublin later that year, a German translation shortly thereafter, and a French version in 1777. As Cook predicted, however, it was the only unauthorized, anonymous book on the voyage, although a fake account also appeared in 1776.

More seriously, Cook had to contend with Johann Forster's expectations of publication. As a condition of joining the voyage, Forster had insisted that he be permitted to publish a book on the expedition, because he was certain it would earn him a great deal of money. The terms of this agreement were never clear, however, and Forster's understanding of them was not the same as the Admiralty's. Sandwich was willing to consider Forster as the Hawkesworth of the second voyage, but wanted to see a description of one part of the voyage first as an example of his literary style. After reading the sample pages, which he found totally unacceptable, he rejected the idea of a Forster-authored account of the voyage. Not surprisingly, Forster was enraged, even going so far as to suggest that Cook's journal contained "inaccuracies" and "vulgar expressions," a strategy hardly calculated to improve his standing with Sandwich, who had long been Cook's patron. Nevertheless, Sandwich was still prepared to consider a joint Cook-Forster venture, with Cook editing his own journal for publication and Forster writing a companion volume on the scientific work. Cook and Forster would share the costs of printing and profits equally; the Admiralty would pay for engraving the plates.

Later, Forster would translate both volumes for French and German publication.

Cook proceeded according to this plan and spent much of the fall of 1775 and early winter of 1776 editing his journal, adding paragraphs, polishing his language, and dividing the text into chapters. He had the help of a clergyman named John Douglas, but Douglas's role was limited to that of editor, correcting grammar and spelling, and suggesting passages that might be changed to comport with the tastes of potential readers—eliminating references to the sailors' liaisons with Tahitian women, for example, and changing the description of the Malekulans' penis wrappers to suggest that the men were more fully clothed; Cook's statement that "the men are naked . . . but the Penis is wraped round with a piece of cloth or a leafe" became "the men go quite naked, except a piece of cloth or leaf used as a wrapper." The engraver of the illustrations for the volume cooperated by providing Hodges's bare-shouldered Malekulan with a length of cloth draped over his shoulder. Cook agreed to most of Douglas's suggestions, even when they entailed changes of substance, urging that the text be made "unexceptionable to the nicest readers."

Forster was at work too, but when Sandwich read Forster's draft in the spring of 1776, he concluded that the volume needed editing to eliminate passages repetitive of Cook's journal. The notion that his work was not perfect as written infuriated Forster, and he refused to allow a word to be changed. Sandwich and his colleagues at the Admiralty held firm, informing Forster that they would have no further dealings with him if he refused to agree to modifications in his manuscript. Forster, never inclined to compromise, became less and less willing to do so as the dispute dragged on, and finally he dropped the project altogether, claiming that the Admiralty had reneged on its agreement. The whole affair angered Cook, who thought Forster had "deceived" him, and, having pulled out of the collaboration, would probably get his account into print first; but Cook recognized that Forster's involvement was not critical to the success of the project and proceeded without him. "I am only sorry my Lord Sandwich has taken so much trouble to serve an undeserving man," he wrote.

Cook's volume, *A Voyage Towards the South Pole and Round the World,* was published in May 1777 and became an instant success. The first printing sold out the day it appeared, and a second edition was printed a few months later. A French translation came out the following year, a third English edition in 1779, and a fourth in 1784.

Forster, meanwhile, went ahead on his own. Chronically short of money, he had no steady source of employment after returning to London and had counted on royalties from publication to sustain himself and his large family. He sold Banks most of his botanical and zoological drawings to raise money, and at the end of 1775 published the first of his scientific works on the voyage, a technical volume titled *Characteres Generum Plantarum,* which described ninety-four new species of plants collected on the voyage. Cook attempted to suppress the printing of this book as a violation of his exclusive right to publish the first account of the voyage, even though it was written in Latin and highly specialized in nature, but Forster successfully persuaded Sandwich that it could not possibly affect the success of Cook's account. Three years later Forster published his most significant contribution to the voyage literature: *Observations Made During a Voyage Round the World, on Physical Geography, Natural History, and Ethic Philosophy.* It was a long and detailed description of the places he had visited, including plant, animal and human life. Two-thirds of the book was devoted to the physical and cultural characteristics of the people encountered on the voyage, and it had a major influence on the developing field of anthropology. Forster was not content with producing specialized scientific works, however. Under the terms of his contract with the Admiralty, he had promised not to publish his own general account of the voyage, but the agreement said nothing about George; so father and son collaborated on a book published under George's name. They managed to get their version out six weeks ahead of Cook.

The Forster volumes created a small storm of controversy among those close to Cook because they were viewed as an obvious attempt to upstage the captain. Wales published a long harangue against the book, which he was convinced was in reality Johann's work because it was written "with so much arrogance, self-consequence, and asperity, and the actions of persons are decided on in so peremtory and dogmatical a manner." Wales accused Johann of publishing his account purely for financial motives and criticized him for attempting to discredit Cook's soon-to-be published journal by implying that the captain would write only about nautical matters. George Forster attempted to defend his father against Wales's accusations and assert his own claim to authorship in a pamphlet rushed into print soon after Wales's appeared.

A comparison of George's volumes with Johann's original journal (only recently discovered and certainly not available to Wales) shows many similarities of content, but the writing in the published account is clearly George's. Nevertheless, Wales and Cook were correct in their assumption that the Forsters produced the book under George's authorship to circumvent Johann's agreement with the Admiralty and to get back at Cook and the officials who supported him. But in the long run the controversy was trivial. The whole business of voyage-publishing was motivated by desire for profit, on Cook's part as well as everyone else's, and assertions to the contrary were self-serving. Such books enjoyed a large market, however, and it would have taken a flood of journals to affect seriously the market for Cook's work. The Forster volumes did not detract from the success of Cook's, and once his own account appeared in print, Cook paid little attention to the affair.

The Forsters' publishing efforts were not enough to support their family, however, and the recognition and financial rewards that Johann had expected as his due following the voyage never materialized. Out of desperation, in 1778 George returned to Germany, where he had good prospects of work as a translator and hoped to find a position for his father as well. The Forsters were well respected in the German states—George was soon offered a position as a professor of natural history in Cassel, and his father corresponded with Frederick the Great on scientific subjects—but both would have preferred to remain in England, where they thought the scientific community was more sophisticated. While George tried to find work for his father in Germany, Johann nursed his grievances against Cook and the Admiralty, writing letters to Sandwich, Banks, and others, trying to gain restitution for the alleged wrongs done him. Meanwhile, he fell so deeply into debt that when the University of Halle offered him a position in 1779, his creditors refused to allow him to leave England. Eventually several German Masonic lodges raised money to pay the debts, and the University agreed to pay his travel expenses. In mid-1780, Johann and the rest of his family moved back to Germany.

At first Johann was bitter about his treatment in England, but gradually he accepted his situation and turned his attention to writing up the rest of his scientific observations from the voyage. He published a book on animal species, a companion volume to the *Characteres Generum Plantarum,* in 1788 and completed a more detailed manuscript on the zoological discoveries, which remained unpublished until after his death because of disputes with the printer over illustrations (Forster wanted elaborate, expensive engravings) and the language of publication (Forster wanted it in Latin, but the printer insisted on German). He also wrote an exhaustive study of exploration from classical to modern times and edited a journal devoted to German translations of scholarly travel literature. The latter, in particular, was a reflection of Johann's increasing interest, first demonstrated on the voyage and in the publication of his *Observations,* in the scientific study of man.

Both father and son remained interested in current exploration as well. George planned to join a Russian-sponsored expedition to the Pacific in 1787, and only his son's strenuous objections prevented Johann from signing on too. War between the Russians and Turks canceled the voyage, and George moved instead to Mainz, where he became librarian to the archbishop there. He also continued to translate accounts of voyages into German, becoming recognized as the ranking German expert on Pacific exploration and culture. Throughout his career, he maintained his commitment to the liberal political ideas he had voiced during the voyage with Cook, supporting the French Revolution and openly criticizing the policies of the German princes, to the consternation of his father and many of his friends. When Mainz was occupied by the French revolutionary army in 1792, George became the leader of the local Jacobins; in 1793 he went to Paris to attend the French National Convention as a delegate from the occupied German territories and urge that they be annexed to the French Republic. He remained in France until his death from pneumonia in 1794. Johann outlived his son by just four years.

While controversy raged over publication of the voyage journals and the Forsters pleaded their case with Admiralty officials, serious plans were underway for the voyage that would return Mai to Tahiti. Refitting the *Resolution* for a new assignment had begun soon after Cook's return in July 1775. The Admiralty ordered the purchase of a second ship in December—the *Adventure* had been assigned elsewhere—and Navy Board officials consulted Cook on the selection of another Whitby-built collier of about three hundred tons, later renamed the *Discovery*. Early the following February, Sandwich invited Cook to dinner to discuss plans for the expedition. Neither he nor anyone else at the Admiralty presumed to ask Cook to undertake another voyage, and Cook himself had indicated, in his request for the Greenwich Hospital position, his decision to retire from active service. But he was clearly the obvious choice, and by the end of the dinner he agreed to command the proposed voyage. Whether Cook volunteered or was persuaded is an open question, but he seemed relieved at the prospect of returning to sea. "It is certain I have quited an easy retirement, for an active, and perhaps Dangerous Voyage," he wrote a friend, but added, "My present disposition is more favourable to the latter than the former, and I embark on as fair a prospect as I can wish."

Of course the voyage was not intended merely to take Mai home, and that is probably why Cook agreed to command it. The English government would hardly have financed so massive an undertaking to deliver a single Pacific islander, although it made an entertaining story for the public. In reality, both the Royal Society and the Admiralty had, for some months before Cook's arrival, been discussing the possibility of another Pacific expedition—this time to the northern Pacific, to search for a northwest passage from its western terminus.

Interest in finding a passage across North America had revived in England around the middle of the eighteenth century, but the failure of earlier efforts to find a passage from the east, reports of Bering's voyages, and the claims of the de Fonte and de Fuca stories all helped shift the focus of discussion from the Atlantic to the Pacific, as indicated in Byron's orders for his voyage in the early 1760s. In 1774 Daines Barrington, a lawyer and member of the Royal Society with a strong interest in Arctic exploration, proposed that ships be sent across the Pacific and up the coast of North America as far as possible in search of a passage across the continent, a plan that won the support of Banks, Solander, and Maskelyne, among others. Barrington and John Pringle, then president of the Royal Society, discussed the proposal with Sandwich, who supported the idea and promised that the voyage would be undertaken when Cook returned. Barrington also helped lobby Parliament to extend a 1745 act offering a reward for the discovery of a northwest passage. That law had cited the commercial advantages of finding a passage; the new act added the importance of its discovery to science.

By the time Cook agreed to command a third voyage, the ships' crews were already being taken on board, even though the expedition would not sail until June. Charles Clerke, originally rumored to be Cook's replacement as captain of the *Resolution,* was appointed to command the *Discovery*. John Gore, Cook's second lieutenant in the *Endeavour,* had not sailed on the second voyage but was back as first lieutenant in the *Resolution.* James Burney, home from his tour of duty off the North American coast, returned as Clerke's first lieutenant. In the *Resolution,* Robert Anderson, William Harvey, William Collett, John Ramsay, and Samuel Gibson were all making their third voyages with Cook. William Peckover, the *Discovery*'s gunner, was also a three-voyage veteran. Others were making their second voyage, among them the nineteen-year-old midshipman George Vancouver, surgeon William Anderson, and William Bayley, astronomer in the *Adventure,* now appointed to the *Discovery.* Altogether, of the *Resolution*'s 112 men, fourteen had sailed with Cook before; ten of the *Discovery*'s seventy were veterans. Midshipman George Gilbert was a newcomer, but his father, Joseph, had been master of the *Resolution* on the second voyage.

The *Resolution*'s second lieutenant, new to Pacific exploration, was a man well versed in science: James King, the twenty-six-year-old son of a Lancashire minister. King had joined the Navy in 1762 as a captain's servant and served in Newfoundland and in the Mediterranean. In 1774 he left the Navy to study science in Paris; from there he went to Oxford, where he worked with Thomas Hornsby, professor of astronomy, who recommended him for the voyage. King's training in astronomy was sufficient for Cook to make him responsible for all astronomical observations aboard the *Resolution,* and he proved to be a thoughtful commentator on natural history as well. Among the others making their first Pacific voyages were William Bligh, the *Resolution*'s master; midshipman James Trevenen; and surgeon's mate David Samwell. Bligh, at age twenty-two, had served in the Navy ten years. As master, he was responsible for navigation and chartmaking, and his skills were much praised by Cook. He was a man of irascible temper, however, and did not get along well with most of the men in the ship. He and King in particular were at loggerheads much of the time. Trevenen, on the other hand, was an exuberant, outgoing youngster of sixteen, fresh from the Naval Academy at Portsmouth. Like Burney on the previous voyage, he was attracted to Cook's expedition by the prospect of adventure and fame, and used some influence to get a berth. He and King became close friends, despite the difference in their ages, and like King, Trevenen proved to be an acute observer of his fellow men. The twenty-eight-year-old Samwell, son of a Welsh vicar, had been in the Navy a relatively short time, but his experience had included a voyage to Greenland in the early 1770s. A man of literary inclinations, he wrote one of the most detailed and perceptive journals produced on the voyage, and later in life gained some reputation as a poet.

The scientific complement was smaller and less distinguished than on previous voyages, although it was supplemented by talented amateurs. No one suggested a return voyage for the Forsters, and Banks was absorbed in his work as director of the Royal Gardens at Kew, a post specially created for him by the King in 1773. Far from allowing the assignment to diminish his interest in exploration, Banks saw it as an opportunity to expand England's botanical collections and centralize them in a single location; he had one of his gardeners at Kew, David Nelson, appointed to the *Discovery* as Bayley's servant, with instructions to collect as many specimens as he could. John Webber, a minor landscape painter, joined the *Resolution* as official ship's artist. He had neither Parkinson's skill at drafting nor Hodges's artistic vision, but with the help of James Cleveley, a carpenter with an aptitude for drawing, he produced a competent record of the voyage.

Cook probably never knew that a more famous supernumerary considered joining the *Resolution*. James Boswell, companion and biographer of Samuel Johnson, met Cook at a dinner party and was deeply impressed, praising him as "a plain, sensible man with an uncommon attention to veracity." When plans were announced for the third voyage, Boswell began to imagine himself going along, although Johnson dumped cold water on the idea, arguing that Boswell would learn very little from the voyage. Boswell was forced to agree; both he and Johnson believed that Europeans' observations in the South Pacific were necessarily limited because of language barriers. Still, the very idea of sailing around the world was appealing, and Boswell's enthusiasm increased when he talked to Cook again, this time at a gathering of Royal Society members. The two men talked about the possibility of sending "men of inquiry" to a few key spots—Boswell suggested Tahiti, New Zealand, and New Caledonia—to stay for three years, learn the languages well, and bring back detailed accounts of their people. Boswell privately fancied himself one of these men, but only if the government would give him "a handsome pension for life." The government was unlikely to undertake such a scheme, however, much less with a promise of pension for life, and Boswell's idea remained a dream.

Provisioning for the voyage proceeded in much the same manner as for the *Resolution*'s first voyage. Cook's conclusions about his men's health were not neglected, and the ships got generous supplies of sauerkraut and malt. A new addition, requested by Cook, was an "apparatus for recovering drowned persons" for each ship. The Board of Longitude returned Kendall's chronometer, which Cook had used in the *Resolution,* for a further test, along with a newer version, also built by Kendall, for the *Discovery*.

The government did not neglect Mai in its preparations, supplying him with, among other things, a suit of armor, a globe, tin soldiers, a hand organ, dishes, kitchen utensils, and the things he wanted most: guns, gunpowder, and port wine. Cook had objected to giving Mai weapons, because he knew the young man would attempt to use them against the "Bolabola men," but Cook was overruled. In addition, the ship's stores included a supply of gifts for Mai to present to the chiefs of the South Pacific islands—feathers, trousers, swords, cut-glass bowls, knives and forks, iron tools and tool chests, spyglasses, chessboards, jewelry, perfume, umbrellas, needles and thread, soap, silver watches, pictures of the king and queen—and more livestock to leave at Tahiti, including cattle, sheep, goats, hogs, rabbits, turkeys, geese, ducks, and two peacocks.

It occurred to some that all the material possessions Mai took back to Tahiti might only serve to make him an object of jealousy among his compatriots. Cook thought he would be viewed briefly as a man of importance, until he had exhausted his tales of strange places, at which point he would revert to his former low status. George Forster thought Mai should take practical goods—more iron tools, for example, rather than fancy, impractical European clothing. Others hoped he would serve as a native missionary and lamented that no one had attempted to instruct him in the principles of Christianity. Granville Sharp, a London philanthropist, made an attempt to teach Mai to read and write, with the goal of instructing him in religion, but gave up after fifteen sessions because the young man's busy social schedule left him little time for more serious pursuits.

Cook boarded the *Resolution* at its berth in the Thames on June 24, 1776, and sailed for Plymouth the following day. The *Discovery* had left ten days earlier, but without Charles Clerke, who was confined in a London debtors' prison. The debts were not his own, but those of his brother John, for whom he had cosigned a loan; since John had recently sailed to the East Indies, Charles was left with the debts. As the days went by and no settlement was reached, Cook and the Admiralty agreed that the *Resolution* should sail anyway, with the *Discovery* following later; the two ships would meet at the Cape of Good Hope. If Clerke was so seriously delayed that Cook had to leave the Cape before his arrival, Cook was to leave instructions for him about a later rendezvous point. These arrangements made, the *Resolution* sailed from Plymouth on July 12. Clerke managed to extricate himself from the law some days later, and the *Discovery* left Plymouth August 1.

IV

THE SEARCH FOR A NORTHWEST PASSAGE

16. The "Friendly Islands"

The *Resolution* reached the Cape of Good Hope in mid-October, three months after leaving Plymouth; the *Discovery* followed three weeks later. Cook and Clerke spent another three weeks taking on additional provisions, including more livestock for Cook's continuing experiment in introducing animal husbandry to the South Pacific, and on December 1 the two ships headed east into the Indian Ocean on a second attempt to locate the lands reportedly discovered by French explorers. Cook was more optimistic about succeeding this time, as he had a chart showing the routes of du Fresne and Kerguelen, which another French explorer, Julien Crozet, had given him when they met at the Cape in March 1775.

The Englishmen found Kerguelen's discovery on Christmas Day: a large, barren, uninhabited island with an excellent harbor but little else of interest. Like Bouvet's "Cape Circumcision," it was shrouded in fog much of the time and thick snow was common even in midsummer, which perhaps explains why an explorer looking for a continent thought he had found one in this isolated bit of rock. Apart from giving the crew a day off to celebrate Christmas and scrounging what grass he could for the livestock, Cook had no reason to linger, and the two ships pressed on toward New Zealand, with a brief stop at Adventure Bay on the coast of Van Dieman's Land to collect more wood, water, and animal feed.

In mid-January 1777, the *Resolution* and *Discovery* anchored in Queen Charlotte Sound, where the local residents once again displayed great reluctance to approach the English. A few canoes visited the ships, but no one would go on board, even though all were men familiar to Cook. The reason for their diffidence was soon clear: they thought the Englishmen had returned to avenge the deaths of the *Adventure*'s crew. Cook attempted to disabuse them of this fear, but Mai undermined his efforts by repeatedly urging him, in the presence of the New Zealanders, to kill the men responsible for the murders. (Cook's persistent refusals mystified Mai, who pointed out that murderers were hanged in England.) Cook eventually managed to regain the New Zealanders' trust, to the point that even the acknowledged instigator of the Grass Cove incident, a man named Kahura, visited the ships with impunity. Cook admired Kahura's courage and was flattered at what he took to be a statement of confidence in himself—"for I have always declared to those who solicited his death that I had always been a friend to them all and would continue to unless they gave me cause to act otherwise," he remarked—although it is more likely that Kahura simply realized that if Cook were going to kill him, he would have done it at his first opportunity. Having been spared then, he could presume to be quite safe—and perhaps even poke fun at men who declined to avenge their friends' deaths.

Within days the New Zealanders were visiting the ships by the dozens, and several erected temporary homes on the shores of Ship Cove, the better to carry on trade with the English. In fact, those members of the crew who had been at Queen Charlotte Sound before were struck by the sizable number of New Zealanders clustering about the ship and by their increasing sophistication about trade. Goods that were sold for nails on the previous voyages now commanded axes or hatchets. Sailors who knew about the "curiosities" available in the South Pacific from the stories of previous voyagers had provided themselves with personal supplies of trade goods and were willing to pay the prices asked, which in turn drew even more canoes to Ship Cove. Burney suspected that the New Zealanders laughed at the English for their readiness to part with valuable goods for native products, viewing the sailors as "dupes to their superior cunning," he wrote.

Trade in women increased too, with the New Zealand men routinely bringing girls on board ship. They proved rather less marketable than material goods, however, partly because the women were not the most attractive to English eyes—Samwell noted that "they like other Courtezans were so lavish of red Paint in daubing their faces & so fragrant of noisome smells that they did not meet with many Admirers"—and partly because the *Adventure* murders had turned the sailors against the New Zealanders, regardless of current friendly relations. The sailors' distaste was reflected in the low prices the women commanded; according to Williamson, women were cheaper than fish. He had no use for them himself, denouncing them as "greasy brown nymphs."

The *Adventure* killings created an undercurrent of tension throughout all dealings between the English and the New Zealanders, despite the facade of cordial trade. The men making their first visit to New Zealand, who had no previous friendly experiences to temper the shock of the murders, were especially prone to denouncing the native people in language riddled with hatred and horror. Gilbert wrote that "the savagness of their dispositions and horid barbarity of their customs is fully expressed in their countenances; which is ferocious and frightful beyond immagination," and Edgar had an unusually vivid impression of the war dance: "There is something in them so uncommonly Savage & terrible their Eyes Appear to be Starting from their Heads, their Tongue, hanging down to their chin, and the Motion of their Body Entirely Corresponding with these in a Manner not to be Described." Anderson and King found the New Zealanders sullen, mistrustful, insolent, and dirty. Even Cook, although persuaded that the New Zealanders would not attempt another attack, took extra security precautions. Shore parties were always armed and accompanied by ten marines; boats traveling any distance from the ship were also heavily armed and commanded by a reliable officer with previous experience of the New Zealanders. The last precaution reflected Cook's belief that the *Adventure*'s men had, at least in part, brought their fate upon themselves by misjudging the New Zealanders' behavior.

About a week after arriving at Queen Charlotte Sound, Cook went to Grass Cove in an attempt to learn more about the *Adventure* incident. The only family living there at the time was headed by a man Cook had met on the previous voyage; he was quite willing to talk. The boat's crew members had been eating on shore, he explained, with one man watching the boat, when a group of New Zealanders snatched some of their bread. The sailors tried to get it back, and in the ensuing scuffle John Rowe shot and killed one man. When another New Zealander called for help, Rowe fired and killed a second man. At this point native reinforcements arrived and set upon the Englishmen, who were unable to defend themselves effectively; except for Rowe's musket, their guns were in the boat.

Later, other New Zealanders told the story with small variations. One version had the native men bringing a hatchet to trade; when the sailors took it and gave nothing in return, the New Zealanders seized their bread. Another claimed that the incident started when a New Zealander attempted to steal something from the Englishmen's boat. John Ledyard heard that a man who had been a persistently troublesome thief on board ship persuaded a few friends to join him in an attack over the objections of most of his compatriots, who looked upon the English as friends. Despite the lack of corroboration for Ledyard's story—he was the only crew member who claimed that the attack was deliberately planned—a number of the Englishmen found his version of events most plausible. Cook and several others, however, persisted in their belief that the incident was unplanned and could have been avoided by more skillful handling of the New Zealanders.

When the Englishmen left Queen Charlotte Sound at the end of March, Cook intended to sail directly to Tahiti for one last, brief stop before heading to the Arctic. He followed a new track as the ships headed northeast, discovering two small islands along the way (Mangaia and Atiu, in the group now known as the Cook Islands). He hoped to replenish his supplies of animal feed there, but high surf and hostile inhabitants made that impossible. At Mangaia a few canoes ventured beyond the breakers, but only one man would go on board the *Resolution*. The English visit made a profound impression on the islanders, however, despite its brevity; for years afterward the Mangaians performed a play, accompanied by music, commemorating the incident. The islanders believed their supreme god Tangaroa had sent a great ship from the sky, filled with godlike men (because of their light skin) whose leader was called Tute. The men spoke strangely and communicated through Mai, whom the Mangaians recognized as akin to themselves. The islanders' hostile intentions were not a figment of English imagination; the chorus in the play chants,

> We come in hundreds,
> Warriors! Hundreds!
> With our spears put up to fight
> We lusty people of Mangaia
> Will destroy this boastful ship.

The crew stopped next at nearby Hervey's Island, discovered on Cook's first voyage, but could find only a small quantity of grass for the animals. At this point Cook realized that they could not make it to Tahiti on their present supplies of animal feed and decided to go first to the Tongan Islands—a sharp detour from his planned course. Tonga was due west and Tahiti northeast, each about equidistant from Hervey's Island, but the prevailing winds blew east to west, making it quicker to get to Tonga. It was already apparent to Cook that the expedition could not possibly make it to the Arctic that summer, and consequently there was no longer any urgency about getting to Tahiti; Cook had the better part of a year to kill in the tropics before heading north the following spring. Under these circumstances, detouring to the Tongan Islands, where he was confident of getting ample provisions for both the animals and themselves, was a reasonable decision. Ironically, once he had changed course the easterly winds died down, and the voyage to Tonga took much longer than Cook had expected. Only an intermediate stop at uninhabited Palmerston Island (also discovered by Cook on his first voyage) saved the livestock from disaster. Finally, on April 30, two months after leaving New Zealand, the ships anchored at Nomuka.

As usual Cook designated certain officers to trade with the islanders and prohibited individual bartering for "curiosities" until a sufficient supply of food was obtained. His orders were largely ignored, however, as sailors rushed to purchase red feathers, remembering (or being told by experienced hands) how valuable they were at Tahiti. The islanders responded to the demand by raising their prices until the feathers became as expensive as what they would purchase in Tahiti. Simple economic reality was no deterrent to the English sailors, however, and the run on feathers continued.

The Tongans' desire for English goods was as strong as the

sailors' for red feathers, and theft soon became a major problem. Even the chiefs stole with impunity. Hoping harsh punishment would serve as a deterrent to others, Cook had one man flogged after catching him with a bolt from a winch, but succeeded only in encouraging chiefs to force men of lower rank to steal for them. Since the chiefs had great power over their subjects and didn't care what the English did to them, further punishment of thieves was useless, and Cook finally gave it up.

After the Englishmen had been on the island about a week, a tall, striking young man named Finau appeared at the ships and presented himself to Cook as "king" of the Tongan Islands. He had come from one of the neighboring islands and, judging from the Nomukans' behavior, was clearly superior in status to the highest-ranking local chief, a man named Topou, who had befriended the English on their previous voyage. Finau appeared to be in his late twenties, handsome and muscular, with flashing eyes and thick, shoulder-length hair. Samwell observed approvingly that he had a high forehead and "aquiline" nose, "a majestic Countenance & a noble and manly Deportment," while Burney thought he had "a degree of wildness in his countenance that well tallied with our idea of an Indian warrior." With his combination of exotic and European-style features, his lively intelligence, and the commanding bearing that marked him as a high-ranking chief, Finau won instant popularity with the English and remained their most important native contact throughout their stay in the Tongan Islands.

It soon became obvious that the English were close to depleting the food supplies available at Nomuka, and Finau suggested that they sail north to the Haapai Islands, where he promised they could have as much food as they needed. Getting there took three days of cautious sailing through waters dotted with tiny islands and reefs, but the results amply repaid the effort, for Finau had not exaggerated the islands' abundance. The ships had scarcely anchored (at a small, fertile island called Lifuka) when canoes surrounded them, bringing hogs, fowl, and fruit to barter. Finau himself, who had sailed ahead, greeted Cook and conducted him ashore to meet the local chiefs. The next day Finau took him to a clearing where two enormous piles of food had been heaped up, with a chief standing in front of each. Cook correctly surmised that the piles represented a sort of tax paid by each man to Finau for the occasion. After Finau presented the gifts to Cook and Mai (one pile was designated for each), the islanders entertained the English with a demonstration of wrestling, which reminded them of the matches they had seen in Tahiti—until the men stood aside and women entered the ring. The "lusty wenches," as Cook called them, fought with as much skill as the men. Nevertheless, one horrified officer tried to break up the match, much to the amusement of the islanders.

Two days after this ritual ceremony of welcome, Finau asked Cook to reciprocate with a demonstration of his country's ceremonies. Cook obliged by ordering the marines to go through their exercises on shore, while a crowd of at least two thousand men and women watched. The islanders, who had obviously prepared for this occasion, followed with a series of dances quite unlike anything the English had seen before. About a hundred dancers, all carefully costumed and carrying short paddles, danced to musical accompaniment "in which every one joined as with one voice," Cook said. Their motions were so perfectly timed that "the whole party moved and acted as one man." King thought the Tongans had planned their show deliberately to upstage the Englishmen's entertainment, and admitted the performance was superior to anything the English could have produced. Apparently the islanders thought so too. They looked with particular disdain at the English musical instruments, including the French horns Cook had taken pains to bring, which had thus far captured no interest at any of the places they visited.

Not to be outdone, however, Cook planned a display of fireworks for that evening. He had the crew bring out every sort of device they had on board, including rockets that exploded under water, with gratifying results. "Such roaring, jumping, & shouting, . . . made us perfectly satisfied that we had gaind a compleat victory in their own minds," King wrote. Unable to imagine how fire could burn under water, the islanders were especially fascinated by the water rockets. Mai took advantage of their reactions to proclaim that the English could destroy not only the earth, but the sea and sky as well, and some of the sailors convinced the men around them that Englishmen had used fireworks to create stars.

The islanders followed the fireworks with more dancing, and the entertainment lasted well into the night. But behind this veneer of hospitality the island's chiefs plotted to murder Cook. According to a sailor who visited Tonga thirty years later and became acquainted with Finau's son, several chiefs conspired to kill Cook, his officers, and the marines during the evening's festivities. Then they intended to seize the ships. Only arguments among the chiefs about exactly how and when to accomplish the deed saved Cook. Finau, who did not initiate the plot but cooperated with those who did, called off the plan when he could not get the others to follow his suggestions.

After about a week at Lifuka, Cook was ready to move on to Tongatapu, but Finau urged him to stay a few days while he went to an island farther north to get a supply of red feathers. Anderson thought Finau was stalling in an effort to keep the English at Lifuka, probably because he enjoyed greater influence there than at Tongatapu; the island's inhabitants, on the other hand, were ready to see the English go and spread a report that a European ship was anchored at Nomuka. Cook guessed the story was a ruse to get rid of him and his men; he was probably right, for they had seriously depleted the island's food supplies, just as they had at Nomuka.

Cook gave up waiting for Finau after five days and headed for Nomuka, confident the chief would catch up with them. On the way south three large sailing canoes approached the ships, and a man named Fatafehi Paulaho, who claimed to be chief of Tongatapu and all the other islands in the area, came on board with a gift of two hogs for Cook. This incident puzzled the Englishmen, since Finau had claimed the same status and obviously commanded the respect of all other chiefs at both Nomuka and Lifuka. When questioned on the point, Paulaho told Cook that Finau was indeed a powerful chief, but not king over all the islands as he had claimed. If Paulaho spoke the truth, then Finau's reluctance to see the English go to Tongatapu became understandable; he wanted to keep them on those islands where he had greatest influence.

Once past the ceremonial exchange of gifts, Paulaho asked why the Englishmen had come to Tonga. Cook produced a map and explained their course; Paulaho in turn showed it to the men with him, demonstrating that he understood clearly what Cook had told him. Both the nature of his questions and his quick grasp of the answers made a deep impression on the English; as King put it, "he shew'd that he was capable of receiving & comprehending subjects which we had consider'd above their faculties in their present state." Paulaho's obvious intelligence and his general demeanor, which King described as "benevolent . . . mix'd with a decent gravity," lent considerable support, in the Englishmen's eyes, to his claim to be principal chief of the islands.

The ships stopped briefly at Nomuka, where Finau appeared the following day, claiming that several canoes loaded with hogs for the English had been destroyed in a storm. Cook did not believe the story, and Anderson's theory was confirmed when they observed Finau and Paulaho together. It was immediately obvious that Paulaho was the superior chief, for Finau observed the same rituals of obeisance in Paulaho's presence as everyone else, although "with such a dejection of Countenance and such apparent depression of Mind (I suppose in consequence of its happening before severall of us to whom he had beking'd himself)" that Clerke felt sorry for him.

Continuing south to Tongatapu, the largest of the Tongan Islands and Paulaho's principal residence, Cook and his officers heard stories of a "king" even more powerful than Paulaho. Having been through this once before, Cook was prepared to take the stories seriously and asked to meet the "king," whose name was Maealiuaki (generally rendered something like "Malliwaggy" by the English). Paulaho demurred at first, saying Maealiuaki had refused to visit the ship because he would have to go below decks, allowing common men to walk over his head; moreover, Paulaho said, anyone who met Maealiuaki must strip to the waist as a gesture of respect. Cook and Clerke replied that their king—who, they informed Paulaho, was a much greater king than Maealiuaki—never required his subjects to strip before him, but only to remove their hats. Paulaho finally gave in, and agreed to introduce the two captains fully clothed.

Maealiuaki was an elderly man, "clearly arriv'd to his second childhood," Clerke thought. Nevertheless, the meeting was a friendly one, and the old chief became a frequent visitor to the ships, despite his earlier reluctance to set foot on board. Shortly after this meeting, Maealiuaki, Finau, and Paulaho all set up temporary residences near the ships, and at least one of them dined with Cook almost every day. Cook enjoyed their company, in part because they kept the lesser chiefs, who had become something of a nuisance with their constant demands for gifts, away. Unlike most Pacific islanders, Maealiuaki took a liking to English food and especially to the Englishmen's wine; "he could drink his bottle as well as most men and was as cheerfull over it," Cook remarked.

Having decided it was impossible to get to the Arctic that summer, Cook was in no hurry to leave the Tongan Islands and stayed a month at Tongatapu. The elaborate entertainments presented at the other islands were repeated here on an even grander scale. Maealiuaki presented Cook with lavish gifts of cloth, red feathers, hogs, coconuts, yams, and fish, and, near the end of the Englishmen's first week on the island, staged what Cook called a "grand *heiva*." The day began with hundreds of the island's residents carrying breadfruit, yams, and fish to Maealiuaki's temporary encampment, where they piled their offerings into two huge heaps. These were intended as presents for Cook and Clerke—gifts of mixed value, Cook observed wryly, since some of the fish had been kept two or three days for the event.

The festivities began late in the morning with an intricate performance involving a chorus of seventy men and four groups of twenty dancers each, accompanied by three men playing what the English called drums, for lack of a better description. These were logs about three to four feet long, hollowed out but with most of the ends left intact. Men beat time on these instruments with sticks about two to three inches in diameter, varying not only the rhythmic pattern but also the tone by beating closer to the center or sides of the instrument, and producing "a rude though loud & powerfull sound." The dancers, using two-foot-long wooden paddles as props, started slowly but quickened their pace as the drummers changed tempo, all the while chanting phrases in response to the chorus. This continued for nearly half an hour, when another group of singers and dancers led by Finau took over. More dancers followed, as the entertainment continued well into the afternoon.

Cook found the performance tiresomely repetitive after a while, although he admitted that "so would, . . . the most of our Country dances to people as unacquainted with them as we were with theirs." Anderson, on the other hand, was absorbed in the intricacies of the dances and recorded them in great detail. He observed that the islanders watching the performance seemed more taken with the language of the songs than with the movements of the dancers—perhaps, he theorized, because the message was a "sentimental" one that appealed to their emotions. Anderson was interested in the movements because of their careful planning and exact timing, and because the overall effect of both singing and dancing was gentle and restrained, in contrast to much of the dancing he had seen earlier. Thinking perhaps of New Zealand, he remarked that Pacific islanders' dancing was usually designed to "raise and nourish many of the most mischievous passions . . . whereas this to appearances inculcates the very reverse in every respect." The message of the performance appeared to be one of love rather than "brutal ferocity," and "every face . . . wore an aspect so placid and benevolent that it is impossible to associate the idea of barbarity along with it."

The next day Cook reciprocated, as he had at Haapai, by directing the marines to go through their maneuvers before an audience of the islanders. Again it was clear that the Tongans thought the English performance a poor second to theirs; they laughed at the clumsiness of some of the exercises. Only the English guns impressed them. Three days later Paulaho made his ceremonial gifts to Cook, an even more lavish spectacle than what Finau and Maealiuaki had offered. Four tall posts were anchored in the ground, defining a space about two feet square; yams were piled on the ground and sticks laid crosswise between the posts to contain the pile. Then the islanders continued to stack up more yams and breadfruit to a height of about thirty feet, capping the display with baked hogs and one live one. "Had our seamen be[en] ordered to do such a thing," Cook remarked, "they would have swore it could not be done without Carpenters and the Carpenters, not without a dozen different sorts of tools, and the expence of at least a hundred weight of nails and after all it would have imployed them as many days as it did these people hours." More dancing followed the completion of this display, and Cook in turn had his men set off fireworks, but they did not have the same effect as at Haapai because water had seeped into them, rendering most useless.

Despite the lavish gifts and entertainment, it became increasingly clear that most Tongans did not harbor benevolent feelings toward the English. The gifts were the work of chiefs competing with each other to impress the English, and to make their offerings they levied heavy taxes of foodstuffs on their subjects. Within a week, wandering Englishmen found themselves no longer welcome in Tongan homes, and theft became a rampant problem. Cook tried flogging thieves, in the hope that such punishment would serve as an example to others, but it proved no more effective here than at Nomuka, and as the days went by he resorted to increasingly severe measures in a vain effort to put a stop to theft. First he confined Paulaho and Finau on board ship as hostages after a turkey and several sheep were stolen, apparently at their direction. The animals were returned within hours, but an alarming number of armed men gathered at the English tents on shore until Paulaho ordered them to disperse. The situation further deteriorated the next day when the theft of two guns so enraged Cook that, disregarding his own frequently stated policy, he fired his musket directly at one of the men responsible, wounding him in the side.

This was the first indication that Cook was beginning to lose his capacity to distinguish major incidents from trivial ones and to exercise the kind of restraint that had kept misunderstandings on both sides from igniting serious conflict on his previous voyages. A more serious example occurred a few days later when one of the islanders stoned a sentry assigned to guard the encampment; in addition to having him flogged, Cook ordered a cross cut into his arm. Such punishment stopped the thieving briefly, but at the expense of generating even greater resentment among the islanders. Even the chiefs, who usually didn't care if Cook punished their subjects, thought such mutilation unnecessarily cruel, as did some of the English officers.

Paulaho patched up his friendship with Cook after his stint as a hostage, but his subjects engaged in a campaign of petty harassment against the sailors assigned to work on shore, hiding in the trees and throwing stones and pieces of wood. Cook was forced to assign sentries to guard the sailors, ordering them to fire with small shot if the islanders became too "insolent." The sailors themselves did nothing to help the situation, often challenging the Tongans to wrestling or boxing matches, which the islanders usually won. Losing made the sailors angry and encouraged the Tongans to step up their taunts. King, who was frequently in charge of the shore parties, feared more serious violence and eventually felt it necessary to take the sailors' pistols and knives away from them.

Men venturing into the interior parts of the island were even more vulnerable than those working on shore. Samwell, on one of his walks, was tricked by his guide and a young girl he met along the way; the girl engaged him in conversation while the guide attacked him from behind and picked his pockets. Two days later, Bligh and Williamson had their guns stolen on an overnight hike. This time Cook refused to take action against the islanders, because the two men had gone off without his permission and he blamed them for failing to guard their possessions properly. He was also annoyed because the fuss the officers made over their stolen guns induced Paulaho and Finau, who feared being taken hostage again, to flee. Through Mai's intercession Cook persuaded them to return to the English encampment, but not before Finau was reassured that he would not be expected to produce the missing weapons.

This incident bore some similarity to the attacks on the Forsters and others during the previous voyage, and like them it created serious dissension about Cook's policies on native depredations. Williamson, like Johann Forster, thought Cook was too easy on the islanders and accused him of acting only when his own possessions were stolen. Given Cook's increasingly harsh responses to theft over the preceding days, this was hardly a valid accusation, but Williamson and others were correct in discerning some inconsistency in Cook's handling of theft. It was not threats to Cook's possessions that spurred him to act, however, but rather threats to his authority, whether over the Tongans or his crew. Bligh and Williamson implicitly flouted his authority by leaving without permission; that their misfortune had disrupted relations with Finau and Paulaho, possibly threatening the trade that provided the Englishmen's food, further angered Cook. But the crew did not make these kinds of distinctions, and saw only that Cook's erratic behavior in dealing with native theft threatened their security at a time when they felt increasingly vulnerable to attack. As Ledyard wrote later, by the time the ships left Tongatapu "we could not go any where into the country upon business or pleasure without danger."

The weeks in the Tongan Islands afforded ample opportunity to fill out the sketchy picture of their culture that Cook and the Forsters had pieced together on the previous voyage. King and Anderson, in particular, made a careful study of the islanders' customs, especially their system of government. Cook and the more experienced officers were by now used to sorting out rivalries and pecking orders among Polynesian chiefs, and it was clear to them that the Tongan system in fundamental outline was similar to that of Tahiti; but this was the first place where they had encountered three different men in rapid succession who claimed to be "king" and the first where the chiefs exerted nearly absolute control over their people.

In questioning the islanders about their political system, the Englishmen learned that "Fatafehi," the name by which Paulaho was sometimes called, was actually a family name and that Paulaho was the fifth of that name to reign as "king" since the last great ships had arrived at these islands. (These ships, the Englishmen surmised, must have been Tasman's.) Paulaho and his forebears, according to tradition, were descended from gods, and each carried the title "Tu'i Tonga," the highest in the Tongan hierarchy of chiefs. Maealiuaki was head of a second great family; Finau was his son. Although they had great authority over the islanders, Paulaho outranked them in social status. The Englishmen used a military analogy in their attempt to categorize the Tongan chiefs in European terms: Maealiuaki had been the "Admiral and General of the forces" under Paulaho's father, and Finau was the current "Generalissimo." As the kingship had been handed down through the Fatafehi line, so military authority had rested for generations with the ancestors of Maealiuaki. Their authority, Bayley theorized, was such that they acted as a check on Paulaho's power, with the capacity to depose him if he became too despotic.

The Englishmen also perceived that each island or small group of islands within the Tongan group had its own chief, and the large islands like Tongatapu were further subdivided into districts, each controlled by a lesser chief. The district chiefs were subject to the island chiefs and they to Paulaho and Maealiuaki, but each had enormous power over the residents of his own district or island. Anderson and King, like Banks in Tahiti, saw parallels between this hierarchical political system and the feudal society of medieval Europe; if Paulaho was "king" and Maealiuaki his chief "general," then the minor chiefs were "Lords or Barons."

The English observers thought the Tongan chiefs exercised an extraordinary degree of control over their subjects, much greater than the chiefs of Tahiti and the other Polynesian islands they had visited, and they speculated about whether this power was absolute. Some thought the chiefs could command offerings of food and condemn men to be punished, even killed, on whim. Others believed their authority was tempered by either law or custom, with the offerings of food given voluntarily or levied in small amounts at a time, as a kind of tax. The existence of private ownership of property, which had so impressed Cook and his men on the second voyage, was also evidence of some limitation on the chiefs' power.

It was hardly surprising that the Englishmen had trouble comprehending Tongan politics, because the system was exceptionally complicated, even more so than that of Tahiti, where the English had had much greater experience. Remarkably, they got the basic outline right. Paulaho, the Tu'i Tonga, was indeed the most influential man in the islands, and his authority was passed down through the family line. Historically, the Tu'i Tonga was the principal ruler of the Tongan islands, with absolute authority over both civil and religious affairs, but by the time Cook arrived the system had evolved such that the position was primarily religious and ceremonial. Maealiuaki was the chief political leader; his position was an elective one, conferred by a group of Tongatapu chiefs and often, although not necessarily, passed from father to son. He was the governing force of the islands, while Paulaho outranked him in social status and moral authority. Thus the English were on the right track in identifying Paulaho as "king" and Maealiuaki as "general," but they failed to understand, at Tonga as at Tahiti, the complex interaction of religious, social, and political authority. In both island groups chiefly status was defined by rigid rules of hereditary descent. The highest-ranking chief enjoyed great prestige and respect but was not necessarily the top political leader, a position based on ability as well as status. The Englishmen were also correct in believing that Tongan chiefs exercised greater power than their Tahitian counterparts. Chiefs controlled agricultural and household production and, as the English had observed, could levy taxes of food to support their families and individuals engaged in specialized production. They could also require their subjects to work on communal projects. In practice, this absolute authority was tempered by the weight of custom; but Cook's visit was a disruptive influence, prompting the chiefs to require larger than normal contributions from their people for the gifts to their visitors.

Cook was ready to leave Tongatapu by the last week in June, but then decided to delay a few more days to observe an eclipse of the sun. The officers responsible for the observations expected to get surprised reactions from the islanders, but in fact the Tongans were familiar with eclipses, which they interpreted as the sun and moon "in the act of coition." In general, the English discovered, the Tongans were quite sophisticated in their knowledge of the heavens, using the moon and stars for navigation and possessing some rudimentary knowledge of the motion of the major planets. They of course had no idea of the purpose of the Englishmen's exercises in astronomy, but concluded after watching the scientific crew at work for some days that the chronometer was a god and the astronomical observations were religious ceremonies dedicated to the sun and moon.

The Tongans' skill at navigation was hardly a surprise, for like the Tahitians they had considerable knowledge of islands in their part of the Pacific, and their sailing canoes were capable of traveling long distances. They told the English about several islands within a few days' sail, notably Fiji—a large, mountainous island about five days' journey northwest of Tongatapu. The Tongans feared the people of Fiji because of their superior weapons and cannibalism, but nevertheless there was peaceful contact between the islands; during their stay at Tongatapu the English met several men from Fiji, who were easily identifiable because they were much shorter and darker than the Tongans.

Notwithstanding such proof of undiscovered islands nearby and the ample time available to explore, Cook showed no interest in searching for Fiji or the other islands described by the Tongans, to the surprise of at least one of his crew members, who remembered his determination to cover as much territory as possible on the two previous voyages. Perhaps Cook thought he had already explored that part of the Pacific adequately, and that one more small island was not worth the trouble to find; but if so, even that attitude marked a sharp change from his earlier behavior, when Forster and others faulted him for being *too* thorough. More likely Cook was tired, and ill—not to the point of impairing his day-to-day functioning significantly, but enough to make him lethargic about pursuing new discoveries and irritable in dealing with the persistent problems of managing a crew and negotiating with native peoples.

He did, however, go to considerable lengths to observe an island ceremony called *inasi* in the brief time remaining before the ships sailed. Normally a harvest festival, this *inasi* was a special occasion honoring Paulaho's son. Cook and Anderson thought it was intended to commemorate the boy's coming of age; more recent writers have speculated that its purpose was to install him as Tu'i Tonga, succeeding his father. Whatever its object, the ceremony was obviously one of unusual sanctity, which the Tongans were reluctant to allow the English to witness.

Paulaho invited Cook to join him and his son for the first part of the *inasi,* when several groups of islanders presented them with offerings of food, but told him he could not attend the next stage, although he could watch from a distance. The chief also asked Cook to keep his crew confined on board ship, as everything on the island would soon be *tabu.* Curious about the preparations underway, Cook left Paulaho and walked toward the *marae* where the second round of festivities was to take place. Several times men urged him to go back, but he persisted until he came within sight of the *marae,* where the assembled crowd protested his presence so strongly that he finally felt compelled to retreat. Not to be deterred easily, however, Cook tried to approach the *marae* later, only to be turned away again. Following Paulaho's instructions, he had to content himself with watching the proceedings from behind a fence, along with a large group of islanders who were also excluded from the proceedings.

The *inasi* continued the following day, and Cook, determined to get closer this time, joined the group at the *marae,* sat down, and refused to leave. After some hurried consultation, the Tongans decided to allow him to remain on condition that he strip to the waist and untie his hair, following their practice. Cook agreed to these conditions, much to the surprise of some of his crew watching from a distance; Williamson complained that such behavior from the captain was undignified. It was, in fact, an unusual move for Cook, and more reminiscent of Banks's enthusiastic participation in Tahitian life, but Cook's willingness to comply with the Tongans' requests accomplished his goal of getting a close-up view of the ceremony.

Soon after the conclusion of the *inasi,* after two and a half months in the Tongan Islands, Cook was finally ready to depart. He left the chiefs several cows, sheep, and horses, and asked Mai to explain that he had brought the animals to the islands at considerable trouble and expense for their benefit. He further instructed Mai to tell the Tongans how to care for the animals and urge them not to kill any until they had time to reproduce themselves. Although Anderson felt the islanders were not impressed with their gifts, Clerke was optimistic that the animals would multiply, because the Tongans, unlike the Tahitians, were not "Children of the Hour," but planned for their future, as their close attention to agriculture showed.

Despite the increasingly severe problems with theft and the obvious hostility of many of the islanders, especially at Tongatapu, the English left the Tongan Islands with positive impressions of their people. All agreed that they were exceptionally attractive physically—Clerke thought them the most beautiful people he had ever met, next to the Marquesans—and laudably industrious. Agriculture was more highly developed there than anywhere else in the South Pacific, and the canoes "neater built, . . . faster, and . . . much better navigated." Equally important, the Tongans were uncommonly generous and benevolent. They would sell anything they had, were cheerful and friendly, and treated their women well—which King, sounding much like George Forster, thought a sign of superior civilization. All except the chiefs appeared to be monogamous, and although there were the usual liaisons between island women and the sailors, even the young single women were less inclined to promiscuity than at most other places they had visited. Anderson remarked that sexual freedom in the Tongan islands did not appear to weaken society, but rather to strengthen it, as he saw no signs of jealousy. Couples eventually formed lasting unions, women showed great affection for their children, and respect for women increased as they grew older.

The Tongans' only vice was the usual one—theft—and most of the officers were unusually tolerant of them even on this point, believing that they stole only out of curiosity. Even Cook tempered his view on the matter by the end of his stay in the islands. Such high praise was remarkable given the undercurrent of native hostility displayed toward the English, especially at Tongatapu. But the rock throwing and petty harassment involved mostly sailors and Tongan commoners. The officers and scientists, who were the ones recording their impressions of the islanders, had contact mostly with chiefs, who did everything they could to impress the English and who, unlike the Tahitian chiefs, exercised a degree of authority over their people that allowed them to control what the Englishmen saw. This very authority itself impressed the English. They remarked on the orderly government of the islands and, although they recognized its despotic nature, thought custom and common sense tempered potential abuses of power. Strong government, meticulous agriculture, well-made ships, kindly behavior, good treatment of women: all added up to a society that the English could understand. As Anderson put it, "the natives of Tonga and the isles around it are upon the whole arrived at as much perfection in their government, at as high a pitch in their agriculture and some other things as any nation whatever under the same circumstances." Except for their lack of metal, "they are in every respect almost as perfectly civiliz'd as it is possible for mankind to be."

17. *Mai Returns*

When the outline of Tahiti came into view on August 12, some of the first-time Pacific voyagers were disappointed. After listening to their more experienced shipmates' stories about the island's extraordinary beauty, the vision they saw—an island so shrouded in clouds that its top was not even visible—hardly lived up to expectations. Once the ships anchored at Vaitepiha Bay and the crew was treated to the usual exuberant Tahitian welcome, however, even the most skeptical admitted first impressions might be mistaken.

The first Tahitians to board the ship, including Mai's brother-in-law, were indifferent to their returning compatriot—until Mai gave his kinsman a handful of red feathers. Instantly the situation changed; the brother-in-law, "who would hardly speak to Mai before, now beged they might be Tyo's and change names." It was not an isolated incident. Although Mai's sister greeted him warmly later that day, the rest of the Tahitians paid him no attention until they saw his stock of feathers.

Cook was not surprised. Throughout the journey he had worried that the Tahitians would take advantage of Mai and his newfound riches, and several times during the weeks before arriving at Tahiti he had tried to talk to Mai about his future, urging him to plan some means of using his unique experiences and the possessions he had accumulated in England to good advantage. But Mai would not listen, insisting over and over that he wanted to destroy the Bora Bora chiefs and restore his native Raiatea to independence. Cook replied that he would not collaborate in these ambitions. Trying to convince Mai of the folly of his plans, he reminded the young man of his low status among the Society Islanders, which succeeded only in reinforcing Mai's inclination to ignore Cook's advice.

Mai's problems soon faded into the background in light of an interesting new development: other Europeans had visited Tahiti since Cook's previous voyage. In the course of several conversations, Cook and his officers deduced that these visitors were Spanish, had sailed from Lima ("Reema," the Tahitians said), remained several weeks, and built a small house not far from where the English ships were now anchored. When the Spaniards went home, they took four Tahitians with them and left four of their own number behind. About ten months later more ships arrived, returning two of the Tahitians (a third died and the fourth decided to stay in Lima) and collecting the Spaniards. The group remained at Tahiti only briefly, promising to return some months later to settle permanently.

According to the Tahitians, the Spaniards had attempted to undercut the Englishmen's reputation on the island, telling them that the king of Lima ruled over the entire universe and that England was a "damn'd, little, dirty, piratical State" that had been destroyed by Spain. The trusting Tahitians believed this story and assumed their English friends must be dead, killed in the great war with the king of Lima. What was potentially more serious, the Spaniards also urged local chiefs not to allow the English to anchor and not to supply them with food if they returned to the island; and the Spanish had earned enough respect among the Tahitians to give the chiefs pause when Cook's men arrived expecting their usual provisions. After conferring among themselves, however, the chiefs agreed to supply the Englishmen with what they needed. Samwell attributed this decision in part to the extraordinary respect the Tahitians felt for Cook, but no doubt the plentiful supply of red feathers helped. (One of the Tahitians who had visited Lima told Clerke he thought it was a poor country because it was devoid of red feathers.) The Spaniards' alleged superiority took a further dive when the islanders realized the extent of Mai's riches. One man who had come back from Lima with three shirts and a suit was chagrined when he saw how much more the English had given Mai; "he cursed the Signior's very heartily & lamented much, that his good fortune had not given him a Trip to England instead of Lima." Mai, for his part, treated the Tahitian visitors to Peru rudely, annoyed that he was not the only world traveler among his countrymen.

The Spaniards' attempts to destroy the Tahitians' loyalty to the English angered Cook and his companions, who thought of themselves as scientific explorers above questions of petty national rivalries. Spanish success in earning the respect of the islanders was also disturbing. Samwell and Anderson, both on their first voyage to the Pacific and fresh from the often hostile encounters at the Tongan Islands, thought the Spaniards had succeeded in making themselves appear superior to the English by treating the Tahitians respectfully and avoiding the casual intimacy that the English encouraged; but Cook pointed out that his return had quickly dispelled whatever credence the Tahitians placed in the Spaniards' claims. In fact the Englishmen probably exaggerated the Tahitians' feelings for their Spanish visitors, interpreting the islanders' acceptance of the stories about Spain conquering England as a sign that they believed England to be weak, when in fact the Tahitians viewed conquest in war much differently than Europeans; they grieved for their English friends, but did not think less of them for their supposed defeat.

Word of the English presence spread quickly, and after a week Tu, disappointed that Cook had not come immediately to Matavai Bay, sent a palm branch via messenger with a "pressing invitation" to visit. Two days later the ships left Vaitepiha Bay for their familiar anchorage at Matavai. Cook called on Tu immediately, with a gift of a Tongan helmet made of red feathers that impressed the Tahitians, King said, as much as a diamond crown would have Europeans. Tu and his family later went on board the *Resolution,* where they expressed pleasure at seeing many of their old friends; but King, who had already formed a negative impression of the Tahitians in his brief stay at Vaitepiha, thought their professions of friendship lacked enthusiasm and sincerity. They seemed excessively concerned about learning the rank of the various Englishmen and "proportioned their Caresses accordingly." Red feathers were the surest route to winning Tahitian favor, and the islanders' eagerness in begging for them King found "disgusting."

Mai met with much the same reaction at Matavai Bay as he had at Vaitepiha. Few people paid attention to him until they discovered his stock of red feathers and other riches; then even Tu took notice. This pleased Cook, who thought Mai should curry favor with Tu, marry his younger sister (Tu himself favored this plan), and settle down at Matavai Bay. Such a scheme, Cook thought, would be the best way for Mai to hold on to his property and preserve his future. Mai, however, was not interested. He didn't like Tu's sister, and he insisted on associating with the lowest-status Tahitians, whose only interest was "to plunder him," which they did quite successfully until Cook intervened. Finally even Tu became annoyed with Mai, putting an end to any lingering prospects of his settling at Matavai.

Three days after the ships anchored in Matavai Bay, a group of Tahitians told Clerke that a Spanish ship had just anchored at another part of the island. Both he and Cook were skeptical of the story, but they sent an officer in one of the boats to investigate. When he found no trace of a ship, the English realized that the Tahitians had concocted the story to see how their visitors would react.

The Spanish visit remained a continuing topic of discussion between the English and the Tahitians, since the English wanted to find out as much as they could about Spanish intentions and the Tahitians were trying to figure out whether the Spanish boasts about their power over the English were true. Cook was annoyed that he couldn't get a clear account from the islanders about the time the Spanish had arrived or the length of their stay, although he deduced from the inscription on a cross they left behind that the ships had arrived in 1774, not long after the *Resolution* left. (Cook had the English claim to prior discovery carved on the back of the cross: "Georgius Tertius Rex 1769-72-74 and 77.") He was puzzled about why four men had remained on the island, however; two were priests, judging from the Tahitians' imitations of them saying grace and reciting the rosary, but if they had attempted to convert the Tahitians to Christianity, their efforts were in vain. The islanders "all unanimously condemn'd [the fl]esh-subduing Dons," Samwell reported, "for that denial which may [be de]emed meritorious in Cells & Cloisters, but will be always looked upon with Contempt by the lovely & beautiful Nymphs of Otaheite." Moreover, according to the Tahitians, only one of the group ever ventured far from their house or made much effort to talk with the islanders.

The Spaniards promised to return to Tahiti, which suggested to Cook that they intended eventually to establish a settlement on the island. Tu was pleased at the prospect, to Cook's dismay, for it was obvious that the chief had no inkling of the negative consequences of European colonization, no sense "that such a step woud at once deprive him of his kingdom and the people of their liberties." Tu's enthusiasm was typical of the friendly and trusting attitude toward strangers that would make the Tahitians vulnerable to European designs on their island—as their good-natured acceptance of repeated English, French, and Spanish visits indicated—but although Cook lamented Tu's naiveté, he did not believe the Tahitians stood in danger of colonization. While the island was an ideal port of call for ships, he could see no reason that Europeans would ever want to establish a permanent settlement there.

Spanish ships had indeed visited Tahiti late in 1774, as Cook deduced, with the goal of establishing a settlement and converting the islanders to Christianity. In his assumption that Europeans could have no motive for long-range colonization, however, Cook reckoned without Spanish paranoia over other nations' forays into the Pacific. The notion of exploration as purely scientific inquiry, so much a part of English and French thinking in the late eighteenth century, seemed to the Spanish nothing more than a cover-up for encroachment on their territory. In an effort to shore up its waning influence in the Pacific, the Spanish government had launched a series of expeditions to Tahiti.

The first, commanded by Don Domingo de Boenechea, sailed from Lima in the fall of 1772 under orders to determine if any foreign settlements had been established there. If not, he was to persuade the inhabitants that Spain had sole rights to their island and begin the work of converting the Tahitians to Christianity. Boenechea and his men spent a month at Vaitepiha Bay, and were no doubt the source of the vague story about "French" ships that Cook had heard on his previous visit. Boenechea returned to Tahiti shortly after Cook left, with orders to take steps toward establishing a settlement. To this end, he had on board two priests, a young marine named Maximo Rodriguez ("Matema" to the Tahitians), a servant boy, and a small wooden building to house them, constructed in sections so it could be easily disassembled and rebuilt. The two returning Tahitians, who had ostensibly become Christians, were to help in converting their fellow Tahitians.

The Spanish ships remained several weeks, again at Vaitepiha Bay, sailing for Lima in January 1775. When they returned the following October with supplies for the little "settlement," the priests insisted on being taken home. They had been ridiculed by the Tahitians; their possessions had been stolen; they feared for their lives; they had converted no one to Christianity. Even the two supposed converts had deserted them, preferring to live with their families. Of course the priests were never in real danger from the Tahitians, but they were objects of ridicule because of their strange customs, and the two men transformed ridicule into hostility. Rodriguez, on the other hand, who learned to speak excellent Tahitian from the islanders who had visited Lima, spent most of his time traveling around the island, living with the people and learning their customs; he compiled the most detailed record of Tahitian life since Joseph Banks. The avowed purpose of the Spanish expeditions to Tahiti was a complete failure, however, and Spain made no further attempt at "settling" the Society Islands.

Throughout the Englishmen's stay, the Tahitian chiefs were preoccupied by their continuing dispute with Moorea. The battle brewing at the time of Cook's last visit had been inconclusive, and Tu's forces, again led by the "admiral," Towha, were in the midst of preparing a new attack. Tu asked for English help in the expedition, but Cook refused, explaining that the people of Moorea had never done him any harm. The next day, Towha sent a messenger to tell Tu that he had killed a man to be sacrificed in a ceremony entreating assistance in the battle against Moorea. Cook had heard rumors that the Tahitians practiced human sacrifice, and Bougainville mentioned it in his journal; seeing an opportunity to confirm the truth of these assertions, he asked Tu's permission to attend the ceremony.

With Anderson, Webber, and Mai, Cook accompanied Tu to the *marae* at Atehuru, southeast of Matavai Bay. According to Mai, human sacrifices were rare, occurring primarily in time of war or famine, but even so Cook and Anderson counted about fifty skulls arranged around the *marae,* none of which showed signs of extended exposure to the elements. When a human sacrifice was deemed necessary, they learned, the priests told the highest chiefs, who settled upon some "useless or ill dispos'd person" as the victim; he would be killed by a quick blow to the head. Cook and Anderson watched the long ceremony with detachment, but afterward remonstrated with the Tahitians about the barbarity of their custom. Cook told Towha that the sacrifice would not please their god, but on the contrary would make him angry and diminish their chances of success against Moorea, and Anderson added that a chief who did such a thing in England would be punished with death, to which "the Tahitian express'd great astonishment and said it was so enormous a crime to kill a chief that he would hear no more of it." Towha was already angry with the English for their refusal to assist in the battle against Moorea—he could not understand why people who called themselves friends would not assist their allies in battle—and he found incomprehensible the English practice, described to him by Mai, of treating all ranks of men alike under the pretense of meting out impartial justice. Cook and Anderson's latest critical reaction to the ceremony at Atehuru only deepened his disgust at English ways. "We left him," Cook observed, "with as great a contempt of our customs as we could possibly have of theirs."

Towha's displeasure did not prevent the English from spending five delightful weeks at Matavai Bay. They had horses on board, and Cook and Clerke took a daily ride along the beach, to the astonishment of the Tahitians, who gathered by the dozens and followed the captains every time they rode. (Mai attempted to impress his countrymen by riding too, but usually fell off his horse before getting very far.) The islanders' fascination with this spectacle did not wear off, even after several days, and Cook thought that the horseback riding did more than anything else to convince them of English superiority and lay to rest Spanish assertions to the contrary. The Tahitians also offered a new form of entertainment to the English: a game remarkably like football, played by young girls using a breadfruit for a ball. The winner of the game paraded naked before the crowd.

Cook further occupied himself in distributing most of the animals he had brought from England and supervising his men in planting a garden. Unloading the livestock was a great relief, for "the trouble and vexation that attended the bringing these Animals thus far is hardly to be conceived." Otherwise, there wasn't a great deal of work to be done—the English were mostly marking time until they could begin their voyage to the Arctic—and the crew members spent much of their time on shore enjoying the pleasures of a tropical spring. Tu entertained Cook frequently, usually at his home at Pare. When Cook complained of a "Rheumatick pain" that lamed one side of his body from hip to foot, Tu's mother, sisters, and eight other women marched on board Cook's boat and insisted upon massaging his body "till they made my bones crack and a perfect Mummy of my flesh." After fifteen minutes, Cook was eager to escape, but the treatment cured his pain, as a similar treatment had cured Wallis on his visit to Tahiti.

The sailors enjoyed their own version of Tahitian hospitality, each with his *taio,* or friend with whom he had exchanged names, and woman. Venereal disease no longer seemed to be the problem that it had been on the second voyage—the Tahitian women had no qualms about sex with infected men, saying that the disease would not harm them—and so there were no constraints on coupling between Englishmen and island women, except the usual rule that *arii* women would cohabit only with officers. According to the women, this prohibition was not because the *arii* placed a high value on chastity (which had always been Cook's explanation) but because they would associate only with men whose rank in English society was the equivalent of theirs in Tahiti. Williamson, an officer, bragged that Cook was wrong about the chastity of *arii* women, and that he had managed to win them over.

Remarkably, theft was only a minor problem, in sharp contrast to all previous visits. Cook attributed this change to the chiefs' desire for a steady trade without interruptions; he also gave Mai credit for getting across to his countrymen the disadvantages of thieving. In fact, Cook's first explanation was more plausible. Repeated visits from Europeans had fueled the Tahitians' desire for iron and woven cloth, and the islanders understood that they could get just about anything they wanted by offering food or their own local products in exchange. It was in Tu's interest to maintain good relations with the English in order to keep up the flow of trade (just as it was in Cook's interest to do the same), and experience had taught him that nothing disrupted good relations more than theft. As Cook realized, chiefs had considerable authority over their subjects' behavior, and the extent of Tu's influence on the island was on the rise, making it easier for him to put a stop to most theft against the English. Tu himself was still not above an occasional trick, however. When Cook gave him a spyglass he didn't want, Tu offered it as a present to Clerke, asking him not to tell Cook. Clerke at first declined the gift—gifts were never proffered without the expectation of return, according to Tahitian custom—but Tu insisted. A few days later, Tu reminded Clerke that he had given him nothing for the spyglass and Clerke, although he did not want the glass, felt compelled to give the chief four axes. Tu said Cook had offered five, so Clerke countered with six. Not until he told Cook about the incident did Clerke realize he had been duped.

Toward the end of September, Tu's forces gathered at Matavai to prepare for the planned attack on Moorea; but before the fleet could get underway, word came that Towha had just returned from the island, where, with a small force, he destroyed several canoes and houses and concluded a truce with the local chief. Tu and Towha had disagreed about strategy for some time, and now the partisans of each man accused the other of undermining his attempts to subdue the neighboring island. Towha blamed Tu for not coming to his assistance sooner; Tu blamed Towha for attacking before his fleet was ready. Cook suspected Tu had delayed his departure because he knew it was Cook's intention to visit Moorea after leaving Tahiti and planned to follow the English ships there, convinced the English presence would intimidate the people of Moorea and aid their cause, even if the Englishmen kept out of the fighting. Thus Tu hoped to achieve what Towha had failed to gain: help from the English in their battle, even if that help was psychological rather than real. Cook also thought Tu was averse to fighting and might simply have been dragging his feet about getting into a battle.

Word circulated that Towha would take his revenge by attacking Tu with a force made up of Vehiatua's subjects. Cook's scruples about becoming involved in island conflicts evaporated when his friend Tu was the potential victim, and he let it be known that he would retaliate against anyone who attacked the chief. The rumors soon faded, however, and Tu joined Towha for a ceremony of Thanksgiving at the Atehuru *marae.* Cook did not feel well enough to attend, but sent King in his place; the two men concluded that the event was not merely a ceremony of thanksgiving, but a confirmation of the truce as well.

By the end of September, about a week after the aborted battle, Cook decided it was time to leave Tahiti. He would have preferred to stay longer, since food supplies were ample and it was still too early in the season to sail north; but Mai remained adamantly opposed to settling at Tahiti, and Cook felt obliged to take him to Huahine or Raiatea. First, however, he wanted to see Moorea.

The ships anchored October 2 in a narrow bay cut between two mountains covered from base to top with thick, deep-green vegetation. Beyond this bay the center of the island rose to a jagged peak with two summits, one slightly lower than the other, which gave the island a distinctive silhouette quite unlike that of other high islands in the South Pacific. Shortly after anchoring, Cook met Maheine, the island chief who had been the target of Towha's attack. Cook judged him to be between forty and fifty years old. He was bald, a condition rare among Society Islanders, and wore a turban most of the time. Cook thought Maheine was reluctant to bare his head not so much because Tahitians were ashamed of baldness, but because they thought Englishmen were, a view that was perfectly understandable given that the English occasionally shaved the heads of thieves as punishment. Indeed, Cook remarked, the Tahitians displayed some suspicion of the bald members of the crew, assuming they, too, were thieves.

Shortly before Cook planned to leave Moorea, a goat put ashore to graze was stolen. Although the animal was hardly essential to the voyage, Cook was determined to get it back, claiming he needed every animal remaining in his cargo to stock other islands. He suspected the theft was not a mere prank but a deliberate act undertaken at Maheine's behest, and he sent several officers to the chief's home to demand the goat's immediate return. Confident no one would dare steal more animals when he had made his displeasure at the first theft so clear, Cook had the remaining livestock put ashore to graze the next morning. At the end of the day, another goat was missing. Several islanders assured Cook that the goat had only wandered off and offered to search for it. Cook accepted their offer of help, but when none of the men returned, he realized they had tricked him into doing nothing until the goat could be taken some distance away. Meanwhile, the officers sent in search of the first goat returned with the animal and the men who had stolen it.

By the next morning most of the islanders had disappeared, and those few remaining reported that Maheine had decamped to the most distant part of the island. Upon learning that the second goat had been taken to the opposite side of the island, Cook sent two officers to retrieve it. When they reached the place where the goat had supposedly been secreted, several islanders put on a show for them, allegedly to pass the time while they waited for other men to bring the animal. This scheme turned out to be another delaying tactic, and when night came the officers returned to the ships empty-handed.

Their failure only fired Cook's determination to retrieve the goat, and what had started out as a reasonable effort to retrieve stolen property using the usual strategies turned into an obsession. Cook felt that backing off at this point would set an unfortunate example "to the people of the other islands we had yet to visit to rob us with impunity"; but in fact more was at stake. Maheine had asked Cook for a pair of goats, a request Cook refused because he believed his few remaining animals were needed more urgently at other islands; not to be so easily deterred, the chief simply took what he wanted. Then, after Cook made his displeasure known, the Mooreans responded not by returning the stolen goat but by taking another one. Finally, most of the islanders had conspired to thwart the Englishmen's efforts to get their animals back. Such actions constituted a sustained assault on Cook's authority, and although it was no different from what he had experienced before, especially on initial visits to new locations, at Moorea it came after an extended period of quiet, orderly trade with the Tahitians. Always prone to anger when he felt himself losing control over a situation, Cook's reaction was even stronger at Moorea in part because of the contrast between what he perceived as the insolent, provocative attitude of the Mooreans and the Tahitians' compliant, respectful behavior. His diminishing tolerance for any action that could be construed as an attack on his authority only made matters worse.

In the futile belief that force would achieve his objective, Cook led thirty-five armed men on foot across the island and directed Williamson to take another group to meet them by boat. When Cook and his party reached the place where the goat was reportedly kept, the people living there denied any knowledge of its whereabouts. One Englishman thought the Mooreans were only pretending ignorance; they made fun of Cook, which so enraged the captain that he threatened to burn their houses and boats if they did not hand over the animal. The threats brought forth more denials, at which Cook ordered his men to set fire to a half-dozen houses and two or three large canoes. Then he led his group back to the beach, burning six more canoes along the way.

Still the goat was not returned, but Cook refused to give up his demand for restitution. After two days of stalemate, he sent word to Maheine that the Englishmen would burn every canoe on the island if the goat was not given up. To drive home his point, he ordered the ships' carpenters to break up three or four canoes lying at the head of the harbor and stow the wood on board to use later in building a house for Mai. Then he and several others walked along the coast, hacking and burning more canoes and houses and shooting hogs. Cook put the total destruction at six or eight canoes, but Bayley claimed the toll was twenty canoes, fifteen to twenty houses, and twenty to thirty hogs; both Clerke and Edgar thought it would take months or even years for the islanders to recover from the damage.

When he returned to the *Resolution* that evening, Cook learned that the goat had just been returned. "Thus this troublesome, and rather unfortunate affair ended, which could not be more regreted on the part of the Natives than it was on mine," he concluded. The next morning, he claimed, "we were again all good friends."

Cook's regrets were hollow and his belief that the islanders had forgiven him naive in light of the destruction he had wrought. The crew rarely questioned his judgment (even though they complained about some of his decisions), but on this occasion even his most loyal officers criticized the use of such extreme tactics for the sake of one stolen goat, arguing that a less violent strategy could have been just as successful. The sailors and marines who had carried out the attack, normally much less sensitive than Cook in their behavior toward native peoples, joined in the rampage reluctantly, taking part only because Cook demanded that his orders be carried out. Gilbert, who recalled Cook's tolerant, humane treatment of other groups they had encountered, was puzzled about the motives for his overreaction at Moorea. The only plausible explanation, Gilbert thought, was Cook's friendship with Tu and the fact that Maheine and his people were Tu's enemies. (In fact Cook had spared the houses of one group of people on the island who were allies of Tu; but Tu paid for English favoritism when Maheine, powerless to do anything against Cook, took revenge by attacking Tu and his people.) Those few who supported Cook's action agreed that strong measures were needed to deter theft and blamed the island's chiefs for bringing destruction on their people by their stubborn refusal to return the goats, but even they were inclined to think the punishment excessively severe. Only Mai enjoyed the assault, attacking canoes and houses with enthusiasm—perhaps as another part of his effort to convince his countrymen of his superiority now that he had seen so much of the world.

Cook decided to leave Moorea immediately after the incident, and the next day the ships anchored at Huahine. The Englishmen's problems with the people of Moorea were not over, however. Just before they anchored at Huahine, a man from Moorea who had sailed with them was caught stealing. In a fit of rage, Cook ordered the ship's barber to shave the man's head and cut off his ears; but after the first part of the sentence was carried out, an officer (probably King), thinking Cook was "only in a Passion," stopped the barber and freed the accused thief, who escaped with one earlobe cut.

Word of the Englishmen's destructiveness, embroidered in frequent retellings, had spread across Huahine even before the ships anchored, which Cook accepted as proof that his actions at Moorea had been justified. On previous voyages the people of Huahine had been the worst thieves in the Society Islands, and he looked forward to a more positive reception this time.

Initially, all went well. Armed with presents, Cook and Mai paid a call on the island's principal chief, a boy about ten years old who had replaced the elderly Ori, now resident at Raiatea. After the usual formalities, Cook outlined his need for supplies and warned the assembled company against stealing from them as they had done on previous visits. Then he asked for a plot of land where Mai could settle. One of the lesser chiefs told Cook that the entire island and everything in it was at his disposal; he could give whatever part he liked to Mai. Cook, however, concerned to make Mai's living arrangements as permanent as possible, asked the men present to designate a specific parcel for him and guarantee that it would be his forever. They assented to this request, and Cook set his carpenters to work building a house for Mai while other members of the crew planted a garden.

For a little over a week the Englishmen experienced no more serious problems than attempting to rid their bread of cockroaches—airing the bread on shore and smoking the ships with gunpowder and brimstone killed roaches by the hundreds, but the next day there seemed to be just as many alive in the hold—until Bayley discovered one of his sextants missing. With Mai,

Cook went ashore, stormed into a crowd watching a theatrical performance, and demanded the return of the sextant, threatening reprisals more severe than those he had inflicted on Moorea if the group would not tell him where it was hidden. Mai grabbed the man he thought had stolen the instrument and threatened to stab him unless he revealed the hiding place. When the man refused to respond, Cook had him seized and confined in irons on board the *Resolution.* After much badgering from Mai, he told his captors where to find the sextant. A search the next morning proved that the prisoner was telling the truth, but because he was, in Cook's opinion, "a hardened Scounderal," Cook had the man's head shaved and his ears cut off before releasing him.

That night the accused thief uprooted everything in Mai's garden; the next morning he accosted his adversary, vowing to kill him and destroy everything he owned. At this, Cook had the man seized and confined on board for the next five days, during which time Mai's house was completed and all his possessions moved into it. At first Mai complained that his new home was too small—the English king had promised him a two-story house, he claimed—but Cook laughed and told him one story was all he needed. Mai also began to see that his English pots and pans and dishes, carefully packed and carried halfway around the world, were of little use in his native surroundings, as he realized "that a baked hog eat better than a boiled one, that a plantain leafe made as good a dish or plate as pewter and that a Cocoanut shell was as good to drink out of as a black-jack." Much to Cook's satisfaction, Mai traded these objects to the ships' crew for hatchets and other items more useful in the Society Islands. Cook also provided Mai with guns and ammunition, against his better judgment, because the young man pressed so urgently for them.

After all was finished, late one night, the sentry assigned to guard the prisoner fell asleep and his charge, having observed where the keys to his irons were kept, freed himself and escaped to shore, with one shackle still locked around his leg. Back on the island, his supporters removed the shackle and sent it to Mai to show that the man he had tormented was free. This gesture was hardly necessary, for the prisoner's disappearance was obvious at daybreak, and Cook unleashed his anger at the men responsible for guarding him. The sentry, who admitted falling asleep, was himself placed in irons and flogged daily for three days, while the officers in charge of the watch were reduced in rank as punishment. Cook offered a reward of twenty hatchets to anyone who brought the thief back to the *Resolution,* without success. A man accused of being the thief's accomplice was captured and confined on board until the ships sailed three days later, but the thief himself remained at large, and Cook had to be content with the Huahine chiefs' promises to protect Mai after the Englishmen departed. As further insurance, Cook gave Mai blue, white, and brown beads with instructions to send one of them to him at Raiatea, where he intended to go next. If all was well, Mai was to send a blue bead; if he was being treated only "tolerably," a white bead; and if he was unhappy, a brown bead. Then, at least, Cook could return to Huahine and try to help if necessary.

When the ships left Huahine on November 2, Mai bade a stoic farewell to his friends, managing to keep his composure until he got to Cook, when he broke down and wept. It was an emotional farewell on all sides. Cook, despite his reluctance to take Mai to England in the first place and his frequent annoyance at the young man's behavior, had developed an affection for his charge that was apparent in his concern to see Mai comfortably settled. He only regretted that Mai lacked the understanding to put his European experiences to some practical use at home; but in that respect, Cook thought, he was no different from the rest of the Society Islanders. After ten years of European visits, they appeared untouched by European influence, and any benefits from European contact would have to come at the instigation of Europeans visiting the islands, not Polynesians visiting Europe. Not that Cook expected the Tahitians to change their ways in response to some perceived superiority of European culture; on the contrary, he hoped Polynesian ways would be preserved with a minimum of European influence. Nevertheless, he wanted the islanders to adopt those practices that could, in his estimation at least, prove useful without undermining native customs. Raising new breeds of animals was the most significant of these, in Cook's mind, and he thought the introduction of English livestock was his most important contribution to Tahiti and its neighbors.

In drawing his conclusions Cook ignored the evidence of his recent visit to Tahiti. He had observed Tu's increasing power, his ambition to conquer Moorea, the islanders' increasing desire for English goods, and their interest in understanding status relations among Europeans—all without recognizing the hand of European influence in shaping these changes. In fact the Tahitians learned a great deal from the English, contrary to what Cook thought, but what they learned was often not what the English thought they should learn. Practical, useful knowledge by English standards (agricultural methods and more sophisticated kinds of tools, for example) was generally of little interest to Tahitians, who had their own, perfectly satisfactory ways of doing things. Rather, the already status-conscious, ceremony-loving islanders were taken with ways in which English friendship and possessions could enhance their standing in their own society. They coveted iron and other English manufactured goods not so much for their practical use, but for the vastly increased status they conferred on those who possessed them—hence the increased power of Tu and his family, who had been the primary beneficiaries of a decade of English attention. An interesting symbolic example was the English flag that Wallis gave the Tahitians in 1767; the men of the *Resolution* and *Discovery* were amused to see that the islanders had transformed it into their own sacred object, decorating it with feathers and displaying it on ceremonial occasions. But the rush to obtain European goods fostered rivalry among the islanders, which in turn spawned theft (virtually unknown before European contact), war, and human sacrifice, which was most often associated with war. Interestingly enough, King criticized the Society Islanders precisely for these practices, accusing previous explorers of exaggerating their organized government and happy, carefree lives, without realizing that he was seeing islands already changed by European contact.

From Huahine an overnight voyage took the Englishmen to Raiatea, where they stayed for another month, still marking time before temperatures in the North Pacific would warm up enough to allow them to proceed. Three men, preferring the idyllic life of the Society Islands to the hardships of an arctic voyage, attempted to desert during that time; one was easily recovered by officers searching the island, but the second two fled to Bora Bora and almost made good their escape. Cook got them back only after taking the principal chief's son and daughter as hostages. The chief himself went to Bora Bora and delivered the deserters to the *Resolution.*

The Englishmen finally took their leave of the Society Islands the first week in December. Shortly before sailing, in an effort to get some idea about possible future stopping places, Cook asked the Raiateans if they knew of any islands to the north or northwest; they replied that they did not. As far as anyone knew, nothing lay between them and the North American continent.

18. To the Arctic

For a little over two weeks, the men of the *Resolution* and *Discovery* sailed through uncharted seas, until Christmas Eve, when they approached a single small, low-lying island. Cook decided to anchor long enough to catch some of the turtles that infested the shallow waters around the island and to observe an eclipse of the sun on December 30.

Christmas Island, as Cook named his discovery, was a turtle hunters' paradise. Although the animals tried to protect themselves by diving into deep holes, they remained visible above the water's surface, and sailors dove in after them, pulling the turtles out by the fins and chasing them. The men looked ridiculous splashing through the water, but it was an effective method of catching the slow-moving creatures; one group caught forty-two in half an hour, and Cook estimated that the crew bagged about three hundred altogether.

In mid-January the Englishmen discovered more islands, a cluster of three this time, and as they drew closer the familiar sight of canoes paddling toward the ships greeted them. When the canoes pulled up alongside, the sailors were astonished to find that these people spoke a language remarkably like Tahitian and looked much like the Society Islanders as well. The canoes were filled with stones and fish; as soon as the islanders discovered that the English were friendly, they threw their stones into the sea and traded their fish for whatever was offered, although they clearly preferred nails. Despite repeated invitations, however, they would not venture on board.

A small group returned the next day and this time accepted the English invitations to visit the ship. "I never saw Indians so much astonished at the entering a ship before," Cook remarked. Based on the men's reaction and his knowledge of previous explorers' voyages, he concluded that these islanders had never before encountered Europeans. Three thousand miles northwest of Tahiti, in a part of the ocean where he had reason to believe no land existed, Cook had stumbled upon an isolated group of islands that would become one of his most famous discoveries: the Hawaiian group, later named by Cook the Sandwich Islands after his patron, the Earl of Sandwich. At this point the ships were cruising off the coast of Kauai, with the smaller island of Niihau in the distance; the third island spotted earlier was Oahu. These islands, the northernmost outpost of Polynesia, were so far from the routes of earlier explorers that they had remained undiscovered by Europeans through two centuries of Pacific voyaging.* Like New Zealand, they had been settled from central Polynesia, most likely either the Society Islands or the Marquesas, around A.D. 1000, and their customs and language retained a remarkable similarity to those of Tahiti and its neighbors.

King took the islanders on a tour of the *Resolution,* offering them beads and mirrors as gifts. The men asked what the beads were used for, and upon being told that they were merely ornamental, handed them back as useless. Later they returned the mirrors for the same reason. They tugged at the Englishmen's cloth to see if it would tear; when it did, they rejected that too. Nails were the only objects the men liked, although they showed great curiosity about other items, especially the ship's sails and

* There is some evidence that Spanish explorers discovered the Hawaiian Islands around the middle of the sixteenth century on one of their regular voyages between Latin America and the Philippines; but, as was the case with many Spanish discoveries, the information was not made public, and knowledge of the islands was effectively lost to future generations. The Hawaiians' immediate interest in iron supports the theory that they had had some earlier contact with Europeans, even though the residents of the island in the eighteenth century had no oral tradition of that contact. The Hawaiians themselves told the Englishmen that they had obtained iron from pieces of driftwood that washed upon their beaches.

china cups, which they thought must be made of wood. (Porcelain or pottery of any kind was unknown to the Polynesians.)

King thought the islanders showed an unusual degree of deference toward the Englishmen, but even so some attempted to steal objects they fancied. One man took the lead and line used for depth soundings, while another made off with the butcher's cleaver. Cook sent Williamson in a boat to intercept the thief, who had fled in a canoe toward the island. As they approached shore, the boat's crew fired at the canoe (against his instructions, Williamson later claimed), and everyone in it jumped overboard and swam to safety. When the Englishmen were close enough to attempt a landing, the men on shore began throwing rocks, forcing Williamson to give up his attempt to retrieve the stolen goods.

Later Cook sent Williamson to search for a suitable landing place on Kauai. Again the lieutenant ordered his crew not to fire on the islanders, a precaution he thought necessary because of "ye great wantonness of the inferior people on board a ship, & ye idea they possess that it is no harm to kill an indian." When the Englishmen came within a few yards of shore, dozens of islanders (King thought at least one hundred) crowded around the boats, trying to jump in. Williamson attempted to get away, with the intent of searching for another landing place where they wouldn't encounter so many people, but the boats were immobilized. One man grabbed the boat hook from Williamson's boat and refused to let go, while others pushed so hard against the sides of the vessel that they nearly capsized it. Williamson hit the man holding the hook with the butt of his gun, with no effect; the sailor clinging to the other end said he couldn't hold on much longer. Although Williamson did not think they were in danger of attack, he feared for the safety of his boats and men, and so he shot and killed the man holding the boat hook.

Having restrained his men from firing on the grounds that they could not be trusted to behave humanely toward "indians," Williamson now justified his behavior out of a need to establish superior force over people he labeled "barbarians." Then, he claimed, "they afterwards readily believe that whatever kindness is then shewn them proceeds from love, whereas otherwise they attribute it to weakness, or cowardice." Not everyone on board agreed with Williamson, however. Cook's often-stated policy was to handle potential native threats by firing small shot or shooting over the heads of adversaries. Even though Cook himself had not always lived up to his own rules in recent months, Williamson knew the captain would not be pleased at the outcome of this first landing attempt and therefore kept the news from him. When Cook finally learned of the incident, after leaving the Hawaiian Islands, he was predictably angry with his lieutenant; and Williamson, pointing out the differences of opinion among the officers in such cases, asked Cook never to send him on another such mission if he was not free to "act from reason & ye dictates of my own Conscience."

When Cook landed on the island the next day, ignorant of the denouement of the previous day's events, several hundred people were gathered on shore to greet him. They dropped to the ground the moment he stepped off the boat, prostrating themselves before him and remaining in that position until he made signs to them to rise. Later, when Cook walked some distance around the island, everyone he encountered behaved in a similar fashion. Nothing of the kind had ever happened in any of Cook's previous encounters with native peoples, and Burney, who knew about the killing of the previous day, attributed such submissive behavior to fear of English power. Cook later concluded, correctly, that it was simply the way they behaved toward their "great chiefs."

The Englishmen had no difficulty getting fresh food at Kauai. Indeed, Clerke noted that provisions were a bargain; a medium-sized nail would buy enough pork to feed his entire ship's company for a day. He attributed the cheapness of food to the islanders' eagerness for iron, but he had forgotten the pattern of trade with other islands; generally, the longer people had been acquainted with Europeans, the higher their prices, as they came to understand their visitors' dependence on local food and their eagerness for local crafts. Unlike other Polynesians, however, the residents of Kauai were remarkably honest in their dealings. After the first day at anchor, there were no further incidents of theft.

Concerned about introducing venereal disease to Kauai, Cook ordered several men who showed symptoms to remain on board ship. He was not optimistic about the prospect of enforcing his edict, however. A similar policy in the Tongan Islands had failed; not only did the diseased men do whatever they could to circumvent Cook's orders, but, he had finally realized, it was impossible to determine precisely who carried the disease. Some men who appeared to be cured were not, and others successfully concealed their symptoms. Nor did the Hawaiian women help the situation, for they were even more eager than most island women for the attentions of English sailors, trying every possible inducement to entice the Englishmen into their homes, sometimes even attempting force. The sailors, of course, cooperated; those who had shore leave smuggled women aboard ship for their comrades confined to their quarters, occasionally going so far as to disguise the women in men's clothing.

After several days at Kauai, a man and woman who were obviously highly respected chiefs—the English predictably dubbed them the "king" and "queen"—visited the *Discovery*. A handsome young couple in their early twenties, they commanded an extraordinary degree of deference from their fellow islanders. As the couple approached the ship, most of the islanders on board jumped into the bay, and the rest fell prostrate on deck, hiding their faces. The young man and woman boarded the *Discovery* but would not immediately go beyond the gangway, apparently because their attendants were concerned about their safety. The gregarious Clerke, eager to assure the "king" of his good intentions, shook his hand and clapped him on the shoulder, only to have the young man politely withdraw his hand and ask not to be touched. His followers, Clerke noted, showed great care in helping the chief in and out of his canoe and into the ship "as tho' a drop of salt water wou'd have destroy'd him." Not even at Tongatapu had the English seen such extreme deference paid to a chief.

The Englishmen resumed their journey north early in February after two weeks at Kauai—ample time to establish that the Englishmen's first impressions were correct: these islanders were clearly related to the Polynesians of the South Pacific. Although darker than the Tahitians and more serious in demeanor, in English eyes, they were otherwise similar in physical features, language, many of their crafts, and their affinity for the sea. Cook and his crew were by now accustomed to finding Polynesians widely scattered across the Pacific, but their latest finding astonished them nevertheless because it was so far distant from all the islands they had seen previously. With his discovery of the Hawaiian Islands, Cook had visited the extremities of Polynesian settlement—Hawaii in the north, New Zealand in the southwest, and Easter Island in the southeast—and, with the exception of Samoa, all the major island groups within it.

Five weeks after leaving Kauai, the Englishmen caught their first sight of the North American coast, at a point that is now northern Oregon. Snow-covered mountains in the distance offered a forbidding prospect compared with the tropical islands they had so recently left, and a spell of stormy weather inspired Cook to name their first landfall Cape Foul Weather. Two days later they passed the spot where several creative mapmakers, inspired by the account of an early seventeenth-century Spanish explorer, had located the entrance to a strait across the continent. Cook, noting that the Spaniard mentioned a river and nothing more, confirmed that the break in the coastline was just that and not the beginning of a northwest passage. His skepticism about finding such a passage in the temperate latitudes mounted as the ships proceeded north, especially since he saw nothing at the latitude where the famous strait "discovered" by Juan de Fuca supposedly lay. (Ironically, there is in fact at that point a broad strait, today named after de Fuca, but the *Resolution* and *Discovery* sailed past it at night.)

Another week—it was nearly two months since the Englishmen had left Hawaii—finally brought the ships to a good harbor, in some ways reminiscent of Queen Charlotte Sound: a large, sheltered bay dotted with small islands, surrounded by hills thick with trees and shrubs. One sailor, taken with the area's fertile land and temperate climate, thought it would be "a very desirable place to form a British Settlement." Although the Englishmen did not know it at the time, the harbor lay on the west coast of a huge island, today known as Vancouver Island after the *Discovery*'s midshipman who would later return in command of his own ship to explore this area more fully. Cook called it Nootka Sound, mistakenly thinking that Nootka was its native name.

Although no houses were visible on the shore, the ships had not been long at anchor when men in canoes paddled toward them. Soon about thirty boats clustered around the ships. By now the Englishmen were accustomed to being greeted by men in canoes, but this welcoming party was altogether different from those of the Polynesian islands. The men were short and slovenly in appearance, their canoes small and simply constructed, their language harsh in tone to English ears. The men were eager to trade, but the goods they offered were not those the Englishmen were used to seeing, and they brought no women with them. Although the men would not board the ships, they showed no surprise or fear but greeted the English with what Clerke described as "a long harangue in the strangest language & accompanied with the strangest gesticulations I ever yet heard or saw." Formalities over, the men proceeded to strike up trade with their visitors.

They offered animal skins, fishhooks, weapons (including clubs shaped much like the New Zealand *patu*), and carved masks of human faces and animal heads, but very little food; early spring in the Pacific Northwest was a time of scarcity. In return, the men at first accepted anything the English offered, but quickly displayed a strong preference for iron. Unlike the other Pacific peoples that the Englishmen had met, the Nootka had iron tools and were well aware of the metal's usefulness. Clerke thought this proved they had been in previous contact with Europeans, or at least with other tribes who had encountered Europeans, which would also explain why they seemed so unconcerned about the English ships' appearance in their bay. In fact, it is unlikely that these men had experienced any direct contact with Europeans, although at least one Spanish explorer had coasted the area and might have stopped briefly in or near Nootka Sound. Farther north and west, however, the Russian fur trade was well established, and the people of Nootka Sound probably knew about Europeans through contacts with other coastal tribes.

Within a day the Nootka were visiting and trading with the English completely without reserve. Language proved no obstacle; "we soon became perfectly acquainted with each others inclinations," Clerke remarked, "tho' our conversation was as perfectly unintelligible to both parties as tho' we had no such faculty as articulation among us." After the first day, the men showed no reluctance about boarding the ships, nor did they hesitate to help themselves to whatever they wanted. "They were as light-fingered as any people we had before met with," Cook remarked, although they were more selective in their theft, taking mostly iron, and were capable of inflicting more serious damage to the ships, since they had knives and could cut the iron implements they wanted from the ropes that fastened them. Among other things, they managed to carry off a fishhook weighing over twenty pounds and stripped the boats of virtually all their iron hardware. But theft did not become a source of serious conflict at Nootka Sound, because the native men were quick to point out the offender when something was missed, and the English were usually able to retrieve stolen objects without too much trouble. More significant, the Englishmen did not set up camp on shore and consequently all contact with the local people was confined to the ships, where their actions were easier to control.

Apart from their skill at theft, the Nootka were altogether unlike the South Pacific islanders. The English found them ugly—small, with flat faces, badly proportioned bodies, and large, bony knees and ankles. Worse yet, they were dirty. The Pacific islanders bathed daily, sometimes more than once, while the Nootka seemed never to bathe, and compounded their natural odors by rubbing themselves all over with animal blubber and ornamenting their faces and bodies with paint and soot, which they laid on carefully, using pieces of polished slate for mirrors. This combination of dirt and ornament made it impossible for the Englishmen to engage in their customary practice of placing all new people they met into a color hierarchy. Eventually, Burney wrote, they found some who were clean enough that they appeared little darker than the "country people" of England.

The Nootka's hair was long, thick, dirty, and infested with lice, which they picked from their scalps and ate. Sometimes they sprinkled their hair with white bird down, which the English likened to Europeans' powdered wigs. Earrings and copper ornaments dangling from strings running through holes in their noses completed their bizarre appearance. They had a curious habit of supplementing their speech by jumping and nodding their heads vigorously while they talked, but even so they appeared lethargic and dull by comparison with the Society Islanders, displaying little interest in the English and their habits and none of the Polynesians' eagerness to please their visitors and adopt them as friends. Samwell remarked of one chief that he appeared "stupid" and "insensible of what we wanted of him, behaving more like a brute animal than a rational Creature, differing in this very much from the South Sea Islanders who easily accomodate themselves to Strangers & betray none of this awkward bashfulness & Timidity." What Samwell described as stupidity, however, was not really lack of intelligence, but lack of interest in the English and their ways.

As much as the English denigrated the Nootka, Cook and his companions admitted they had some striking qualities as well. Their music was surprisingly pleasing; they often sang while paddling around the ships, beating time by striking their paddles against the sides of the canoes. Their carved wooden masks displayed extraordinary artistry and craftmanship, and they knew how to work iron into useful implements without benefit of heat. The Nootka were also skillful traders, quickly developing a ritual for their dealings with the English. Each day, several canoes would approach the ships and paddle rapidly around them; then one man, presumably a chief, would rise with a spear or other weapon in hand, often covering his face with a mask, to make a speech. Sometimes a song followed; then the trading could begin. The individuals who made up the trading party changed, but the procedure did not. The Nootka drove hard bargains, usually demanding iron or brass in payment, and were not above using tricks to command higher prices. Samwell recorded an instance when one man offered a bird mask for sale, while another used a small whistle to imitate bird calls in such a way that they seemed to be coming from the mask. The buyer was taken in, paying more than the usual price. The men also expected payment for what the English had been accustomed to take free, including wood, water, and grass for the animals. "I have no where met with Indians who had such high notions of every thing the Country produced being their exclusive property as these," Cook wrote. Clerke took to calling the Nootka "our Landlords."

Most significant, the men who had first greeted the English were determined to monopolize trade for themselves. Five days after the ships arrived in Nootka Sound, Cook observed a group of men arming themselves near the spot where a shore party was gathering wood and water. Cook ordered his men to retreat and take up their weapons; observing these moves, the Nootka men immediately made signs to indicate that they had no hostile intentions against the English, but were protecting themselves against native enemies. Later that day, a dozen large canoes entered the bay, and the two groups engaged in what appeared to be violent and angry speeches. These newcomers seemed to want to approach the English ships, but the Nootka Sound men would not allow it. Neither side resorted to violence, however, and after several hours of haranguing each other the visitors left. After this affray, Cook and others recalled similar incidents when strangers had entered the bay and been sent away by the local residents; "our first friends . . . seemed determined to ingross us intirely to themselves," he remarked. In fact, over the next several days, the Nootka permitted other tribes to trade with the English, but only when one of their own number acted as intermediary. In this fashion, Cook observed, they prevented the English from bidding one group against another, thus keeping the price of native goods high while the value of English goods decreased steadily. Cook also believed that the Nootka men resold English goods to other tribes; they were sometimes gone four or five days at a time, returning with a fresh supply of skins and artifacts.

After a week at Nootka Sound the needed repairs to the ships were almost completed, when the carpenters discovered that part of the *Resolution*'s mainmast had rotted. Four days later, the mizzenmast broke in a storm. Fortunately there were ample supplies of wood available, but it meant remaining several days longer while repairs were completed.

The delay allowed an opportunity to learn more about the Nootka. Up to this time, the Englishmen had seen none of the local women, although a few sailors had tried to communicate their willingness to pay for sex. Eventually a group of men brought three women to the ships. The women were just as ugly and dirty as the men, in the Englishmen's estimation, and had ornamented themselves with an extra measure of greasy red paint especially for this occasion. Not to be put off by mere appearances, some of the younger officers procured washtub, soap, and water, and scrubbed their visitors thoroughly, to the young women's astonishment. Few of the sailors were interested in the women even when clean, but those who were traded their pewter plates, the usual asking price for a night's pleasure, until some of the men had no more utensils for their food. (All the sailors had long since used up their personal supplies of hatchets and iron tools in the South Pacific.) The women were "modest and timid," obviously submitting to this prostitution only because they were forced by the men who had brought them to the ships.

The Englishmen assumed the Nootka were offering their daughters or sisters, but recent scholars have suggested that the women were in fact servants or slaves; Nootkan sexual attitudes were much more restrictive than those of the Polynesians, and it is unlikely that men would have permitted family members to engage in prostitution. Indeed, it seems clear that they would not have offered women in trade at all had not the English persistently asked for them. King astutely thought the Nootka had slaves, presumably captives taken in battle, and remarked that they occasionally tried to sell children to the English. But no one connected the possible existence of slavery with the prostitution of women, although Bayley remarked that only the "lower sort" of women were involved.

One sunny day, the first in two weeks, Cook explored the Sound with several of his officers, stopping at a village where they recognized a number of their trading partners and received a cordial welcome. Because the ships were anchored at some distance from the local villages, this was Cook's only opportunity to see the Nootka at home. He visited several houses, which were little more than crude huts, smoky and smelly inside, and found most of the women sewing clothing from a fabric remarkably like the cloth made in New Zealand. Cook also watched the men smoking fish, the staple of the Nootka diet.

In their excursion around the Sound, the Englishmen visited five villages with a total population of about 2,600, according to King's estimates. Based on these brief visits and the meetings on board ship, King felt reasonably certain that the Nootka lived in independent communities, with rivalry among them that sometimes led to war. They appeared to observe little distinction in rank, although some men were clearly chiefs. Beyond this, King could say little. He was cautious in his description because he believed it difficult to describe cultures very different from one's own; as he explained, "few of us are capable of seperating the invariable and constant springs by which we are all mov'd, & what depends on education & fashion." This problem, compounded by the language barrier, meant he could describe the Nootka only on the basis of their outward actions, and who could say whether their behavior in the presence of the English was typical? King, more than his colleagues, understood that the very presence of the English might shape native behavior; he observed that the Nootka men had ceased their normal, everyday pursuits when the English arrived. King also understood that similar behavior could elicit different responses from different men: one observer might describe the Nootka, for example, as "Sullen, Obstinate & Mistrustful" and another as "docile, good natured & unsuspicious." Most agreed, however, that the Nootka were courteous, brave, fair (except for the occasional theft) and firm in defending their rights. Usually calm, they could be quick to anger if they thought they had been insulted, and they showed no fear of the English or their weapons. For these reasons, the Englishmen respected them, and the two groups got along remarkably well during the four-week stay. Even theft ceased to be a problem after the first day or two. If there was none of the affection that the English felt for the Tahitians, there were fewer problems too. As one officer remarked, "we sailed out of this place with less mischief and less things stolen from us by the natives than Ever we have done before at any place."

From Nootka Sound Cook continued north, attempting to stay close enough to shore to chart the coastline, although bad weather forced the ships out to sea for days at a time. He was using as guides two books on Russian discoveries in the North Pacific: an account of Bering's voyages by Gerhardt Müller, translated into English by the geographer Thomas Jefferys, and a description of more recent Russian voyages in the Arctic region by Jacob von Stählin, translated by Matthew Maty, secretary of the Royal Society. Unfortunately, the two accounts agreed in few respects. Stählin's book included a map of what he called "the New Northern Archipelago," showing Alaska as a large island between the North American and Asian continents with a series of smaller islands to the southwest. Müller's map, on the other hand, showed a great bulge in the North American continent stretching almost to Asia. Although the map was ostensibly based on Bering's own observations, his landfalls along this coast had been far enough apart to cast some doubt on the chart's accuracy. Cook hoped Stählin's map was correct (it was, after all, based on more recent information than Müller's), since it showed a wide passage into Arctic waters between North America and Alaska, but little of what he could observe, when it was clear enough to see anything at all, matched either chart; and the disjunction between charts and reality became more serious as the ships continued north.

When the Englishmen came upon a broad inlet around the middle of May, Cook hoped it would prove to be Stählin's passage. The timing was fortunate, as the *Resolution* needed repairs to its hull, and Cook decided to anchor long enough to complete the work and take on a supply of wood and water. The fog was much too dense to permit much investigation of the inlet, however, and when the skies finally cleared, five days later, it was immediately obvious that the Englishmen were not at the entrance to a transcontinental strait but in the midst of an enormous sound surrounded by mountains. Cook sent boats to explore two arms of the sound on the chance that they might lead to a passage inland, but given the appearance of the land and the nature of the tides, he was not surprised when both parties returned without success. After naming the area Sandwich Sound, Cook directed the ships to proceed along the coast in search of a more promising opening to the north.

Sandwich Sound (now known as Prince William Sound) lies on the south-central coast of Alaska, southeast of the modern city of Anchorage. Up to that point, Cook and his crew had been sailing generally northwest; beyond Prince William Sound, the coastline shifts abruptly, trending southwest. This change in direction was discouraging to Cook and his men, who expected to follow the coast continuously north until they found a northwest passage or reached the Arctic. Instead, it became increasingly clear as the days went by that the North American continent extended much farther west than Stählin's map indicated, diminishing the chances of finding a passage through it. The innumerable inlets and islands along the coast did offer some hope to those who were determined that a passage must exist, however, aided by the almost constant fog, which obscured the massive mountains inland and permitted the optimistic to think that they might be coasting along a series of islands rather than the edge of a continent.

The most promising sign of a northwest passage came at the end of May, when the ships reached a break in the coastline several miles wide. Exploring further, the Englishmen could see what appeared to be a series of islands to the north, and Cook thought it possible that they might find a passage through them. But another day proved that the weather had played tricks again. The "islands" were actually mountain peaks obscured by fog; the vista revealed when the skies cleared "had every other appearence of being part of a great Continent."

Cook was convinced the inlet offered no hope of a passage across the continent, but, pressured by some of his officers and aware of the criticism he would draw if he left any promising possibilities unexplored, he devoted another six days to sailing north through the ever-narrowing waterway. Eventually the ships reached a point where the inlet branched into several smaller ones, and Cook sent Bligh in one of the small boats to explore the most promising. Bligh returned with disappointing news: the passage narrowed sharply just a few miles north and the water became brackish, suggesting that it had no outlet. His findings and the extent of the mountains in the distance destroyed any lingering hopes Cook harbored that the land between Prince William Sound and his present location was a series of islands. Obviously Stählin's map was wrong and Bering's reports more nearly accurate; the North American continent did in fact extend much farther west than he had been led to believe. (Cook's analysis of his situation at this point was quite accurate. The inlet, known today as Cook Inlet, has no outlet; Anchorage lies at its northern end.)

For the next month the Englishmen continued their damp, chilly, monotonous course along the Alaskan coast. An encounter with several native men who paddled out to the ships broke the tedium briefly; the group boarded the *Discovery,* bowed to the crew in the European manner, and gave Clerke a small box containing a paper written in Russian. Some speculated that it was a message from shipwrecked Russian sailors (no one on board could read Russian, so theories abounded) but Cook thought it more likely a message for the next Russian ship passing by. The incident made it clear that the Englishmen were now within Russian fur-trading territory; more encounters with native men over the next several days confirmed this view, as the men showed no surprise at the sight of large ships and "had acquired a degree of politeness uncommon to Indians."

By the end of June the Englishmen found themselves coasting along a series of islands, and Cook directed the ships north through a narrow passage between two of them. Headwinds and strong tides made it difficult going, but the crew was rewarded after a day's work by the sight of the open expanse of the Bering Sea ahead of them. They anchored briefly on the north side of an island called by its inhabitants Unalaska (one of the islands in the Aleutian chain), where they found more signs of Russian influence—which was all to the good, in the Englishmen's opinion. The Unalaskans were handsome, clean, polite, and scrupulously honest in trade; judging from the crosses around their necks, some had been converted to Christianity. "They are to the last Degree obliging and civil, as intrinsically so, as any degree whatever of civilization, could render them," Clerke thought, and King gave credit to the Russians, "who seem'd to have thought it worth the trouble to Correct their passions, & make them better members of society."

Samwell, however, hired a local guide to take him to a village some distance from the ship, where he saw a side of native life more like that at Nootka Sound. The houses were tiny, fetid, hollowed-out spaces below ground, entered by ladder. "They are excessively nasty & stink most abominably of rotton Fish & other Filth," he wrote; "they appear on the outside like a round Dunghill & the whole Town stinks worse than a Tanner's yard," a consequence of the supplies of fish and whale blubber lying about. But the people were as friendly and honest as those who had visited the ships, and Samwell and his companions were quick to take advantage of their hospitality; "having been used to many strange Scenes since we left England, we spent no time in staring about us with vacant astonishment but immediately made love to the handsomest woman in Company." These women, unlike the Nootka, had no objections to coupling with strangers.

From Unalaska Cook directed the ships east along the northern side of the Alaskan peninsula and then north along the mainland coast. The expedition had followed this course for nearly a month when Cook decided to strike out to the northwest, into the Bering Sea. During the previous weeks William Anderson, the surgeon and naturalist, had become seriously ill. Both he and Clerke had displayed symptoms of tuberculosis for more than a year—Clerke apparently contracted the disease while in debtors' prison in London—and although Clerke remained well enough to continue his command, Anderson declined steadily during the weeks of cold and fog. On August 3 he died. Later that day the Englishmen discovered a large island, which Cook named after him.

Five days after Anderson's death they reached the Bering Strait, which separates the Bering Sea from the Arctic Sea. After sailing across the Strait to the Asian mainland and stopping briefly at a coastal village, they pushed on through the passage and into the Arctic Sea. Cook and his officers were still confused by the discrepancies between their maps and direct observation; as a result they were still not certain whether the strait separated North America and the "island" of Alaska as pictured by Stählin, or Alaska and Asia. The issue remained unresolved for the moment as the expedition sailed confidently north, encouraged at the sight of open sea in all directions. Their optimism was short-lived, however, for six days later the men on the *Resolution* observed "an extraordinary white appearance in the sky near the horizon." The veterans of the second voyage recognized the signs of approaching field ice. Their less experienced companions did not want to believe them, but by afternoon none could deny that they had reached an impenetrable mass of ice.

For the moment the crew diverted themselves hunting walruses or "sea horses," as they labeled the huge animals, which clustered by the dozens on the ice floes. Since the ships were short of provisions, Cook sent boat parties out to kill some of the animals, which weighed as much as twelve hundred pounds by the sailors' estimates. The task proved more difficult than Cook had anticipated, however. Although the walruses appeared to be asleep, at least one on each iceberg was awake, as if designated to warn the others of danger, and when the boats got too close they all splashed into the water. This group spirit was apparent in other ways too. When one was wounded, others came to its aid, and mothers defended their young until they were themselves killed, even to the point of trying to upset a boat that had taken a baby. King was disturbed at the killing of animals with such human affections; it was, he thought, "a cruel sport."

Preparing walrus meat was a complicated business. First the carcasses had to hang for a day to let the blood drain; then they were towed overboard for twelve hours, boiled four hours, cut into streaks, and fried. "Even then," Gilbert said, "it was too rank both in smell and tast to make use of except with plenty of peper and salt." Apart from Cook and Clerke, who professed to like the meat, the crew ate it only because they were ravenously hungry and sick of salt provisions; Cook made certain that the squeamish fell into line by cutting off the daily ration of salt meat. A few men who got sick after eating walrus steaks survived on bread and water until Cook relented and restored their other rations.

Unwilling to give up trying to sail farther north, Cook adopted a strategy he had used several times in the Antarctic, sailing in a zigzag course in an attempt to get around the ice. But when the ships were almost trapped in another mass of field ice six days later, he decided it was too late in the summer (it was now the end of August) to continue the effort. Instead he would find a suitable place to spend the winter "so as to make some improvement to Geography and Navigation" and return to the Arctic the following season, getting an earlier start next time.

The days in the Arctic Sea fairly well convinced Cook that the land west of the Bering Strait was part of the coast of Asia, as described in Müller's account of Bering's voyage, but he was still unclear about whether the land to the east was Stählin's island of Alaska or the North American mainland. On the voyage south from the Arctic Sea he coasted enough of Alaska's west coast to convince himself that it was indeed part of the North American continent, and that Stählin's map was wrong.

Cook learned more about just how inaccurate the map was on the voyage south, when he anchored at Unalaska once again. During this visit the Englishmen met several Russians resident on the island; their leader, a man named Ismailov, studied Cook's maps, pointing out several errors, and gave Cook his own, more up-to-date charts to copy. Cook in return corrected the latitude of some of the islands shown on Ismailov's charts and gave him a quadrant with instructions for its use. The combination of his own observations and Ismailov's information made the multiple defects of Stählin's map so obvious that Cook was convinced Stählin himself must have known of its weaknesses, and he railed at the audacity that had allowed it to be published at all: "What could induce him to publish so erroneous a Map? in which many of these islands are jumbled in regular confusion without the least regard to truth and yet he is pleased to call it a very accurate little Map? A Map that the most illiterate of his illiterate Sea-faring men would have been ashamed to put his name to."

When planning his voyage, Cook had intended to winter at Kamchatka, a Russian settlement on the Asian mainland, but while cruising along the Alaskan coast he decided to sail back to Hawaii instead, in part because of the uncertainty of getting adequate provisions at Kamchatka but more significantly because he did not wish to remain idle six or seven months; in the Hawaiian Islands his men could pass a pleasant winter and undertake a thorough survey of the islands at the same time. With that goal in mind, toward the end of October he and his men left Unalaska on their voyage south, with the pleasurable prospect of a tropical interlude ahead of them.

19. Kealakekua Bay

The *Resolution* and *Discovery* approached the Hawaiian Islands toward the end of November, and morale soared as temperatures warmed and the crew contemplated a winter of plentiful food and willing women. Several miles east of Kauai the group discovered another island, known as Maui; when the local residents, including an obviously high-ranking chief named Kalaniopuu, paddled their canoes out to the ships, it became clear that word of the English visit of the previous winter had spread.

Cook decided not to anchor at Maui, but barter for provisions with the men who visited the ship as it coasted offshore. Still, he took his usual precautions against introducing venereal disease, examining his men for symptoms and forbidding women to board the ships. His orders were useless against the nimble young women, however, and some of the officers who tried to carry out the spirit of Cook's orders by keeping the women at arm's length drew ridicule for their pains. Several island men showed symptoms of disease and appealed to the Englishmen for help in curing themselves, which King took as a sign that they blamed the English for infecting them. Others on board, however, thought it unlikely that the illness could have spread from Kauai to Maui in so short a time, even assuming the English had infected women at Kauai; they concluded that venereal disease must be indigenous to the Hawaiian Islands, and perhaps to other South Pacific islands as well. It was a self-serving (and false) argument that convinced neither Cook nor King. Hawaiians traveled freely among their islands, just as the people of the Society Islands did, and nine months was more than a sufficient incubation period for gonorrhea and syphilis.

As the ships sailed slowly east off Maui's coast the crew spotted another, much larger island; on December 2 they came within canoe range of "Owhyhee" on Hawaii. As at Maui, the local residents brought provisions to trade, and Cook decided to continue sailing offshore, bartering with the islanders as he went along, instead of looking for a place to anchor. Following this plan, he hoped to scout as much of the island as possible, while avoiding the problem they had often encountered at other islands, where the local people showered them with food during the first few days at anchor but then traded erratically. In the Tongan Islands, Cook had been forced to move from island to island to keep his crew supplied with food; now, facing as much as three months in the Hawaiian Islands, he wanted to ensure a steady flow of provisions. The best way to accomplish this goal, he thought, was to remain offshore and establish a carefully regulated trade with the Hawaiians. (Maintaining a steady supply of food had not been a problem on Cook's earlier voyages, because, with the exception of the first visit to Tahiti, he had never stayed long enough in one place to diminish native food supplies severely.)

As a means of regulating trade, the plan worked well. Food was available whenever it was needed, and the Hawaiians traded fairly, sending their wares up to the ships and then going on board to bargain over price rather than insisting on payment in advance, the usual practice elsewhere. They never attempted to cheat on sales or, even more remarkably, commit theft, although they bargained vigorously and would refuse to sell rather than accept a price they thought was too low.

But Cook's strategy was disastrous for morale. His men resented their confinement on board when they were so close to shore, and they claimed that the closely regulated trade meant shorter rations than they were accustomed to in the tropics. When supplies of beer ran out and Cook proposed alternating the daily ration of grog with a "beer" brewed from the local sugar cane, the crew rebelled, sending Cook a petition claiming that the beer was "injurious to their healths" and the quantity of provisions substandard for the tropics. Outraged that his judgment on matters of sailors' health should be questioned, Cook called his men on deck and informed them that he had, up to that time, heard nothing about the alleged shortness of provisions, and would rectify it if true. But, he went on, it was "something extraordinary" for them to believe the sugar-cane beer bad for their health when they were perfectly willing to eat the cane raw. His only concern, he told them, was to preserve the grog supply for the coming months in the Arctic, when they would stand in greater need of it; he certainly would not force anyone to consume the beer, but if they refused, they would have nothing to drink but water. Moreover, he considered their letter a "very mutinous Proceeding," and in the future they could expect no more "indulgence" from him. The men agreed to think over Cook's threat, eventually deciding they preferred water to the sugar-cane beer.

Cook blamed the conflict on the narrow-mindedness of his crew, who resisted "every innovation whatever tho ever so much to their advantage," and went on at length about his attention to his men's health; but in fact the accusations and counteraccusations were a reflection of deepening stress on both sides, fostered by two and a half difficult years at sea. About a year earlier, when Cook had proposed eliminating the daily grog ration until reaching the Arctic in order to preserve supplies, the crew agreed without protest. Their resistance now indicated their displeasure at the thought of enduring another Arctic summer when the original plans had called for only one such stint. Cook's angry reaction and subsequent self-justification were further signs that the judgment, balance, and sense of fairness for which he was often justly praised had seriously diminished over the course of the voyage. Whether he was dealing with native peoples or his own crew, the sympathy and sensitivity that set him apart from most sea captains increasingly gave way to demands for unquestioning obedience and rage at anything that thwarted his will. Burning houses and canoes at Moorea, inflicting harsh punishments at Huahine, forcing men to eat walrus meat even after it made them sick, now persisting in an unpopular policy of remaining offshore and cutting off grog besides—all of these decisions contrasted sharply with Cook's temperate, considerate behavior on previous voyages.

Cook had never fully recovered his health after the severe illness of his second voyage. One scholar who has studied the medical evidence suggests that it was caused by a roundworm infestation, which could interfere with the absorption of vitamins, especially the B group, and ultimately lead to serious vitamin B deficiency. The symptoms include persistent ill health, fatigue, loss of interest and concentration, irritability, and depression, all of which Cook displayed to some extent throughout the third voyage.

In addition, there were strong psychological reasons for Cook's increasingly erratic behavior. The voyage had started later than planned, which radically altered its subsequent course. Unable to make it to the Arctic by the summer of 1777, Cook lost the better part of a year in reaching his most important destination; he spent that time cruising the South Pacific in waters that were, for the most part, well known to him. Cook hated to be idle, and long stays at islands he had already visited, however pleasant for his crew, were to him a waste of time. That they created problems in supplying his ships and occasional conflict with native residents only deepened his frustration. Having finally reached Arctic waters in summer 1778, Cook found his maps wildly inaccurate and realized that prospects of finding a northwest passage were slim. In his usual thorough manner, he decided to return for another attempt the following summer, but he had spent enough time coasting ice fields in the Antarctic to know it was extremely unlikely he would find a way through the ice that had stopped him north of the Bering Strait. In short, after two dramatically successful voyages, Cook faced the prospect of failure.

Despite his crew's protests, Cook stuck to his plan of sailing around Hawaii, trading with the islanders as he went along. The Hawaiians were eager to trade—some who didn't want to wait their turn to paddle their canoes up to the ships jumped overboard and swam carrying their goods—and the women were equally eager to sell their favors, despite Cook's efforts to keep them out of the ships. Eventually he relaxed his restrictions, and by the time the ships reached the southern coast of the island, so many women had taken up residence on board that Cook complained they interfered with sailing the ship.

After six weeks of this routine, Bligh returned from one of his missions exploring the coastline at closer range to report that he had found a good harbor, and on January 17 the ships sailed into a broad, shallow bay on the southwestern coast of the island. Cliffs rose sharply from the water's edge at the north end of the bay, tapering off to the south, where a stretch of flat, fertile land offered promise of safe landing, water, and food. A few houses clustered near the shore, and the high land beyond the beach was home to a great many more people, if the hundreds of canoes approaching the ships were any indication. By the time they anchored, King estimated, fifteen hundred canoes holding about nine thousand people surrounded the *Resolution* and *Discovery*. At least three hundred more women and boys swam to the ships, and a few men paddled out on planks. They crowded on board the ships, singing, jumping, obviously delighted with their visitors; the sailors, "jaded & very heartily tir'd" of the weeks aboard ship, were no less pleased with the meeting.

The sheer numbers of Hawaiians posed some problems, however. Dozens who could not get aboard the *Discovery* clung to its side, causing the ship to heel dangerously, and several men took advantage of the confusion to steal until a chief named Parea, who had visited the *Resolution* while it was cruising offshore, proved able to control the crowds. After he calmed the crowds on the *Resolution,* King sent him over to the *Discovery* to get the situation there under control while another chief, Kanina, boarded the *Resolution* to help maintain order. When the decks became so jammed that it was impossible to move, at a signal from Kanina all the Hawaiians jumped overboard. A third chief, Koa, also visited the ship to pay his respects to Cook, bringing gifts of red cloth and a pig.

Later in the day Cook, King, and Bayley went ashore with Koa and Parea. A small group met them on the beach, carrying wands tipped with dog hair and repeating the same sentence over and over. King could recognize only one word: "Erono," the name the people at Kauai had used for Cook. The Hawaiians led their visitors to the south end of the beach, where twelve carved wooden images were ranged in a semicircle around a cluster of hogs, sugar cane, breadfruit, and coconuts on a platform between two small buildings. Koa draped several yards of red cloth around Cook, presented him with a hog, and led him to each of the images in turn. The circuit completed, Koa prostrated himself before the center image, kissed it, and instructed Cook to do the same—"a long, & rather tiresome ceremony," in King's estimation. A ceremonial feast followed. Afterward Cook and King walked through the village, where the inhabitants fell to the ground as Cook approached, just as they had at Kauai, and performed a simpler version of the original welcoming ceremony at each house he visited.

Throughout these ceremonies the Hawaiians continued to address Cook as "Erono." Samwell thought the label was a title or rank, the highest the Hawaiians observed and one that "is looked upon by them as partaking something of divinity." The purpose of the welcoming ceremony, he believed, was to confer this title on Cook. In fact "Erono" was not a title, but the name of a god, Lono; the Hawaiians believed Cook was his incarnation. The rites directed by Koa, at a sacred place called a *heiau* (equivalent to the Tahitian *marae*), were intended to introduce Cook to other gods, represented by the wooden figures, and acknowledge his authority.

The Hawaiians continued to show obeisance to Cook whenever he stepped ashore, and as the crowds got larger the spectacle became ludicrous to the Englishmen. Whenever Cook approached, everyone fell to the ground; as soon as he passed, they picked themselves up and followed him. But if he turned around, or even looked over his shoulder, they instantly fell prostrate again; apparently it was *tabu* (*kapu* in Hawaiian) to look upon his face. Some who didn't react fast enough but stayed on the ground a few seconds too long were trampled by others rushing to keep up with Cook. Eventually the Hawaiians adopted a compromise system of walking on hands and feet, which proved easier than constantly shifting from a prone to an upright position. Clerke, as captain of the *Discovery*, got similar treatment at first, but as he later explained, "I disliked exceedingly putting so many people to such a confounded inconvenience, and by application to the Arees I got this troublesome ceremony taken off." Cook, on the other hand, enjoyed it and even went so far as to have some of his crew carry him on their shoulders when he visited the village.

The rest of the Englishmen enjoyed no such marks of respect, and the Hawaiians continued to crowd the ships unceremoniously every day. When they became too numerous or too rambunctious, Parea would be pressed into service to clear out the ship, which he accomplished by throwing stones at the canoes as they approached. The islanders were generally well behaved, however, and theft was only rarely a problem, with one exception: the Hawaiians were so eager to get iron that they sometimes pulled nails from the ships' bottoms. The sailors, who learned quickly that nails were the best way to buy women, also took nails surreptitiously, as supplies of trade goods were running low. Between the islanders and the crew, the two ships might have suffered the problems of the *Dolphin* had not the officers learned to keep close watch. For those stationed on shore, however, the situation was quite different. The Englishmen, with the Hawaiians' permission, had selected a spot for their observatory that was considered sacred ground, and no one except the priests, directed by a man named Kao, would come near them at their waterside encampment. (The Englishmen in time dubbed Kao "the bishop.") Some of the sailors tried to bribe a priest to let women come to their tents at night, but he refused, saying that any women caught on sacred land would be killed.

A week after the ships anchored in this bay, called Kealakekua by the Hawaiians, a double canoe about sixty feet long, one of the largest the Englishmen had ever seen, came toward the ships. Two smaller canoes followed it. In the larger canoe they recognized Kalaniopuu, the elderly chief who had greeted them off the coast of Maui, with several attendants, including his young nephew Kamehameha. All were dressed in long cloaks and helmetlike caps made of red and yellow feathers and carried poles topped with feather ornaments called *kahili.* (The Englishmen referred to them as "fly-flaps.") After presenting Cook with gifts of hogs and vegetables, they invited him to go ashore with them. On the beach Kalaniopuu removed his cloak and placed it over Cook's shoulders; then he piled up half a dozen more at Cook's feet.

The splendor of Kalaniopuu's dress, the size of his entourage, the solemnity of his visit to the ships, the lavishness of his gifts to Cook, and the obvious deference paid him by the Hawaiians all made it clear to the English that this "old immaciated infirm man," as King described him, was the highest-ranking chief of the island. Like the chief who visited the *Discovery* at Kauai, Kalaniopuu bided his time before calling on the Englishmen and orchestrated his visit with great pomp and dignity. In doing so, both chiefs displayed some ambivalence about the strangers in their midst; the Englishmen were potential rivals, and yet, as gods, they were sources of *mana,* or sacredness. They were also enormously popular with the Hawaiian commoners. It was therefore imperative for the chiefs to greet Cook, both to take charge of the situation and maintain status among their own people, and also to share in the Englishmen's *mana.*

The prestige Kalaniopuu enjoyed among his subjects and the

control over them exercised by lesser chiefs like Parea and Koa made a deep impression on the Englishmen. Clearly the Hawaiian social and political system was similar to those of the Society and Tongan Islands, in that the "aree" (*alii*) constituted the highest class, with a small group within that class (*alii nui*) holding highest status, much like the *arii rahi* in Tahiti; below the *alii* was a group of intermediate rank (*konohiki*), who governed local districts, and below them, a large class of commoners (*makaainana*). The English thought the Hawaiians also had a class of servants or slaves, which was probably true; a group of people called *kauwa* were outcasts, a status that was hereditary, like all other ranks in Hawaiian society. As in the Tongan Islands, the highest chiefs could trace their lineage back several generations. Kalaniopuu, King learned, was the fifth of his lineage. But the authority of the Hawaiian chiefs—as indicated especially in their ability to control the behavior of their people and their freedom to commandeer food as gifts for the English—appeared even greater than that of the Tongan rulers, and the English gave them credit for keeping the weeks at Kealakekua Bay untroubled by major theft or other hostilities.

The Englishmen were correct in their perception that the Hawaiian chiefs exercised extraordinary power—the Hawaiian Islands were the most highly stratified of all the Polynesian societies—but their social system was far more complex than the English imagined. The *alii nui* controlled all land on the island, which was divided into districts under the control of lesser *alii*. They in turn divided their lands into tracts managed by the *konohiki,* who parceled out small holdings to the commoners. This hierarchy controlled both land and the right to use the ocean up to one and one-half miles from shore and the water generated by an elaborate irrigation system. Men at any level could call upon individuals below them to contribute labor for communal projects, such as building houses, canoes, and irrigation ditches; they could also dispossess lower-ranking men from their land, although in practice dispossession was not done arbitrarily, but as punishment for refusal to use the land productively or contribute labor. Similarly, food and other goods needed for gifts and ceremonial occasions were collected from the commoners by the *konohiki* and then passed up the hierarchy.

The Hawaiians' respect for Englishmen (and for English property), their chiefs' ability to keep them in line, and their exuberant spirits and good-natured generosity endeared them to the English, even though they appeared less attractive physically than the Tahitians (being generally smaller and darker) and had an unfortunate addiction to *kava,* an intoxicating drink made from the root of a plant of the same name. (*Kava* was consumed all over the South Pacific, but nowhere in such great quantities as in Hawaii.) King, especially, admired the Hawaiians' agricultural skills and their capacity to grasp ideas that were foreign to them; they asked cogent questions about English civilization, and several watched the armorer at work with a degree of attention that showed they were trying to understand the processes by which he shaped iron. One man was later seen heating a piece of iron over a large fire, using two stones as an anvil and hammer.

Despite the Hawaiians' generosity to the English (or perhaps because of it), after about two weeks Kalaniopuu pointedly inquired of Cook how much longer he and his men planned to stay. He seemed pleased to hear that Cook intended to leave soon, but shortly afterward asked him to leave King at Hawaii. (The Hawaiians assumed that King was Cook's son.) Others spoke to King, promising that he would be a "great man" among them. King was not surprised that the Hawaiians wanted him to stay when they were so eager for the rest to be gone; similar efforts to recruit particular individuals as permanent residents had been made on other islands too, "often from no better motive than what Activates Children, to be possess'd of a curious play thing." He might well have added that the islanders' motives were similar to those of Englishmen who wanted to take a Polynesian home with them. The Hawaiians had another motive, too, however; by keeping at least one of the Englishmen among them, particularly one that they perceived to hold high rank, they would share in the benefits of his *mana.*

The Englishmen left Hawaii during the first week of February, but had barely reached the open sea when a sudden storm blew up, seriously damaging the *Resolution*'s masts. Cook debated whether he should return to Kealakekua Bay or trust to finding a suitable harbor in the islands farther west; the former alternative seemed the more prudent, and a week after taking their leave, the Englishmen sailed back into Kealakekua Bay.

The Hawaiians were not pleased to see them. Indeed, few even bothered to greet the ships as they returned, which "in some measure hurt our Vanity," King remarked, "as we expected them to flock about us, & to be rejoiced at our return." English feelings were mollified somewhat when they learned that Kalaniopuu had ordered that no one could go to the ships until he had visited, and after he made his appearance others quickly followed. But prices for provisions and other goods went up substantially; Kamehameha visited the ships wearing an elegant feather cloak that he was willing to sell only for the exorbitant price of nine iron daggers. Cook thought the Hawaiians feared that the English intended to settle on their island, and he tried to reassure them by pointing to the broken mast. Heinrich Zimmermann, one of the sailors on the *Discovery,* had a different explanation; shortly before leaving Kealakekua Bay, Cook had taken poles marking off a sacred area to use as firewood. He obtained the chiefs' permission first, but Zimmermann thought the commoners had been resentful of Cook's action. In addition, William Watman, an elderly seaman, died just before the ship's departure, which Zimmermann thought must have destroyed the Hawaiians' belief in the Englishmen's sacredness.

Subsequent events suggested there was some truth in Zimmermann's explanation. An officer sent ashore to get water reported that he paid several islanders to help, only to have a chief stop them; instead of helping, they pelted the watering crew with stones. Clerke complained that theft, a minor nuisance during the preceding weeks, was becoming a serious problem, such that "every day produc'd more numerous and more audacious depredation." In one of the more serious incidents, a Hawaiian on board the *Discovery* stole the armorer's iron tongs, jumped overboard, and made for shore in a canoe. Clerke ordered muskets fired over the thief's head and sent Edgar and Vancouver in pursuit. The sound of musket fire alerted Cook and King, who were working at the observatory on shore; seeing a canoe paddling furiously toward the beach followed by the *Discovery*'s boat, they guessed what had happened and, taking two marines with them, went tearing off—Cook ran so fast that King had trouble keeping up with him—in an attempt to intercept the canoe when it hit the beach. At about the same time, one of the *Resolution*'s small boats also joined the chase.

The thief had escaped by the time the two boats reached shore, but another man handed the tongs to Edgar, who then decided to seize the canoe in retaliation despite the recovery of the stolen goods. The Hawaiians resisted, throwing stones at Edgar and

Vancouver until Parea forced them to stop. When Parea turned away, however, the mob knocked Vancouver down. The boat's crew, having left the ships in such haste that they had not brought guns with them, were forced to retreat to the ships—with some difficulty, since all but one of their oars had been broken in the fray. Cook and King, meanwhile, lost sight of the canoe, but met a group of Hawaiians who offered to guide them to the thief. After about three miles it became clear that the men were deliberately leading them astray. Cook stopped occasionally to inquire about the thief, threatening to have one of his marines shoot if the Hawaiians did not turn over the culprit immediately; after he tried this tactic a few times, however, the islanders merely laughed at the threat.

Back on board ship, Cook was incensed when he learned that his men had attempted to seize a canoe unarmed. Such an incident only increased the Hawaiians' confidence, he thought, and now he would have to display force to keep the upper hand. As a start, he ordered all the islanders on board to leave; later, when the thief was finally apprehended, Cook had him flogged. Such measures had no deterrent effect, however. The next morning, one of the *Discovery*'s boats was missing. This was an extremely serious matter, especially coming as it did on the heels of the previous day's theft and the generally provocative behavior of the Hawaiians. Cook decided to employ simultaneously two of his favorite strategies for dealing with native theft: he sent several boats to the entrance of the bay to detain any canoes attempting to leave and then went ashore accompanied by several marines to find Kalaniopuu, with the intention of persuading him to return to the *Resolution* as a hostage. If he found that the islanders had fled, Cook planned to retaliate by burning houses and seizing the large canoes on the beach. Clerke remained on board the *Discovery*, too ill to accompany Cook ashore; King, on Cook's orders, went to the observatory, where the priests, aware of the seriousness of the situation, asked anxiously what Cook was planning to do. King reassured them, saying that Cook was angry about the theft of the boat but that no one would be hurt.

When Cook and the marines landed on the beach, dozens of Hawaiians flocked around them as usual, either unaware of or unconcerned about the theft. Kalaniopuu's sons offered to escort him to their father, who agreed after some hesitation to accompany Cook to the *Resolution*. As they prepared to walk back to the beach, two Hawaiians arrived with the news that men in the *Discovery*'s boats had killed a chief on the opposite side of the bay. This intelligence angered the crowd gathered around Cook, which by this time amounted to several hundred people, and some began collecting stones. Kalaniopuu still appeared willing to return to the ship, however, and the two men walked hand in hand to the beach after Cook ordered the marines to march ahead to the water's edge so the Hawaiians couldn't cut off their retreat to the boats.

When the entire group arrived at the bay, a woman ran up to Kalaniopuu, crying and pleading with him not to go to the ship, telling him he would surely be killed. He was clearly shaken, although he made no attempt to leave Cook's side, and the crowd grew more restive. Phillips, the leader of the marines, observed another chief hiding a dagger under his cloak. Fearing that the man intended to stab one of the Englishmen, Phillips asked Cook for permission to shoot, which Cook—"ever too tender of the Lives of Indians," Samwell thought—denied. Then another Hawaiian raised his arm as if to throw a stone. Phillips, convinced by now that their lives were in danger, urged Cook to retreat, but instead Cook fired at the man and then knocked him down with the butt of a musket. When a second man threatened to throw a spear, Cook shot at him and missed, killing instead the man standing next to him. At this show of force the Hawaiians retreated slightly, but rather than dispersing as Cook expected them to do, they began throwing stones and drawing their daggers. The marines fired, followed by the men in the boats waiting offshore, but Cook gave orders to stop. Then he signaled the boats to come in closer, as he and the marines waded into the shallow water to meet them.

When Cook turned his back to the crowd, a Hawaiian hit him on the back of the head with a club and then struck him in the neck with a dagger. He fell into the knee-deep water, and several people forced his head down, beating him with clubs and stones, while others ganged up on the marines. The men in the smaller of the two nearby boats rowed closer and pulled in Phillips and the other three marines who had managed to swim toward them; then Phillips jumped back into the water to rescue a fifth man, who was badly wounded. The second boat, commanded by Williamson, had mistaken Cook's signal for an order to move away; by the time Williamson realized his mistake, he decided it was hopeless to attempt any assistance. The crew of the smaller boat managed to drive the Hawaiians back with their gunfire, but by then it was too late. Cook and four marines were dead. The small boat, overcrowded with the survivors, returned to the *Resolution.* Williamson thought about trying to recover the bodies, decided against it, and returned to the ships also. The Hawaiians dispersed, taking the five bodies with them.

From the *Discovery,* Clerke heard the shots and watched the attack through his spyglass, but he was too far away to see exactly what had happened. Not until Williamson came on board did he know that Cook was dead. On the *Resolution,* according to Gilbert, "a general silence ensued throughout the ship, for the space of near half an hour: it appearing to us somewhat like a Dream that we could not reconcile ourselves to for some time. Greif was visible in evry Countenance . . . for as all our hopes centred in him; our loss became irreparable and the sense of it was so deeply Impressed upon our minds as not to be forgot."

Despite Cook's increasingly rigid and distant demeanor during the voyage, most of his men felt a deep respect and affection for him, and his death was a devastating blow. The crew regarded him as stern, reserved, often distant, and subject to occasional violent fits of anger, jokingly known among the crew as "heivas" because Cook in his rages reminded them of the more vigorous of the Pacific islanders' dances. The slightest contradiction from an officer could bring forth one of these displays. But Cook's men also respected him for his ability and courage. His navigational sense was so acute that he sometimes ordered the crew to change course before they noticed anything amiss; when danger could not be avoided, Cook not only was impervious to fear himself, but, by the force of his example, maintained calm throughout the ship.

Perhaps more important, he conveyed a genuine sense of regard for his men, whatever their rank. George Forster thought Cook gained the affection of his men in part because he did not allow himself any comforts denied to his crew—when rations were short, Cook and the officers ate no more and no better than the sailors—and in part because of his obvious concern for their health and well-being, which did not go unnoticed among men who had served under less enlightened captains. Such gestures as allowing an occasional party or giving up his private quarters when the weather was bad so the sailmaker could work more comfortably "won for him the hearts of these rough and tough fellows, who had seldom been treated like this before." Forster also noted that Cook never exposed his men to dangers that he was unwilling to face himself, and was always in the first group to go ashore at a new, untested location. And there were occasions when Cook let down his reserve. Trevenen remembered fondly the few times when he and other young officers accompanied Cook on an exploring party in one of the small boats; the captain would sometimes relax and "converse familiarly with us."

Both feared and loved, Cook was a kind of father figure to his crew and the guiding spirit of the expedition. "We all felt we had lost a father," Zimmermann wrote; ". . . the spirit of discovery, the decision, and the indomitable courage, were gone." And Trevenen remarked that "I, as well as the others, had been so used to look up to him as our good genius, our safe conductor and as a kind of superior being, that I could not suffer myself—I could not dare to think—he would fall by the hands of the Indians, over whose minds and bodies he had been used to rule with absolute sway." Trevenen pointed to the irony of Cook's death: he had enjoyed extraordinary popularity among Pacific islanders, and had established a reputation for fair dealing with them. That he might be killed at their hands was unthinkable. That Hawaiians, who had shown more reverence for Cook than any other people they had encountered, should become his murderers was the final irony.

The people Cook encountered on his voyages revered him for much the same reasons that his crew did: he got across his sincere affection for them, treated them with respect, and was never condescending. He had an "instinctive knowledge" of how to please them, according to one crew member, or, as George Forster put it, "his mind, which was free from the fetters of prejudice and his respect for human rights inclined him towards mercy and forbearance"—in contrast to the "contemptuous zeal" of many of his companions (and predecessors among explorers), who viewed native peoples as savages to be suppressed with force at the least sign of resistance to English wishes.

But with the Pacific islanders as with his crew, Cook's behavior was often contradictory, mixing sympathy and respect with demands for unquestioning obedience and fits of anger. The most distinctive features of Cook's personality, mentioned over and over by those who knew him, were his respect for his fellow men and his overpowering determination and persistence. Both, it may be argued, were rooted in his exceptional self-confidence, the sort of confidence that allowed him to write that he had gone as far as any man would go in the Antarctic and had "done with the Southern Pacific Ocean" after the second voyage. Convinced of his own worth and of the value of his work, he had no need to profess superiority over others and no hesitation about pushing himself and his crew to extreme limits in pursuit of his goals. Such conviction and sense of purpose made possible feats like charting the east coast of Australia and circumnavigating the Antarctic, but they also fostered a cocksure attitude that brooked no questioning of his decisions and handled challenges or setbacks with ill grace. Most of the time Cook's concern for others and sound judgment kept his boundless self-assurance and legendary hot temper in check; King wrote perceptively that "His temper might perhaps have been justly blamed, as subject to hastiness and passion, had not these been disarmed by a disposition the most benevolent and humane." But on the third voyage, benevolence increasingly gave way to anger and impatience, as a result of fatigue, ill health, and mounting evidence that this voyage would fail to achieve its objectives. Even so, Cook's bouts of temper remained aberrations that did nothing to diminish the depth of his men's grief at his death.

Clerke took command of the *Resolution* immediately, turning over the *Discovery* to John Gore, the *Resolution*'s first lieutenant. Then he sent Bligh to evacuate the men working at the observatory on shore, a task accomplished under a shower of stones. None of the Englishmen was hurt, but their retaliatory gunfire killed eight Hawaiians.

Although eyewitnesses to Cook's death disagreed on details, all believed that the attack had not been premeditated, but was rather the result of an unfortunate combination of circumstances: the news of a chief's killing, announced at an inopportune moment; the Hawaiians' reluctance to see Kalaniopuu go to the ship; Cook's killing of two men; and his failure to retreat to the boats sooner. According to Phillips, just before Cook fired, the Hawaiians at the water's edge had moved aside to allow the marines to pass; if they had returned to the boats at that point, he thought violence could have been avoided. But instead Cook shot two men, angering the crowd, and the marines' fire failed to disperse them as it had in other threatening situations and as Cook expected it would again.

The entire episode was another example of Cook's occasionally erratic behavior over the past several months. He persisted in his efforts to get Kalaniopuu to return with him to the *Resolution* in the face of mounting opposition from the Hawaiians; he ignored Phillips's repeated warnings to retreat; and after telling Phillips not to fire, did so himself in an impulsive fit of anger against a man who threatened him. This time, Cook's determination to accomplish his objective—forcing the return of the *Discovery*'s boat by taking Kalaniopuu hostage—was ill-advised in light of the potential dangers, and his angry lashing out at the Hawaiians was very likely the spark that set off their attack.

In trying to understand the reasons for Cook's murder, the Englishmen concentrated on the events of February 14, without considering how they fit into the context of their three weeks at Kealakekua Bay. An English ship captain who visited Hawaii fourteen years later was told that Cook's decision to replenish the ships' wood supply with stakes from a sacred area had been his undoing; his story corroborated Zimmermann's theory that the commoners were angered by the action, even though the chiefs had given their permission. The priests disapproved too, and had predicted that Cook would be killed as a result. This story may not have been true in every detail—the captain who heard it thought the chiefs and priests had been looking for an excuse to kill Cook and used the incident of the stakes to incite their subjects to violence—but in fact Cook's fate was bound up with Hawaiian religious belief, although in a more complex way.

That the Hawaiians viewed Cook as the incarnation of the god Lono is undeniable: they addressed him as Lono, they paid him the homage due a god, and they welcomed him to their *heiau* as Lono. The Englishmen's arrival at Kealakekua Bay coincided with an annual festival, known as the *makahiki,* dedicated to Lono. According to Hawaiian tradition, Lono had been the dominant god of the island until another god, Ku, came from a distant mythical land, built temples, overthrew the chief, and installed a new one loyal to him. Lono was a benevolent and peaceful god; Ku was violent and warlike, and the temples he built were dedicated to the practice of human sacrifice.

Kalaniopuu traced his lineage to the chief installed by Ku, and worship of Ku dominated Hawaiian religious practice at the time of Cook's visit, but the cult of Lono remained an integral part of the island's culture, and many of the priests at Kealakekua Bay were his followers. For several weeks each winter, from about mid-November to early February, the Hawaiians observed the *makahiki* festival, when ceremonies associated with Ku, notably the practice of human sacrifice, were suspended and replaced by the rites of Lono. At the end of the festival, Ku returned, ritually reconquered Lono, and resumed his dominant position until the next year's rites.

The Englishmen happened to arrive off Maui about the time that the *makahiki* festival began. Part of the ritual involved a procession around the island of Hawaii, and Cook's ships proceeded to make a circuit of the island, in the proper direction. Then they anchored at Kealakekua Bay, the center of the Lono priests. The Hawaiians did not make sharp distinctions between god and human, heaven and earth, or the supernatural and natural, and they believed that gods sometimes reappeared in human form. Indeed, legend held that Lono had appeared on the island at other times in the past. So Cook's timely arrival made him appear to be another incarnation of Lono, and his behavior—his acceptance of the Hawaiians' ceremonies honoring him—fit perfectly into their belief.

Kalaniopuu's association with Ku made him a rival to Cook, but as long as Cook played out the role of Lono as temporary visitor, all would go well. And, to the Hawaiians, this seemed to be exactly what happened. The end of the *makahiki* festival and the return of Ku were traditionally marked by a human sacrifice; on February 1 William Watman died. The Hawaiians asked that he be buried in the *heiau,* and Cook concurred. (Again Zimmermann had the right idea but the wrong details in seeing some significance in Watman's death.) Then they began asking when the Englishmen planned to leave. The ships sailed February 4, right on schedule. Cook's promise to return the following year also fit neatly into the ritual.

But Lono was not supposed to return two weeks later, and when he did, he posed an immediate threat to Kalaniopuu. Ku had come from a distant, legendary place, by boat, to usurp Lono's power; now it appeared that Lono was perhaps trying to do the same thing. The encounter between Kalaniopuu and Cook on February 14 was a confrontation between the embodiment of Ku and the embodiment of Lono. In this context, Cook's appropriation of the sacred stakes may have taken on significance—by taking them he violated a *kapu* and could therefore be sacrificed.

Moreover, according to Hawaiian religious belief, the sacrificial victim became incorporated into the god to whom he was sacrificed. By killing Cook, Kalaniopuu appropriated Cook's *mana* or sacredness, in effect combining the qualities of Lono and Ku. Thus while Cook's murder was not premeditated in the usual sense of the term and might have been avoided by more careful handling of the situation, it was not accidental or sparked merely by the conflicts of February 13 or 14. Rather, without knowing it, Cook and the Englishmen had become caught up in the centuries-old legends that defined Hawaiians' religious culture.

The Englishmen had their own rituals to play out: recovering Cook's body and avenging his murder were their immediate concerns. It was obvious to the cooler heads on board that whatever action might be taken against the Hawaiians would have to wait until the bodies were recovered. Because King was a favorite with the Hawaiians, Clerke sent him to attempt this task. King had one boat halt a little distance offshore while he went closer to the beach in a smaller boat, flying a white flag. The Hawaiians on shore had no difficulty understanding this conciliatory gesture; they took off the feather mats that they wore as a kind of armor and sat down, while Koa swam out to the boat carrying a piece of white cloth. With tears in his eyes, he threw his arms around King; but he carried a dagger, and King, who had never trusted Koa, remarked, "I had the caution to hold the point while he was embracing me." Koa promised to bring Cook's body immediately, but disappeared for some time, while other chiefs pressured King to go ashore, meet Kalaniopuu, and fetch the body himself. Meanwhile, the second boat moved close enough to shore that the sailors could talk to the Hawaiians on the beach, who claimed that Cook's body had been cut up. "They all appear'd very much pleased at seeing us come in such an humble manner, to sue for the body of our Chief," Edgar observed. One man told Vancouver, who understood the Hawaiian language reasonably well, that the body had been carried a long way off but would be returned the next day; others said "the god Cook is not dead, but sleeps in the bush, and in the morning he will return."

Koa's behavior and the obvious lack of remorse among the Hawaiians convinced King that friendly persuasion would not be effective in retrieving the bodies. He sent Vancouver back to the

Resolution to report to Clerke, who responded with orders to return to the ships after informing the Hawaiians that the English would destroy their village if they did not return Cook's body by the next day. This threat brought only displays of contempt from the people on the beach, which so angered King that he was about to shoot into the crowd until Burney stopped him, arguing that they should not destroy what had been accomplished so far by a needless display of violence. King acceded to Burney's suggestion, mostly because he thought an attack would violate the spirit of their flag of truce, but he feared their conciliatory behavior only encouraged more insolence, since courage appeared to be the most valued trait among the Hawaiians, "as in most unciviliz'd states." The men in the boats restrained themselves with reluctance as they watched some of the Hawaiians parading around in the dead men's clothes. One even dared to carry Cook's sword.

This issue of conciliation versus retaliation fostered intense debate among the Englishmen for the next several days. On the one hand, they wanted to do nothing that would endanger the return of Cook's body; many thought the best strategy would be to maintain a posture of friendliness until they recovered the body and then retaliate just before sailing. Others, pointing out that it would be unfair to act as if all were forgiven and then mount an attack, wanted to act immediately if the body was not returned, while a third group argued for a sincere attempt at reconciliation on the grounds that Cook's murder was unpremeditated and, in any case, further violence would accomplish no good purpose. The debate turned partly on attitudes toward the Hawaiians, but also on a question of strategy: some thought friendly behavior was the best way to get Cook's body back, while others were equally convinced that the Hawaiians would respond only to force. Many feared Cook's body was already irrevocably lost, perhaps even eaten, although the Englishmen had no evidence to indicate that the Hawaiians were cannibals.

The next morning Koa came to the ships with a pig as a gift for King—since the Hawaiians thought King was Cook's son, they assumed he was now in command—and lame excuses about Cook's body. King was so angry at Koa, whom he called a "hypocritical dog," that he wanted to refuse the pig, but Clerke thought it best not to reject any offers of peace. After some debate among the officers, they decided not to carry out their threat of attack, but to continue working to prepare the ships for departure while renewing their attempts to get Cook's body back by diplomatic means. When they were ready to leave, they would take their revenge.

After dark that evening, two priests came to the ship with a piece of flesh, claiming it was part of Cook's thigh. The body had been cut up into many pieces and distributed around the island, they reported, and they warned the Englishmen not to go on shore, for the Hawaiians' professions of friendliness were merely a ruse. The Englishmen asked if any part of Cook's body had been eaten; some thought the priests said his flesh would be consumed that night, while others were convinced they had denied that any such thing would take place, a confusion created by faulty understanding of the language coupled with a tendency to cling to preconceived ideas. Like the men on shore the previous day, the priests seemed convinced that Cook was not really dead; they asked repeatedly when Lono would return and what he would do to the Hawaiians when he did. More Hawaiians asked the same question in the ensuing days, while others asserted that Lono would reappear in about two months' time.

In fact, the Hawaiians accorded Cook's remains the treatment reserved for their highest chiefs, consistent with their view of Cook as Lono. Chiefs taken in battle were treated as sacrifices, their bones removed from the flesh and distributed to the followers of the conquering chief. The long bones were reserved for the chief himself, while the skull was an offering to the god. Although not literally captured in battle, Cook was treated as a sacrifice to the god Ku; the Hawaiians who said Cook's bones had been taken all over the island were telling the truth. The questions about when Lono would return, mystifying to the English, indicated the Hawaiians' belief that Cook, as Lono, had been vanquished by Ku but would reappear. But despite this respectful treatment of Cook's remains, the Englishmen were not mistaken in their perception that most of the Hawaiians held their visitors in contempt and felt no guilt over Cook's death. The deification of Cook was largely the work of chiefs and priests; the common people, who had borne the economic burden of supporting "Lono" and his entourage, viewed Cook's death as a cause for celebration rather than as a matter of solemn ritual.

The day after the priests' visit, Clerke sent a group of sailors to get water, accompanied by a heavily armed contingent of marines. Most of the Hawaiians fled when the Englishmen landed, but others threw stones from hiding places behind nearby houses, while more gathered on top of a hill and rolled boulders down to the beach. Later that day the watering party returned to complete their task, this time with orders to burn the houses around the well so the Hawaiians couldn't use them as cover for further stone-throwing. This order was all the men needed to go on a rampage. They did not stop with the houses immediately around the well, but burned most of the village—between 50 and 150 buildings in all. Most of the residents fled in panic, but a few remained to defend their homes, and about half a dozen were killed by the English sailors. Still not satisfied, some of the sailors beheaded two corpses and stuck the heads on poles in the bows of their boats. The atrocities undoubtedly would have continued, Edgar thought, if the officers on shore had not forcibly stopped them.

Late that afternoon a crowd of Hawaiians, led by a young priest who had been particularly friendly to the English, marched down the hill toward the bay, carrying white flags and green branches. Some had hogs and baskets of fruit as well. Although the English fired on them, they kept walking toward the beach after hiding briefly in a patch of high grass. Afterward the priest went on board the *Resolution* and begged Clerke not to kill any more of his people—he said about fifty had been killed in the various encounters with the English—promising that the islanders would cease molesting the English.

True to their word, after this incident the Hawaiians stayed well away from the sailors collecting water, and white flags could be seen flying in the hills. Two days later, a priest named Hiapo came to the ships on behalf of Kalaniopuu, bringing gifts and the message that his chief wanted peace with the English. Clerke and Gore agreed to a truce, provided the Hawaiians returned Cook's bones and gun. A day later Hiapo appeared with what he claimed were some of Cook's bones; the day after that he brought part of the captain's head and his gun. Many of the Englishmen remained suspicious of the Hawaiians' motives, however, believing that Kalaniopuu simply wanted to get what more he could in the way of English goods before they sailed; and indeed, trade between the sailors and Hawaiians resumed briskly after peace was declared.

The Englishmen burned Cook's remains at sea, and the next day, February 23, they left Kealakekua Bay.

* * *

Because conditions at Hawaii had made it impossible for the Englishmen to get all the water and provisions they needed for their voyage north, Clerke directed his course west to look for likely landing places at the neighboring islands. After searching in vain at Maui and the adjacent islands of Lanai and Molokai, he went on to Oahu, anchoring briefly at Waimea Bay on the northwest coast. The inhabitants were friendly, but the only water available was brackish, so he moved on to Kauai, where he knew from previous experience that water and food should be available.

At Kauai, however, the islanders were distinctly unfriendly, apparently resentful of the Englishmen's return visit. Several thousand men, obviously in an ugly mood, gathered around the sailors dispatched to shore for water. A few attempted to steal the cooper's tools, while others tried to seize the marines' guns. The men who remained on board the ships watched helplessly, unwilling to fire on the islanders because they knew from bitter experience that a crowd of such size could overwhelm the small band of Englishmen despite their superior weapons. Some of those on shore were convinced they would never get back to the ships alive, but their fear was inspired more by memories of Kealakekua Bay than by a realistic consideration of their present danger. King, who led the group, said later he thought they had been faced with a "tumultuous mob," potentially dangerous to be sure, but not organized by anyone in authority or intending to do them harm.

With the help of several chiefs, relations with the islanders settled down after the first visit ashore, and the Englishmen remained at Kauai a week replenishing their water supplies. Shortly before they left, one of the chiefs tried to recruit several sailors to help fight against rivals at Niihau—the two islands had been at war since the first English visit, battling over the goats Cook had left at Niihau, according to one version of the story—but desertion was no longer a potential problem aboard the *Resolution* and *Discovery*. "That idea of turning Indian which was once so prevalent among them . . . is now quite subsided," Clerke wrote, "and you could not inflict a greater punishment upon those who were the warmest advocates for this curious innovation in life than to oblige them to take that step which 16 or 18 Months ago seemed to be the ultimate wish of their Hearts."

In mid-March, their reprovisioning completed, the Englishmen took leave of the Hawaiian Islands and sailed north to begin their second attempt at finding a northwest passage.

To complement the work of the previous summer, Clerke intended to explore the Asian side of the Bering Sea, stopping at the Russian fur-trading settlement on the Kamchatka peninsula. After a cold, trying six weeks—the ships were badly in need of repair and the men's clothing nearly worn out—the Englishmen reached the harbor of St. Peter and Paul, where the Russians maintained the settlement that served as headquarters for their Arctic fur-trading operation. Apart from the brief stop at Unalaska the year before, this was the Englishmen's first encounter with Europeans in three years.

The Russians living in the village at the edge of the harbor were suspicious of the Englishmen and questioned them closely about their mission, apparently suspecting that they were traders bent on invading the Russians' fur markets. Later the Englishmen learned that the outwardly gracious and hospitable Ismailov had written to his superiors at Kamchatka describing the English as fortune-hunters, perhaps even pirates. He had also misrepresented the English ships, describing them as small packet boats, so the Russians' apprehensions increased when they saw the size of the *Resolution* and *Discovery*. When the governor of Kamchatka, who lived in a town some miles inland, sent two emissaries to greet the English and learn their business, the men refused to go on board the *Resolution* unless two English sailors remained on shore as hostages for the Russians' safety.

Once the governor's representatives arrived on board, however, the misunderstandings quickly cleared up, and the two men agreed to supply the Englishmen with whatever provisions they needed. One of the officials at Kamchatka was scheduled to go to St. Petersburg soon, so Clerke gave him copies of Cook's journal, his own continuation of it, and Bayley's and King's astronomical observations, with a request that the Russian send them on to London as a precaution, in case the expedition should meet with disaster during the remainder of the voyage.

The Englishmen left Kamchatka in mid-June, passed through the Bering Strait about three weeks later, and ran into closely packed icebergs shortly afterward. For another three weeks they sailed a tortuous course through the ice, changing direction frequently to avoid massive bergs and move farther north; but at the end of July, Clerke, by now seriously ill and barely able to manage the day-to-day command, decided it was useless to continue.

It took almost a month to get back to the coast of Kamchatka. Clerke's health deteriorated steadily, and for the last ten days of the voyage he was unable to leave his cabin. On August 21, the crew caught sight of the harbor of St. Peter and Paul; the next day Clerke died. At his request, he was buried in the Russian village by the harbor. Gore took over command of the *Resolution,* and King moved to the *Discovery*. After a month of repairing and resupplying the ships, they sailed south, skirting the coast of Japan and anchoring at Macao, a Portuguese colony on the coast of China, in early December. Here they learned to their surprise that the revolt of the American colonies had escalated into a war between England and France. Having survived the dangers of an Arctic voyage, they now faced the dangers of French attack and capture on the last leg home—or so they thought.

Gore had the ships' stanchions and railings strengthened and the sides built up to allow protection to gunners; he also bought more guns. But in fact the French government had issued orders that Cook's ships were not to be molested because of the scientific importance of their mission, and Benjamin Franklin, the American colonies' agent in France, had drafted similar orders to commanders of American ships. Franklin asked American ship captains to treat Cook and his crew with "Civility and kindness . . . as common Friends to Mankind" because of the importance of their work to all nations. The Continental Congress did not see fit to adopt Franklin's proposed orders, but most American ships operated close to their own shores and posed much less of a threat to the *Resolution* and *Discovery* than French ships.

The Englishmen also discovered that the sea otter skins they had bought for trifles on the Alaskan coast fetched high prices at Macao, and many of the seamen left the colony with unexpected cash in their pockets. In fact, so many of the crew wanted to go back to Alaska for more furs that King and Gore had a near-mutiny on their hands when they refused.

After six weeks at Macao, the Englishmen sailed for the Cape of Good Hope, arriving in mid-April. Letters from the Admiralty awaited them there, announcing among other things the French proclamation protecting them and Spain's entry into the war. (The

Spanish government also issued orders protecting the expedition.) They left the Cape early in May and reached English shores three months later, but contrary winds precluded sailing up the Channel and into the Thames. Instead the ships proceeded around the western side of England and anchored at the Orkney Islands northeast of Scotland. King went to Aberdeen with the charts and journals to be conveyed to the Admiralty, but the rest of the crew remained in the Orkneys for a month, to their great frustration. They finally made it to London early in October, two months after first sighting England. The voyage had lasted four years and three months in all, the longest by far of the eighteenth-century circumnavigations of the globe.

20. *Cook the Hero*

Clerke's letter informing the Admiralty of Cook's death reached London on January 10, 1780. Sandwich immediately wrote Banks with the news, and the next day the newspapers spread it all over London. The accounts were brief, noting only that Cook had been killed at the hands of Hawaiian natives and that Clerke intended to make another try at finding a northwest passage after replenishing his supplies at Kamchatka.

The few newspapers that went beyond these basic facts praised Cook not for his contributions to world geography, but for his achievement in revolutionizing shipboard hygiene. "It is almost incredible," the *London Evening Post* stated, that Cook had established a system of "diet and cleanliness" so effective that he lost only one man from illness on the second voyage. In a country with an economy heavily dependent on oceangoing trade, reducing mortality on long voyages from appalling proportions to nearly zero was an accomplishment of major significance. As one ship captain put it, Cook "would deserve the gratitude of his country, of every maritime people, and of humanity at large, if his discoveries had been confined even to those improvements in the interior government of ships and their crews."

Other writers praised Cook's humanitarian behavior toward the people he encountered and remarked on the irony of his murder at the hands of Pacific island natives. The *Gentleman's Magazine* for January 1780 drew an analogy between Cook's career and that of Magellan, who was also killed by native peoples toward the end of his circumnavigation. Magellan, the writer pointed out, had been the aggressor in that confrontation, while Cook never deliberately harmed the people he met on his voyages; rather, he "always studied to benefit the savages whom he visited." A eulogy in verse form written some months later went so far as to state that an unselfish concern for the good of mankind had been Cook's motive for giving up the comforts of home and risking his life exploring the world. In the poet's opinion, his voyages were a means of uniting "savage" and "civilized" people. These writers, and others who praised Cook's humanitarianism, did not know the number of cases in which Cook had in fact adopted an aggressive posture toward Pacific peoples; such incidents were glossed over or omitted from published accounts.

By the time the *Resolution* and *Discovery* reached London in October 1780, the voyage itself was an anticlimax. On the face of it, the expedition had been a failure, having neither discovered a northwest passage nor conclusively disproved its existence in the way that the second voyage had laid to rest speculation about the southern continent. Little new had been added to previous knowledge of islands or people, apart from the discovery of the Sandwich Islands, and no one could predict in 1780 how important those islands would eventually become. And there was no personality to help focus attention on the voyage—no Cook, no Banks, no Mai, not even Charles Clerke. The methodical John Gore and the intellectual James King were no match for their predecessors in the public imagination. Consequently, the ships' arrival went almost unnoticed in the press, and public interest in the voyage focused almost entirely on Cook's death.

Not that the voyage was ignored in official circles. Several officers were promoted: Gore and King to post-captain and Phillips, the lieutenant of marines who had survived the attack on Cook, to captain. Gore was named to the post at Greenwich Hospital that Cook had resigned to undertake the voyage. Some of the younger officers, who were well aware that service with Cook had been an important path to promotion for others like them, looked forward to action on the American front. And the Lords of the

Admiralty began the task of publishing the official account of the voyage, which was a good deal more complicated than publishing the second, since Cook, Clerke, and Anderson were dead and King shipped out to the West Indies late in 1781. The Admiralty again employed John Douglas, who had helped Cook prepare his account of the second voyage, and called upon Banks to select illustrations, arrange for their engraving, and oversee the entire production. They also asked Cook's old rival Alexander Dalrymple to supervise the engraving of the maps.

Douglas used Cook's journal as the basis for his work, with extensive additions from Anderson's. By his own admission, he made more changes in the original manuscripts than he had in preparing the account of the second voyage, perhaps because Cook was not there to object; but unlike Hawkesworth, Douglas did not attempt to romanticize or draw moral precepts. When King returned from the West Indies in 1782, the Admiralty asked him to compile an account of the voyage after Cook's death. King received credit as coauthor of the volume with Cook; Douglas was an anonymous ghostwriter, as he had been on the previous project.

Hampered by King's absence and the number of people involved, the project proceeded slowly. Preparing the engravings presented the most serious problem; it was difficult to find the right quality of paper, and disputes arose about who should pay for them—the Admiralty or those who would profit from the publication. The profit question itself was a delicate one, given the many individuals involved and the fact that some of the most important players were no longer around to speak for their own interests. The whole matter was complicated by political changes at the Admiralty, which resulted in Sandwich leaving the government. Late in 1782, more than a year after work had begun on the project, when neither the text nor the engravings were completed and the question of royalties was still up in the air, Sandwich asked Banks to intercede with the Lords of the Admiralty to expedite production of the engravings and settle the financial issues.

Banks proposed an arrangement under which Cook's heirs would get about half the proceeds from the volumes, on the grounds that Cook had commanded the voyage for about half its length; Clerke's heirs and Gore would divide one quarter in proportion to the time they spent in command; and King would get the remaining one quarter for his work in writing the final section. A year later, however, when King wrote to Sandwich inquiring politely about his share, the issue still had not been settled. Eventually, Banks's scheme was adopted with certain changes. Clerke's heirs got one eighth and Gore nothing; instead, Bligh got one eighth with a small payment deducted for Anderson's heirs. By that time King too was dead (also of tuberculosis), and his share went to his heirs.

Nothing was said about royalties for Douglas; presumably he got a flat fee. This arrangement apparently satisfied him, since he never complained about his own remuneration, although he did run into conflict with Banks's management of the project in other ways. Douglas was especially upset when Banks agreed to send the proof sheets to a French bookseller who intended to bring out a French edition soon after, or even before, the English volumes. Banks claimed this was part of a deal to get the paper needed for the engravings. Douglas also thought Mrs. Cook and King were getting shortchanged, remarking that "the only Person who was a great gainer was the Bookseller who Published the work." In fact Mrs. Cook did quite well. Although she felt compelled to write Sandwich in June 1781 asking (in extremely deferential tones) for a portion of the profits, and to repeat her appeal in December 1783, she eventually did get her half share.

The nearly four years required to produce the volume meant, inevitably, that it was scooped by unofficial accounts. The first was an anonymous volume published in 1781, written by John Rickman, a lieutenant in the *Discovery,* and printed by the same bookseller who had published Marra's unofficial account of the second voyage. It went through two editions quickly, plus a pirated Dublin printing, and was summarized in some of the London magazines. Johann Forster, who got a copy of Rickman's journal from a friend in England, rushed a German translation into print late that same year. William Ellis, the *Discovery*'s surgeon's mate, published a two-volume narrative in 1782. Heinrich Zimmermann returned to Germany immediately after the voyage and published his account in Mannheim in 1781; the American John Ledyard, who also went home to his native country after the voyage, brought out his version in Hartford in 1783.

None of these books constituted serious competition for the official account when it finally appeared in June 1784. It was a handsome edition, three quarto volumes with a fourth folio volume of the engravings that had caused so much trouble. At a cost of nearly £5, it was beyond the reach of the average Englishman, but still the first edition sold out in three days and potential buyers offered up to ten guineas for a set. Second and third editions, as well as a French edition, appeared in 1785. George Forster translated the book into German. By early 1786, the profits amounted to over £2,700. The book continued to sell steadily, if slowly, and in 1801 the publisher reported total profits on the volumes of about £4,000. By that time, several smaller and cheaper editions had appeared, reducing demand for the lavish official version but also making the work available to many more people.

The continuing popularity of the volumes on Cook's third voyage were part of a general public interest in travel literature that had remained undiminished since the early seventeenth century. Hawkesworth's volumes were the books most frequently borrowed from at least one major library in the 1770s and early 1780s; they went into a seventh edition in 1789. In the last twenty-five years of the century, more than three hundred new books on travels to exotic places were published in England, in addition to the accounts of voyages that appeared regularly in periodicals such as the *Gentleman's Magazine* and the *Annual Register*.

This broad readership for the accounts of Cook's voyages was only one sign of the enduring public fascination with the captain and his achievements. Over the course of the 1780s and 1790s his reputation was celebrated in print, on the stage, and on canvas, building up an image of Cook as a national hero.

The Royal Society itself led the way. Early in 1780 the members decided to strike a medal depicting a portrait of Cook as a tangible way of displaying their respect for his accomplishments and their "Zeal for perpetuating the Memory of so valuable and eminent a man." They received nineteen proposals for the design of the medal, including entries from Johann Forster and Nevil Maskelyne. The winning design suggested the importance of Cook's discoveries for the British empire, with an image of Britannia holding a globe on one side of the medal and a profile of Cook on the other. Other proposals took inspiration from Cook's achievements in science and navigation; one showed him leaning on a rudder and pointing to a globe, with a quadrant and compass on one side, while another emphasized his success in reducing mortality at sea by using an image of Hygeia, goddess of health, to symbolize his concern for the welfare of his crew as "one of the most shining parts of his character." But it was a third designer who hit upon the most compelling reason for Cook's popularity when he wrote, "The most important consequence of all his voyages, and the characteristic mark which has distinguished his Labours from those of any other navigators, was their tendency, . . . to civilize and unite all mankind, and to spread & extend Arts, Science, & Commerce." Cook became a hero to the average Englishman not because he mapped the Pacific, demonstrated how to determine longitude at sea, disproved the myth of terra australis, or collected enough sketches and specimens of plant and animal life to keep scientists busy for decades—the sort of accomplishments that Cook himself valued—but because he discovered new lands without the violence that had marked earlier periods of exploration (and without the horrifying mortality rates of earlier long-distance voyages), opening vast regions of the globe to the spread of European civilization by peaceful means.

A 1791 article in the *Gentleman's Magazine* summed up Cook's image as it had developed in the decade after his death: he rose from poverty to distinction through his own merit; he experienced the hardships of a sailor's life and did everything he could to ease them; he was strong and brave, yet "possessed all the tenderness which was requisite to diffuse benevolence amongst untutored minds." It was tragic that he "fell a sacrifice to that nobleness of spirit"—yet the manner of his death added to his luster. Cook was not only a hero for his humanitarianism, but a martyr to it.

The first biography of Cook, published in 1788 by the Unitarian minister and amateur scientist Andrew Kippis, extended and enshrined this interpretation. Kippis went further than most writers in transforming Cook's voyages from scientific missions with a multiplicity of purposes to expeditions motivated wholly by the urge to civilize and improve the world. He was not entirely naive about the results of Cook's discoveries, recognizing that many thoughtful observers questioned the value of European contact for Pacific peoples, but concluded that Cook's humanitarian motives outweighed any possible negative consequences. European agriculture and tools would make the islanders' lives easier, he believed, and continued contact with Europeans would spark their curiosity and improve their "infant minds." Most important, he credited Cook's voyages with "spreading . . . the blessings of civilization among the numerous tribes of the South Pacific Ocean, and preparing them for holding an honourable rank among the nations of the earth"—a view that Cook himself and the men who helped organize his expeditions would have found exaggerated, if not totally false. Nevertheless, it became the most compelling justification both for Cook's voyages and for continued European involvement in the Pacific in years to come.

The English were not alone in idealizing Cook in the years after his death. Accounts of his voyage were translated and widely distributed on the continent, especially in France and the German states. The Elector of Bavaria spent hours talking with Zimmermann and gave him a pension; Louis XVI was so taken with the accounts he read of Cook's voyages that he had a special edition prepared for his son and initiated plans for a new French expedition modeled on Cook's third voyage. To be headed by J. F. Galoup de La Pérouse, it was intended to focus especially on the scientific study of native peoples. Several continental writers published essays eulogizing Cook; like their English counterparts, they praised him as a self-made man and "man of peace," who brought agriculture instead of war to the Pacific and opened the world to peaceful navigation. In their most extreme flights of fancy, these writers compared Cook favorably with classical heroes, especially Ulysses, for his courage, and even—because of his humble background—to Christ.

Eulogies published in newspapers and magazines provided the most explicit portraits of Cook as hero, but stage plays and paintings, the former in the guise of entertainment, offered symbolic statements of the same beliefs. English, French, and Spanish playwrights all produced dramas on Cook's death. *The Death of Captain Cook,* published in 1789, dispenses with facts and turns Cook's death into a romantic tragedy in which he dies as a consequence of his selfless, humanitarian actions, defending one chief against the attacks of another in a dispute over the affections of a beautiful young woman. (The "good" chief in the play is Parea, who had befriended the English during their stay at Kealakekua Bay; the "bad" chief is the untrustworthy Koa.) Cook intervenes in a battle between the two chiefs and their supporters, rescuing Parea from imminent defeat; later, in a show of magnanimity, he saves the captured Koa from execution at the hands of Parea's supporters. Instead of showing gratitude, however, Koa slinks off, returning later to stab Cook in the back. But Cook has his revenge; in his dying moments he shoots and kills Koa. The final scene shows a stately burial, with Parea, his wife, and the other Hawaiians weeping over Cook's tomb.

The most popular English play about Cook resurrected Mai as a principal character. *Omai, or a Trip Round the World* was a pantomime created by the painter and stage designer Philippe Jacques de Loutherbourg, with John Webber, the artist on Cook's third voyage, as consultant on costumes and scenery. The play opened at Covent Garden to rave reviews in December 1785, and ran for fifty performances that season; it was revived during the next two winters as well.

Omai weaves the principal Tahitians who befriended the English and the events of the second and third voyages into a wholly fabricated plot. Mai is presented as the son of Tu and a candidate to become "king" of Tahiti; Hitihiti, supported by Purea, is his rival. Towha, a supporter of Tu and Mai, proposes that Mai go to England to woo "Londina," daughter of "Britannia." When Mai arrives in England he finds, of course, a rival for the lady's hand—a Spaniard. Mai and Londina fall in love at first sight, but her father is opposed to the match, and Mai spirits her away, with the father's representatives in pursuit. The couple set off around the world, from Kamchatka to Tonga to Hawaii (where they encounter Purea, who tries to kill Mai) to Tahiti, where Mai is crowned king in triumph.

Throughout the play, Mai is the focus; Cook does not even appear as a character. But the ending makes it clear that the underlying theme is Cook's achievement in discovering widely disparate people previously unknown to Europeans and in conveying to them an image of European benevolence. The play concludes with an elaborate procession of representatives from most of the places Cook visited, including New Zealand, Tanna, the Marquesas, the Tongan islands, Hawaii, Easter Island, Kamchatka, Unalaska, and Nootka Sound. An English captain presents Mai with a sword, a prophet predicts eternal friendship between England and Tahiti, and finally an enormous painting, which depicts Cook floating up to heaven into the arms of Britannia and Fame, descends from the ceiling as the entire cast sings a song of praise to him.

De Loutherbourg's drawing—a more detailed version, entitled *The Apotheosis of Captain Cook,* was later printed—was only one of many visual paeans to Cook. Several artists painted the scene of Cook's death, including John Webber and John Cleveley, brother of James Cleveley, who had been a carpenter on the *Resolution.* Webber may actually have witnessed the murder, and both he and Cleveley had access to eyewitness accounts; their paintings attempted to illustrate the incident accurately, and are similar in their major features. Both show Cook standing at the water's edge with dozens of Hawaiians crowding around him, one poised to plunge a dagger into his back. Cook gestures to the boats just offshore, his outstretched palm seemingly telling them to stay where they are and hold their fire, an interpretation consistent with the popular image of Cook as a benevolent man martyred by ungrateful natives.

Another painting, by George Carter, fancifully shows Cook fighting back, the butt of his gun raised against one of his attackers while another lies dead at his feet. Significantly, although this picture shows a more aggressive Cook, he uses his weapon as a club rather than firing on the Hawaiians, and even that much is done clearly in a posture of self-defense. A fourth painting, by D. P. Dodd, is the least-known, and not only because it is the weakest artistically; Dodd shows Cook in defeat, lying face down while the Hawaiians struggle among themselves for the weapon that killed him—hardly a suitable pose for a hero.

The most famous rendering of Cook's death was painted by Johann Zoffany, who would have joined the second voyage had Banks not backed out. Zoffany was an accomplished artist, one of the best-known of his day, and in painting Cook's death he invoked a timeless theme: the death of a martyr. Zoffany stripped the scene of all extraneous details—the offshore boats, the Hawaiian landscape—to make the dying Cook, surrounded by his murderers, the centerpiece of the painting. Cook lies fallen, but propped on one elbow and facing his killer, who stands above him, dagger in hand. In the background three marines battle a swarm of nearly naked, muscled Hawaiians, painted to look like Greek warriors. As historical representation, the picture is entirely false; when Cook was stabbed, he fell face down in shallow water, and the surviving marines scrambled for the boats. But for Zoffany's purpose the composition was far more effective than Webber's and Cleveley's more accurate rendering, conveying a dramatic sense of struggle while focusing on Cook's dignity and resignation in the face of death. To accomplish his end, Zoffany employed a pose used by earlier artists to depict the martyrdom of sacred and classical heroes and, more recently, by a new generation of artists painting secular heroes and contemporary events—notably Benjamin West, whose popular painting of the death of General James Wolfe (killed in battle at Québec in 1759) would have been well known to Zoffany.

The most extreme examples of the urge to idealize Cook's death were not the paintings, however, but the drawings showing his apotheosis—de Loutherbourg's version and another by Johann Ramberg. Ramberg's drawing, titled *Neptune Raising Captain Cook to Immortality,* was published in 1787 as the frontispiece to Thomas Banke's *New System of Geography;* the engraved version of de Loutherbourg's drawing was published in 1794. Both show Cook floating on a cloud midway between earth and heaven, looking uncomfortable, even skeptical, as undoubtedly he would have, could he have known about such excessive hero worship. Ramberg shows Cook welcomed by angels with Neptune at his feet, while Britannia receives tribute from the four continents below. De Loutherbourg also shows Cook surrounded by trumpeting angels carrying him into heaven, although he seems to be pushing them away with one hand while clutching a sextant with the other.

The paintings, engravings, and accounts of the voyages were all reprinted many times over in the decades following Cook's death, creating and maintaining Cook's image as a national hero. Subsequent generations added their own paeans to Cook, in the form of pictures, biographies, novels, moral tales for children, and physical monuments of all sorts—there are over two hundred around the world today—from the commanding statue near Admiralty Arch in London to the Captain Cook Memorial waterjet that regularly spews a stream of water 150 meters skyward from a huge lake in the center of Canberra.

Among his contemporaries, Cook won praise above all else for his humane treatment of Pacific peoples, but humanitarianism in the abstract was not the stuff of hero worship. What appealed to the European imagination was the belief that discovery of new lands would lead to the peaceful expansion of European civilization. Instead of subjugating newly discovered peoples, as the Spanish and Portuguese had done in South America (English commentators ignored their nation's policies in North America), Europeans would improve Pacific islanders' lives by offering the benefits of their presumedly superior knowledge and culture. Cook's effort to introduce English-style agriculture was only a beginning. The more practical-minded talked of gaining new markets and new sites for colonies, but tempered their eagerness for the economic and political advantages of Pacific discoveries by asserting that those markets and colonies would become a means of conveying European civilization to the Pacific; both the discoverers and the discovered would benefit, and their mutual profit would be a means to achieving peace and unity in the world. Ramberg's *Apotheosis* symbolized this hope by showing the four continents paying tribute to Britannia; whether in gratitude for opening the world to the benefits of civilization or as payment for the material goods that went with it may be left to the interpretation of the viewer.

Although few in the 1780s and 1790s disputed the potential value of European influence in the Pacific, an occasional voice doubted that the denizens of distant islands would benefit from their "discovery" by Europeans. George Forster had suggested, in his account of Cook's second voyage, that European contact with Pacific cultures would ultimately destroy them. And one of the many pieces of bad verse written about Cook's voyages did not praise his discoveries but criticized the damage he had wrought. *The Injured Islanders* made the obvious point that Europeans had introduced venereal disease to the Pacific islands, but also argued, more significantly, that English "toys" created a desire for more material possessions and promoted distinctions of wealth in a society where all had existed equally. English iron was fashioned into weapons that made war more deadly, and jealous rivalries encouraged by desire for material goods made war more frequent. The author ignored the fact that class distinctions had always existed in Tahiti, but was perceptive in his charge that desire for English goods would exacerbate rivalries among Tahitian chiefs. Written in the form of a letter from Purea to Wallis lamenting the loss of her power, the poem attributed the Tahitian chiefs' attack on Purea to their jealousy over the wealth she had gained from the English during Wallis's visit. It was a reasonable interpretation. Certainly by the time of Cook's visits, English favoritism and material gifts, especially iron, played a critical role in the balance of power among Tahitian chiefs.

Cook himself was ambivalent on the issue. He often deplored the introduction of venereal disease, the most immediate and obvious negative consequence of European exploration in the Pacific. His occasional reflections on the Polynesians' and Australians' contentment with their lives suggest his recognition that advanced civilization was not necessarily an unmixed blessing. And yet he was determined to establish English agriculture in the Pacific, because this was a way of conferring tangible benefits without imposing alien values on the people he visited. He did not see that English agriculture itself implied a value judgment about the business of raising food—a preference for orderly cultivation over merely gathering what grew wild or could be grown with a minimum of effort—nor could he have predicted that his animals might upset the balance of nature, as they did in New Zealand, where goats eventually multiplied to the point that they became a nuisance, or in Hawaii, where cows created a similar problem. As Cook explored the world in the name of science, wishing to take nothing from the Pacific except what would add to the world's knowledge, so he wanted to leave nothing among the natives of the Pacific except what might clearly help them improve their lives. And while he had some inkling that England might eventually establish settlements in the Pacific, he expected them to be way stations for future ships, and the sites he favored were sparsely populated. Cook did not share George Forster's prescience; indeed, few men did. It was left for the next generation of explorers—notably two men trained by Cook, William Bligh, and George Vancouver—to return to the Pacific and discover just what changes the English explorers had wrought.

V

EXPLORATION AND EMPIRE

The Legacy of Cook's Voyages

21. *Exploiting the Pacific: Convicts, Furs, and Breadfruit*

Among Cook's many apprentices who returned to the Pacific after his death, William Bligh was undoubtedly the most famous. Master of the *Resolution* on the third voyage, praised by Cook and others as a skilled navigator and chartmaker, he ensured his place in history when, a decade after the third voyage, he became the victim of the most spectacular mutiny in British naval history. His mission on that voyage was a peculiar one: the British government dispatched him to gather a quantity of breadfruit plants from Tahiti and carry them to the West Indies, where, it was hoped, they would be transplanted and eventually serve as a major food source for the islands' slaves. Bligh took his assignment seriously, for it was, in his view, the first effort to put the discoveries of Cook's voyages to practical use. As he explained it, "The object of all former voyages to the South Seas, undertaken by the command of his present Majesty, has been the advancement of science, and the increase of knowledge. This voyage may be reckoned the first, the intention of which has been to derive benefit from those distant discoveries."

Bligh's remark was not altogether accurate, however. As he made preparations for his voyage during the fall of 1787, a fleet of eleven ships was already on its way to establish the first English colony on the Australian continent, a settlement that began as a dumping ground for convicts but developed into a cornerstone of the British empire. The fleet arrived at Botany Bay in January 1788, just a few weeks after Bligh left on his voyage to Tahiti. And by that time, dozens of ships had been collecting furs at Nootka Sound and other ports along the North American coast discovered by Cook. The traders sold their pelts on the coast of China, often stopping for provisions at the Hawaiian Islands.

Indeed, the striking point about British activity in the Pacific at the end of the eighteenth century was not that Bligh's voyage was the first attempt to exploit Pacific discoveries for practical ends, but rather how many different forms of exploitation were underway less than a decade after Cook's death. They included both state-supported and privately sponsored ventures pursuing a variety of goals—commercial, political, scientific, sometimes a mix of all three. Nearly all had at least two points in common, however: in their organization, outfitting, and attention to scientific observation of the places they visited, the late-eigheenth-century expeditions were modeled on Cook's voyages; and in their planning they showed the guiding influence of Joseph Banks. In addition, many were commanded by men who had begun their careers under Cook's tutelage.

Although Banks's career as a Pacific explorer ended when he withdrew from Cook's second voyage, he remained deeply involved in the publication of the voyages and in the distribution of artifacts collected by Cook and others. After Cook's death he became even more influential, extending his activities to include advising government officials on subsequent voyages and selecting scientists for them. Banks's election as president of the Royal Society in 1778 (a post he held until his death in 1820) cemented his position as England's premier scientific organizer; he became, as one historian has put it, "the general director of exploration," a role that gave him ample scope to exercise his interest in putting scientific discoveries to practical use.

In the spring of 1779, some months before the *Resolution* and *Discovery* returned to England, Banks recommended to a committee of the House of Commons that they take steps to establish a penal colony at Botany Bay, the harbor on the southeastern coast of New Holland that he had explored with Cook in the *Endeavor*. The committee was charged with considering the creation of a new convict colony, an issue that had become increasingly pressing since the beginning of hostilities with the American colonies.

The English courts had for many years sentenced certain criminals to "transportation"—banishment, in effect—and the government had shipped them to the southern American colonies, where they were sold as laborers for the term of their sentences. The practice was an essential component of the English criminal justice system. Long-term imprisonment was not considered an acceptable form of punishment because of its expense, and the death penalty, although prescribed for about two hundred different offenses, could not be used in every case in which the law permitted it without invoking public outrage. Banishment was often the preferred alternative, and between 1769 and 1775 the English government sent nearly a thousand felons a year, on the average, to the American colonies.

Beginning in 1775, however, the colonies refused to accept convicts, and as a partial solution to the problem, the government purchased derelict ships, or "hulks," as they were called, in the Thames and in seaports along the southern coast to house felons sentenced to transportation. By the end of the 1770s these hulks were seriously overcrowded; a crime wave in the early 1780s, fed by recession, increasing food prices, and demobilization of troops after the American Revolution, exacerbated the problem. Escapes became frequent, and the potential for epidemic disease loomed as a serious threat. The convict problem became a major public issue, particularly in areas near the prison ships, where fear of escapes and disease reached near-panic levels. As a result, pressure to find a new location for transported criminals mounted.

Banks's urging of Botany Bay was based on his recollection of the nine days he spent there in 1770. The area had a mild climate, and although much of the land had appeared barren, there was enough rich soil to support a substantial population, in his opinion. Grass suitable for grazing livestock, timber, and fish flourished in abundance, and the indigenous population should pose no threat, since the area was thinly populated. The few people Banks had seen appeared "treacherous" but "extremely cowardly." Faced with English settlers, Banks believed they would simply run away. When questioned by the committee members about whether a colony might offer benefits to England beyond serving as a location for transporting convicts, he responded that it should provide a market for English goods as it increased in size, and that a continent as large as New South Wales would surely produce other riches in time.

But the prospect of starting an entirely new colony on a largely unexplored continent twelve thousand miles from England was more than the House of Commons was prepared to contemplate in 1779. The committee recommended that the law authorizing the government to ship convicts to North America be revised to permit transportation to other parts of the globe, but stopped short of suggesting alternative destinations. Government officials considered several possible sites over the next several years, mostly in Africa or the East Indies, and found them all wanting because of poor land or climate conditions. In the meantime, convicts sentenced to transportation continued to crowd the hulks.

Four years after Banks's testimony, another veteran of the *Endeavour,* former midshipment James Matra, took up the cause of colonizing Botany Bay. In July 1783 he wrote to Banks saying he had heard that the government planned to establish two settlements in the South Pacific—such rumors were rife in London at the time—and that one of them, in the vicinity of Botany Bay, was to be headed by Banks. Matra, whose letter suggested that his career had been all downhill since the *Endeavor*'s voyage, said he would be interested in taking part in any such scheme.

Banks's reply to this letter has been lost, but it must have been encouraging, as Matra wrote soon afterward to Evan Nepean, Under Secretary of State for Home and Plantation Affairs, with a proposal to colonize New South Wales as a means of compensating for the loss of the American colonies. He was not interested in identifying a new site for transported felons, however, but in relocating Americans who had been loyal to Britain during the Revolution—not surprisingly, since he had grown up in New York and his father was among the loyalists there. The climate and soil of the area around Botany Bay were suitable for growing tea, coffee, silk, cotton, indigo, and tobacco, Matra argued, which would give England the means of creating "a revolution in the whole system of European commerce." Not only would the colony eventually produce valuable trade goods, but from the beginning it could serve as a port for ships involved in the fur trade on the northwest coast of America and in trade with the Dutch East Indian ports recently opened to England under the terms of the Treaty of Versailles. Botany Bay also had the potential to become an entrepôt for shipping English wool to Japan and initiating trade with Korea, a nation as yet untapped by European merchants. New Zealand's high-quality flax and timber, both valuable for England's shipbuilding industry, would be a relatively short distance away, and in time of war a settlement in New South Wales would provide a strategically useful base in the Pacific.

Matra's proposal drew no immediate response; the government was seeking a home for convicts, not loyalist Americans. In April 1784, Lord Sydney, recently appointed Home Secretary, granted Matra an interview and told him as much, leading Matra to revise his plans for Botany Bay to point out its value as a prison settlement. Stressing its potential as a place of reformation, he urged that convicts be given land immediately without being reminded constantly of their crimes. Under such a plan it was likely that felons could become useful, even "moral" members of society, an argument that owed much to contemporary social reformers' theories about how to reduce crime and rehabilitate criminals.

The ministry ignored Matra's plan because of the enormous cost involved in transporting felons to New South Wales—the voyage would take at least six or seven months and cost up to six times as much as the voyage to North America—and pursued instead the search for sites closer to home, in Canada, the West Indies, Central America, Africa, India, and Sumatra. Efforts to ship convicts to Honduras and Newfoundland in 1784 failed, and other possible sites were ruled out for a variety of reasons; Canadian residents objected to the idea of felons in their midst, the West Indies had no need for convict labor because of its slave population, India also had a large supply of cheap labor, and the climate in much of Africa was considered lethal for Europeans.

In April 1785 a Parliamentary committee charged with reviewing potential sites focused its attention on several locations in Africa as well as two new proposals touting Botany Bay from Sir George Young, a naval officer, and John Call, a former civil servant in India. The two plans were similar, and like Matra's, linked the search for a new convict colony with England's need to recover from the loss of its American colonies. Invoking Cook's name, Young urged that England seek some practical use from the captain's discoveries. "This Spirit of Discovery, the Genius for Trade, and skill in navigation, for which we have always been, and *still* are famous, will soon be transferred," he wrote; ". . . certain it is, Cook's Discoveries were not given in vain; and a very small beginning on this Plan, is sufficient to promote the general good of the humane race." Both Banks and Matra testified in favor of Young's proposal, but distance remained an insurmountable obstacle, in the opinion of the committee, and the members recommended instead a site at Das Voltas Bay, on the African coast between Angola and Capetown.

In September 1785 the Admiralty dispatched an expedition to investigate Das Voltas Bay. It returned in mid-1786 to report that the entire coast of Africa was unsuitable because of its poor soil, harsh climate, and hostile inhabitants. About a month later, in an abrupt change of direction, Sydney announced that the government had selected Botany Bay as the site for a new convict colony, and that preparations for its settlement would begin immediately. Although he borrowed some of Matra's, Call's, and Young's points about the commercial and strategic value of a British presence in New South Wales, it was clearly the government's last choice. There was simply no other place to send the mushrooming numbers of convicts jammed into prison ships on the Thames.

In May 1787, just nine months after announcing its decision to inaugurate a convict colony on the shores of New South Wales, the British government sent eleven ships with 759 convicts, both men and women, and two hundred marines, some accompanied by wives and children, to Botany Bay—"a reckless act on the part of a desperate ministry," in the words of one historian. Commanded by Arthur Phillip, a forty-eight-year-old career Navy officer, the fleet reached its destination on January 20, 1788, where the settlers quickly discovered that Joseph Banks's optimistic reports about climate and soil were wrong. Banks had visited Botany Bay in the autumn, when the temperature was mild, rainfall reasonably frequent, and vegetation at its peak. Phillip and his group, by contrast, arrived in the southern hemisphere's summer to find searingly hot temperatures and parched, sandy soil devoid of all but the hardiest plants. Banks had been right on only one point: the native men posed no threat, although the settlers soon ended up competing with them for fish, the one source of food the area offered in abundance.

Less than a week after the fleet arrived, Phillip moved his company a few miles north to Port Jackson, a harbor Cook had noted but not investigated. Phillip thought it "the finest Harbour in the World, in which a thousand Sail of the line may ride in the most perfect security." In one of its many coves, he and his marines and convicts laid out a village, which Phillip named Sydney in honor of the Home Secretary. Soon after this move, he sent twenty-four men to establish a settlement on Norfolk Island, where Cook and his companions had reported excellent-quality flax and timber, as a first attempt to fulfill the government's hope of realizing some commercial gain from its new venture. (The official plan for the Botany Bay colony noted optimistically that flax and most "Asiatic productions" might be cultivated and timber harvested from New Zealand.) Such an effort was premature, however, as the settlers had all they could do to survive in their first two years.

English government officials looked upon the settlement of New South Wales as an expedient, a last-ditch means of ridding England of its rapidly growing convict population. The possibility of reaping some trade benefits from the colony was largely an afterthought. The men who had proposed the settlement, however, as well as those who pushed for colonization of other sites, hoped to take advantage of the broadening of the world's frontiers for their own personal gain, and framed their proposals to stress political and commercial advantages to the nation in the hope of getting public subsidy for their ventures. Similarly, those men who set their sights on the North Pacific, mounting fur-trading expeditions to the west coast of North America, were motivated by a desire for profit but spoke also of the need for further exploration in the North Pacific and the area's potential strategic advantages to the British empire. Their strategies for gaining government support indicated that scientific and commercial motives, always mingled in eighteenth-century exploration, became increasingly difficult to separate by the end of the century. Not until Spain actually threatened British trade along the North American coast, however, did the government become directly involved in further voyages to the North Pacific. In both North and South Pacific, the impetus for exploiting Cook's discoveries came from individual entrepreneurs, who attempted to link their interests to those of the British government.

By the time the first fleet of convicts reached New South Wales, dozens of ships had sailed back and forth across the North Pacific, collecting furs along the west coast of North America and selling them in Canton, often with reprovisioning stops in the Hawaiian Islands. Cook's crew had discovered the value of North American furs when they reached China on their return voyage and sold their Alaskan souvenirs for enormous sums. James King, in his part of the published account of the voyage, noted the high prices paid for furs that the crew had obtained rather casually—many of them skins of indifferent quality, previously used by the Alaskans—and proposed that the East India Company undertake an expedition combining further exploration of the Asian coast with systematic purchase of furs for profit.

London merchants moved quickly on King's suggestion. In May 1785, a group of merchants formed the King George's Sound Company (King George's Sound had been Cook's original name for Nootka Sound) to sponsor fur-trading voyages to North America. The investors, who had Banks's backing, easily mixed private and public, commercial and scientific motives in promoting their venture. In seeking government permission for their first voyage, they pointed to potential benefits as diverse as opening trade with Japan, establishing a fur-trading base on the Alaskan coast, and exploring areas left uncharted by Cook, even possibly discovering a northwest passage. Despite its lofty pretensions, the company's goal was trading in furs, and in September the investors sent two ships to the North American coast for that purpose. The captains were two of Cook's men: Nathaniel Portlock, master's mate in the *Discovery* on the third voyage, was commander of the expedition and captain of the *King George,* the larger of the ships; George Dixon, formerly the armorer in the *Discovery,* was captain of the *Queen Charlotte.*

Portlock, like Bligh, was conscious of the manner in which Cook's pursuit of discovery in the name of science was being put to practical use. At the beginning of his account of the voyage, he observed that although Cook had not succeeded in the objective of his third voyage—the discovery of a northwest passage—he nevertheless "furnished philosophy with many additional facts, and he opened to commerce several extensive prospects." Indeed, Portlock believed, the disinterested scientific nature of Cook's voyages, and the fact that their results were published to the world, made it possible for all nations to benefit from them, whether "for the cultivation of science, or for the advantage of traffic."

In May 1786, the *King George* and *Queen Charlotte* anchored in Kealakekua Bay, the first European ships to visit the Hawaiian Islands after Cook's murder. As the Englishmen approached the island, fires burned along the coast to signal their arrival to the inhabitants. Canoes soon clustered around the ships, and many people inquired about James King, but to Portlock the islanders appeared reticent, even fearful, compared to those he remembered from his previous voyage; they seemed to think the Englishmen had returned to avenge Cook's death. Even so, dozens of men and women clamored to board the ships. Portlock thought their behavior "insolent," and in the absence of any powerful chief who could maintain order, the Hawaiians continued to create trouble for the Englishmen throughout the day.

One chief among the crowd told Portlock that Kalaniopuu was dead and that his nephew Kamehameha ruled the island. Portlock remembered Kamehameha as a rather minor chief, and was puzzled about how the young man had gained so much power in so short a time; but he was not to learn the answer, since Kamehameha did not visit the ships, and when the unruly Hawaiians continued to harass the Englishmen the following day, Portlock, fearing violence if he stayed, decided to sail elsewhere in search of provisions. After firing cannon to force the islanders to disperse, he and Dixon sailed their ships out of the bay, headed toward Oahu.

Anchoring in a well-sheltered bay on the southeast side of the island, they spent four days replenishing their water and food supplies with the help of a chief named Kahekili, who, unlike his compatriots at Kealakekua Bay, kept his people under close control. From him Portlock learned something about the political changes that had taken place in the seven years since Cook's death. Kamehameha's rise to power on Hawaii (achieved by killing Kalaniopuu's son in battle after the old chief's death) was only part of a broad shift in political alignments among the islands. Of the principal chiefs in 1778, only Kahekili was still alive; formerly chief of Molokai, now he claimed to rule Oahu and Maui as well. His half brother Kaeo ruled over Kauai and Niihau, and at the time of Portlock's visit the two were trying to negotiate an alliance that would give them control over the entire island group except Hawaii. This represented a significant consolidation of power since the time of Cook's visits, when the islands were divided into four kingdoms ruled by rival chiefs. Portlock saw evidence of the battles that produced these political changes in the large assortment of weapons on Oahu and the number of young men skilled in using them. The islanders importuned him to make more weapons, which he refused to do, fearing they might be used against future European visitors.

After a brief stop at Niihau to get a supply of yams, Portlock headed north, anchoring at Cook Inlet on the coast of Alaska in mid-July. His expedition returned to the Hawaiian Islands in the fall of 1786, when it became too cold to remain comfortably on the Alaskan coast, and then again in the fall of 1787 after their second season of collecting furs. Portlock continued to stop at Oahu, Kauai, and Niihau, although he avoided Kealakekua Bay after his unpleasant experience there. With this routine—summers on the northwest coast, winter visits to Hawaii to refresh the crew and gather fresh provisions—Portlock established a pattern that most subsequent vessels engaged in the fur trade would follow.

Portlock and Dixon were among the first Europeans (apart from the Russians) to sail to the northwest coast of America in search of furs, but they were soon followed by a number of other expeditions, mostly initiated by Englishmen at first, but later by French, Spanish, and American merchants as well. When the two captains reached Prince William Sound on the coast of Alaska in April 1787, they met one of their competitors in a sorry state, his ship frozen in the ice and more than half his crew dead of exposure and scurvy. He was John Meares, a merchant employed by the East India Company, who had left India the previous spring en route to Nootka Sound. There he learned that Portlock and Dixon had preceded him, depleting the available supply of furs, so he went on to Prince William Sound, where he made the mistake of attempting to remain through the winter. The hardships he and his crew endured brought them little reward, for as Meares told Portlock and Dixon, several ships had been plying the Alaskan coast during the previous months, bidding up the price of furs. His account persuaded Portlock and Dixon to separate to improve their chances of obtaining skins at reasonable prices. Meares, meanwhile, made his way back to China after stopping in the Hawaiian Islands, where he quarreled with the residents of Kauai and picked up a chief named Kaiana, who wanted to go to England. Portlock and Dixon paid the price for Meares's poor treatment of the people at Kauai when they arrived a few weeks later to find the islanders mistrustful and reluctant to barter food.

Portlock's, Dixon's, and Meares's experiences meeting or narrowly missing each other on the Alaskan coast and in the Hawaiian Islands was typical of fur traders plying the North Pacific in the 1780s and 1790s, a sign of the increasing number of ships involved in trade and the limited number of ports where they could seek cargoes or provision. The King George's Sound Company sent out its second expedition early in 1787, about a year and a half before Portlock and Dixon returned to England. Again they selected a veteran of Cook's voyages as commander: James Colnett, who had been a midshipman in the *Resolution* on Cook's second voyage. At Joseph Banks's urging they also appointed Archibald Menzies as surgeon. Trained in botany, Menzies would add a scientific dimension to an otherwise purely commercial voyage.

Colnett arrived at Nootka Sound in July; Dixon, now operating independently of Portlock, appeared a month later. After two seasons on the northwest coast, punctuated by a winter in Hawaii, Colnett sailed for Canton, where he met Meares, recently returned from his second voyage. The two men decided to join forces, forming a fleet of four ships. They left China early in 1789; in July, Spanish vessels captured them, charging them with trespassing in Spain's territorial waters. Meares, who was not authorized by any English trading company and flew Portuguese colors as a way of avoiding English licensing requirements, was released, but Colnett, operating with English papers, was taken to Mexico and imprisoned for a year. England did not recognize Spain's claim to the northwest coast of North America, and Colnett's capture nearly led to war between the two nations.

By the time Meares and Colnett sailed, English merchants operating on the northwest coast faced challenges not only from Spain, trying to enforce its claim to possession of the entire North American coast, but from France—which in 1785 had sent out a major exploring expedition commanded by La Pérouse in an effort to keep pace with English territorial and commercial expansion in the Pacific—and from the United States, which had no expansionist ambitions but whose merchants plunged aggressively into the fur trade. In reality, Spain was too weak to enforce its claims; Colnett's capture was a last desperate effort to enforce its disputed sovereignty. La Pérouse's expedition ended in disaster, wrecked near Vanikoro in the Santa Cruz Islands, and the few subsequent French attempts to stake out a share of the fur trade ended with the Napoleonic Wars. The Americans, unencumbered by the licensing requirements imposed on British merchants by the East India Company and accustomed to long, risky voyages as an essential part of their economy, proved the most serious competitors, eventually dominating the fur trade.

As competition for furs increased, prices rose; and as early as the late 1780s, it was impossible for a ship to collect enough furs in a single season to show a profit. Even Portlock and Dixon spent two seasons collecting their cargo, and later ships often needed more. As a result, the Hawaiian Islands became crucial as a winter rest and pre-provisioning stop. (Portlock thought they would be useful for providing fresh provisions to the new colony in New South Wales too, but that role fell to Tahiti, which was much closer.) The people of Kauai, who had received visits from Meares. Dixon, and Portlock in rapid succession in the fall of 1787, were seeing only the beginning of a constant stream of European and American ships, all looking for fresh food, water, and women. Far from deploring this invasion of their islands, the Hawaiian chiefs proved to be shrewd traders themselves, using their visitors to advance their own ambitions.

In trading with South Pacific islanders, Cook had always tried to work through chiefs to ensure a steady supply of provisions at reasonable prices and keep the island people under control. This strategy was most successful in the Hawaiian and Tongan islands, where society was highly stratified and chiefly control extremely strong. But Cook and his successors, who followed his example, were not alone in preferring this method of operating; the island chiefs also found it worked much to their advantage, because it allowed them to monopolize the riches of their European visitors. During Cook's time, chiefs throughout the South Pacific had occasionally placed a *tabu* or *kapu* on hogs and other provisions when they were angry with the Englishmen; in the 1780s and 1790s, Hawaiian chiefs increasingly employed this tactic, not to display anger but to control trade by restricting the flow of goods to Europeans. In addition, they invoked the chiefly prerogative of requisitioning foodstuffs from their subjects, which they then used in barter with their visitors. Through this combination of tactics, Hawaiian chiefs managed to maintain high prices and keep the largest share of prized European objects for themselves.

Weapons and ammunition increasingly became the chiefs' preferred form of payment. The men of Oahu had badgered Portlock for iron daggers; two years later, at Kauai, Kaeo dragged his feet on returning an anchor stolen from Colnett's ship, hoping to ransom it for weapons. By the early 1790s, less scrupulous visitors discovered that guns, no matter what their condition, bought far more food than any quantity of iron. As a result, fur traders began carrying supplies of old guns, many of them rusted and all but useless, solely for purposes of trade, and chiefs often refused to trade food for anything except weapons and ammunition. Commoners had no use for guns, preferring iron tools in payment, but the chiefs' control over trade meant that their subjects' wants took second place.

Few Hawaiians knew how to use guns, however, and European or American visitors came to be valued as instructors in the use of weapons and as mercenaries in the interisland wars that consumed much of the chiefs' attention in the 1780s and 1790s. In addition, the chiefs discovered that former sailors and marines were useful as agents in conducting trade with European and American ships. In 1787, John Mackay, surgeon on board the English ship *Imperial Eagle,* received permission to stay behind when his ship sailed to China, making him the first European to take up residence in the Hawaiian Islands. When Colnett left Kauai in March 1788, Kaeo urged him to leave some of his men to help fend off an anticipated attack from Oahu and Hawaii. Colnett refused, although he thought the request was justified, since the other islands' hostility toward Kauai, in his opinion, stemmed largely from its contact with Europeans. Because Kauai and Niihau were the most frequently visited Hawaiian Islands at that time, their chiefs had greater opportunity to accumulate European goods, creating jealousy among the chiefs of the other islands. Colnett's sense of guilt led him to teach some of the Kauai warriors how to use the muskets they had acquired, although he stopped short of giving them more.

Later ship captains were less concerned about keeping their full complement of men, and at least ten men were living in the islands by 1790, most on Hawaii but a few on Oahu. By the end of the century their numbers had increased about tenfold, inaugurating a legend about the idyllic life of the South Pacific, free from the constraints of Western civilization. In reality these adopted islanders were, especially in the eighteenth century, hard-bitten, nearly illiterate, sometimes criminal men trying to escape the horrors of life on long sea voyages. But Hawaiian chiefs found them useful, and gave their favorites land and houses in exchange for their services as intermediaries with trading ships and as mercenaries in war.

Oddly enough, Tahiti, the South Pacific island most frequently visited by eighteenth-century explorers, became less popular as a port of call in the generation after Cook, as ships headed to New

South Wales sailed across the Indian Ocean by way of the Cape of Good Hope, and fur-trading voyages found it more convenient to use the Hawaiian Islands as a base for reprovisioning. The notion of importing food for the Port Jackson colony from Tahiti, occasionally discussed in official circles, did not prove practical in the colony's early years. As a consequence, the regular contact with Europeans that helped shape Hawaiian development from the late 1780s on did not become a significant feature of Tahitian life until the expansion of the whaling industry in the 1820s. But Tahiti was not ignored; European exploitation of its resources simply took a different form, one unique to the history of European-Pacific relations.

Losing the American colonies left the English government without its chief dumping ground for convicts; it also left the planters of the British West Indies without their primary source of food for slaves, since most food supplies had been imported from the American mainland, and all American–West Indian trade was prohibited by Parliament after the Revolutionary War. Illegal trade continued, but the cost of food went up substantially. This situation dramatized what had long been acknowledged as a weakness in the West Indies—their almost complete dependence on sugar as a cash crop—and sparked new efforts to diversify the islands' economy.

The Royal Society of Arts, Manufactures, and Commerce, an organization dedicated to applying scientific knowledge to practical purposes, had initiated the search for a wide range of crops suitable to the West Indies in 1760 when it announced a prize to encourage cinnamon production. In subsequent years the Society offered premiums to promote a variety of other crops, including silk, cotton, cloves, and coffee, as well as one for anyone in the islands who could establish a botanical garden devoted to "useful plants." These programs reflected the strong interest among eighteenth-century botanists in applying their scientific knowledge to practical uses—specifically, in experimenting with transplanting commercially valuable tropical crops from their native habitats in the East Indies and elsewhere to areas under British control, with the ultimate goal of reducing England's dependence on rival nations for these goods. This was exactly the sort of undertaking that appealed to Joseph Banks, with his passion for botany and his imperialistic tendencies. He became a member of the Society in 1761 and by the early 1770s was active in directing its policies; it is likely that he was among those who first suggested the idea of transplanting breadfruit from Tahiti to the West Indies.

Such a plan was a major departure from earlier schemes for diversifying West Indian agriculture, which had focused on cultivating luxury products for the benefit of the mother country. Banks corresponded with West Indian planters, however, and knew that they were more interested in finding a new source of cheap food for their slaves than in raising more cash crops. In 1775 the Society for West India Merchants offered to finance any venture that would import an inexpensive food-producing plant. Recalling his Tahitian experience, Banks suggested breadfruit, and soon afterward both the Royal Society of Arts and the Royal Society promised prizes to the first person who transplanted breadfruit plants successfully.

Interest in such a project intensified after the American Revolution, and early in 1787 Banks and Lord Sydney began discussing the possibility of importing the plants from Tahiti. Banks at first proposed linking a breadfruit expedition with the next shipment of prisoners to New South Wales; after delivering their convicts, the ships would proceed to Tahiti to collect plants and then carry them to the West Indies before returning to England. He scrapped the idea as too complicated, however, and suggested instead a separate voyage devoted exclusively to the task of obtaining and delivering breadfruit. Government officials agreed to his plan, and selected another Cook protégé, William Bligh, master of the *Resolution* on Cook's third voyage, to command the expedition. Banks recommended David Nelson, one of his gardeners at Kew and also a former member of the *Resolution*'s crew, for the post of botanist and gardener. Their ship, named the *Bounty* by the voyage's organizers, featured Banks's second attempt at ship remodeling, this one more appropriate and more successful than the first. He designed a false floor for the great cabin with holes in which pots containing the plants would rest securely for their long voyage from Tahiti to the West Indies.

The *Bounty* sailed two days before Christmas 1787, arriving at Tahiti the following April. The islanders mobbed the ship even before it had anchored. "In less than ten minutes," Bligh recalled, "the deck was so full that I could scarce find my own people." The Tahitians wanted to know if the ship was from Lima or "Pretanee," and upon learning that it was English, asked about Cook and Banks. The *Bounty* was only the second European ship to visit

Tahiti after the *Resolution* and *Discovery* departed for the last time ten years before, and Cook's memory had taken on heroic proportions in the intervening decade. (The first visitor, the *Lady Penrhyn,* had stopped briefly a few months earlier on its way back to England after delivering prisoners to New South Wales.) Tu carried a portrait of Cook with him everywhere. A young man who showed Bligh the picture called it "Toote Earee no Otaheite" ("Cook, chief of Tahiti"), and said Cook had told Tu always to show the picture to visiting Englishmen as a way of securing their friendship. Under the circumstances, Bligh thought it prudent to let the Tahitians think Cook still lived, and ordered his men not to tell them otherwise.

Once anchored in Matavai Bay, Bligh set about establishing a shore routine in much the same way that Cook had. He appointed Peckover, the gunner, to manage trade, sent Nelson and his assistant to scout the supply of breadfruit trees, and established from talking to the islanders visiting the ship that Tu was still the principal chief of the region. Later Bligh learned that Tu had changed his name to Tynah after the birth of his son, who assumed his father's name and title according to Tahitian custom. The practice puzzled Bligh, who remarked that such frequent name-changing made it difficult for Europeans to keep everyone's identity straight.

"The Otaheiteans appear to me to be much altered for the better since 1777," was Bligh's initial judgment. He marveled at the absence of theft and the islanders' "good sense" in adopting English tools in place of their own. Only one minor incident, the theft of a tin pot, marred the first few days at anchor, and at that a chief "flew into a rage" and told Bligh to tie up and flog any islander caught stealing. Astonished at such behavior from a Tahitian chief, Bligh agreed to the punishment. Over the next few weeks, however, Bligh took a lenient attitude toward native theft, much more so than Cook ever had. When a buoy was stolen, Tynah expected the usual sort of retaliation and stayed away from the *Bounty;* Bligh, more concerned about maintaining friendship than recovering stolen property, sent a delegation to the chief's home at Pare to invite him back to Matavai Bay.

Given Bligh's mission, it was imperative that he remain on a friendly footing with Tynah and his subjects, but his relaxed attitude about Tahitian stealing also reflected his belief that the

Bounty's crew deserved much of the blame for the occasional incidents of theft because of their carelessness in guarding the ship and its cargo. When one of the boat's rudders was stolen practically from under the nose of the sailor assigned to guard it, Bligh decided to make an example of the man and had him flogged. On another occasion he punished the butcher for allowing his cleaver to be stolen. These incidents were early indications of Bligh's growing conviction that his men were lazy, stupid, even willfully destructive, an attitude that contrasted strikingly with his genuine respect and affection for the Tahitians.

Once the initial flurry of activity subsided, the Englishmen discovered that more had changed at Tahiti than the islanders' propensity to theft. The large houses and all but three of the huge double canoes that Bligh remembered were gone; the residents of Matavai Bay lived in small sheds. The animals Cook had left were gone too. About five years after Cook had left for the last time—sixty-three moons, according to the Tahitian method of recording the passage of time—the people of Moorea had joined with those of Atehuru, on the west side of Tahiti, to attack Pare. They destroyed houses and canoes, killed most of the cattle, and took the rest of the animals back to Moorea. Tynah and his family fled inland, leading Bligh to conclude that the chief was neither very powerful nor very courageous, an opinion, he claimed, that all the men of the *Resolution* except Cook had shared. (Bligh, like most Europeans, did not understand the Polynesian view of warfare, which saw no value in pursuing a losing cause and advocated flight when defeat became inevitable.) He did recognize, however, that the Mooreans had attacked in retaliation for Cook's rampage there, and that Tynah's district was the target because of his favored position with the English.

Bligh's low opinion of Tynah was reinforced by the chief's excessive eagerness for gifts. The first time they met, Tynah questioned him closely about English ships. When Bligh told him about huge ships carrying as many as a hundred guns, even drawing a picture for the disbelieving chief, Tynah's response was to ask that such a vessel be sent to Tahiti laden with English goods, including beds and high-backed chairs like those he had seen on visiting ships. Disgusted at Tynah's greed for items so unnecessary in Tahiti, Bligh nevertheless played along with him for the sake of maintaining good relations. Later, Tynah's list expanded to include axes, saws, cloth, hats, more furniture, guns, and ammunition; "in short," Bligh remarked, "every thing he could think of mentioning."

Tynah also wanted Bligh to store the chief's newly acquired possessions on board the *Bounty,* because he had no place of his own where they would be safe from theft. (This was a new problem for Tynah, brought on by his own greed for European goods; Cook and his contemporaries had often observed that the Tahitians rarely stole from one another.) Bligh set aside a large chest in his own cabin, but after about two weeks he noticed that its contents were diminishing even though he kept up a steady flow of gifts to the chief. He soon discovered that Tynah's wife had her own storage area in one of the officer's cabins and occasionally transferred some of her husband's possessions there, fearing Bligh would stop giving presents when Tynah's chest was full. Eventually, at Tynah's request, Bligh had the ship's carpenters make him a huge trunk, large enough for the chief and his wife to sleep on, for that was the only way they could secure their gifts.

Tynah was not alone in his greed for English goods. Other chiefs were equally eager to share in Bligh's largesse, and kept close track of who received what from the captain. Bligh deplored such behavior, but blamed it entirely on Europeans. "Whatever good we may have done them," he wrote, "we have given them a Taste for luxury and indolence, indeed I fear it is already too much the case among our good friends of Otaheite."

Within a week of the *Bounty*'s arrival, reciprocal exchange of gifts with the Tahitian chiefs replaced the system of trade that Bligh had attempted to establish as his principal means of obtaining provisions. It was a more expensive way of feeding his men than trade with the general population of the island, but given Bligh's dependence on the goodwill of the chiefs to accomplish his mission, he had little choice. This arrangement was a variation on the system of chiefly control over trade that Cook had encouraged in all the islands and the fur traders had found highly developed in Hawaii, and its purpose was the same: to reserve the lion's share of English trade for the chiefs. By now Tynah and his compatriots had enough experience with Europeans to know exactly how much they could expect from their visitors and by what means. Even the reduction in theft was related to the chiefs' increasing control over European trade. Cook had always believed that the islands' rulers, if not directly complicit in theft, could put a stop to it if they chose, while the chiefs, for their part, knew that nothing incurred English displeasure or cut off the flow of gifts faster than a serious theft. During Bligh's visit they clamped down on their people and professed their abhorrence of theft, which was, of course, exactly what Englishmen liked to hear.

Bligh played along with the chiefs because his need for breadfruit plants in huge quantity was likely to be seen as peculiar, perhaps even suspicious, by the Tahitians. Fearing the islanders would raise the price of the plants if they knew why the Englishmen wanted them, he ordered his men not to talk about the real purpose of their voyage; but Bligh cleverly managed to get the plants free by telling Tynah that all his presents were really gifts from King George and suggesting that the chief might like to send something to the king in return. Tynah rattled off a list of possible gifts, including breadfruit, and Bligh seized his opportunity. Within a few days, while the Tahitians dug up breadfruit plants by the dozen and the Englishmen potted them for eventual transfer to the ship, Bligh had convinced the chiefs that he was doing them a favor by taking the plants to "the *Earee Rahie no Britanee*"—the highest chief of England.

Even with the Tahitians' full cooperation, collecting and potting the plants and tending them until they were large enough to withstand the voyage to the West Indies required several weeks. Most of the ship's crew had little to do during this time, and lazy days on the beaches and in the villages of Tahiti were a remarkable contrast to the rigors of life aboard ship. Bligh spent some time expanding upon Cook's agricultural experiments, collecting the few animals that had survived the attack from Moorea in a single location where he hoped they would mate and instructing Nelson to plant a huge garden near the beach. (It proved no more successful than earlier efforts; within a day after Nelson finished his labors, Tahitians walked all over the young plants.) Most of the time, however, Bligh and his officers were free to visit Tynah and other chiefs, partaking of the banquets and witnessing the *heiva*s for which the island had become famous. They had more time on the island, and more leisure, than any previous English visitors to Tahiti.

Only an occasional problem marred the pleasures of island life. When stormy weather in early December made the *Bounty*'s anchorage in the poorly sheltered Matavai Bay precarious, Bligh considered a move to Moorea, but Tynah begged him not to go;

as an alternative, Bligh investigated the harbor at Pare and discovered it to be far superior to Matavai Bay. On Christmas day he moved the *Bounty* and its crew, establishing a precedent for subsequent European visitors to Tahiti. The harbor at Pare later became the island's major port and the site of its largest city, Papeete.

A few days after the Englishmen had settled into their new base of operations, Bligh discovered one boat and three men missing: Charles Churchill, the corporal of marines, and two seamen. Several chiefs told Bligh that the deserters had ditched the boat at Matavai and were on their way by canoe to the tiny island of Tetiaroa, about twenty miles north of Tahiti. At Bligh's request, Tynah promised to help get the deserters back, and several other Tahitian men returned the boat—in striking contrast to Cook's day, when the islanders were more likely to harbor deserters. Meanwhile, Bligh railed at his own men for allowing the desertion in the first place. He ascribed primary blame to the mate on watch, turning him before the mast as punishment, but also used the occasion to vent his wrath at all the petty officers, the most "neglectful and worthless" bunch he had ever seen. He even threatened to resort to flogging, unheard of as punishment for officers, but for the moment contented himself with lecturing them on the Articles of War. Some days later, with the help of several chiefs, Bligh discovered the deserters; he had them flogged and then dropped the matter.

The incident appeared minor to Bligh, marring only briefly the pleasures of the daily routine at Pare. After eight months, the Tahitians were used to the Englishmen's presence, and they no longer crowded the ship every day, but came and went a few at a time. The constant exchange of presents with chiefs and the rivalry between Tynah and the other chiefs for Bligh's attention faded, and Peckover resumed his initial assignment of managing trade on shore. For diversion, about an hour before sunset every evening, the Tahitians gathered on the beach and paraded up and down until dark, "to see and be seen." "I believe no ship was ever in so happy a situation," Bligh remarked. Relations with the chiefs couldn't have been better, which Bligh attributed to his decision never to take chiefs hostage, even under the most extreme provocation. They had expected such retaliation when Churchill and his companions deserted, and Bligh believed that his "moderation" in not confining them had increased their confidence.

This peaceful idyll was broken in the first week of February when, after a windy night, the crew discovered that the cable holding the main anchor had been cut. Only a single strand of rope kept the ship from tearing loose and running aground in shallow water near the beach. Close examination showed that the cable had been cut deliberately. Bligh was stunned, unable to believe that the Tahitians would commit such a "malicious act," knowing as they must that it would result in serious, perhaps irreparable, damage to the ship. He concluded that someone from another part of the island must be responsible, but still he took a harsh tone with Tynah, demanding that the chief produce the culprit. Many of the local chiefs left Pare at this unusual display of anger, but Tynah and his wife stayed, protesting to Bligh that they were not responsible and might not be able to find the person who was. Tynah suspected that the motive for cutting the cable was ill will toward himself, not the English, and Bligh was inclined to agree, although he didn't say so to Tynah, since he thought the chief was lazy and would seize any excuse to avoid trying to find the culprit.

The mystery was never resolved. Later Bligh thought perhaps one of his own men had cut the cable in a desperate attempt to damage the ship so badly that the English would be forced to remain at Tahiti.

The consequences of the Englishmen's presence at Tahiti and their close relations with Tynah and his family became increasingly clear to Bligh during his weeks on the island. He had been quick to recognize that the attack from Moorea had most likely been in retaliation for Cook's rampage there; later, around the time of the cable-cutting incident, Tynah told Bligh that he feared another attack from his enemies as soon as the English left Tahiti, and Bligh fully agreed. "There is a great deal due from England to this Man and his Family," he wrote. "By our connections with him and them we have brought him numberless Enemies." Unlike Cook, Bligh recognized that English ties to the chiefs around Matavai Bay had altered the balance of power in Tahiti, creating jealousy among other chiefs. Only fear of English retaliation kept the rival chiefs in check, and any hope Tynah had of maintaining his position of power would evaporate when Bligh departed. For this reason Bligh decided to give Tynah the guns and ammunition that he had repeatedly requested. Bligh also believed that English ships sailing the Pacific in the future had an obligation to check up on their friends at Matavai Bay: "If . . . these good and friendly people are to be destroyed from our intercourse with them, unless they have timely assistance, I think it is the business of any of his Majestys Ships that may come here to punish any such attempt." Only if the Tahitians believed that the English would avenge injury to their friends would Tynah and his followers be safe from attack during the periods when English ships were not there to protect them.

The Tahitians had become dependent on Europeans in other ways as well. They used iron knives and hatchets in preference to their own tools, made fishhooks from pins, and were adept with needles and scissors. European clothing was also in great favor, regardless of its condition; they "wear with great pride an old shoe and an old Stocking, altho it is with the utmost difficulty they can keep the shoe on their foot, and render them the most laughable Objects existing." If the spectacle of Tahitians wearing European clothing was amusing, their dependence on European tools was potentially serious. Bligh thought it demonstrated their good sense, but as James King had pointed out after Cook's last voyage, by the time the Tahitian's European tools wore out, they might well have forgotten how to make their own.

By the middle of March 1789, the breadfruit plants were well established and Bligh began to make preparations to leave. Tynah and his wife had often expressed an interest in going to England, and now the chief became importunate on the subject. He wanted to visit King George and assumed his fellow *arii rahi* would welcome him; he also wanted to see the world, and in fact it seemed to Bligh that Tynah's principal motive was his "frantick" desire to visit other countries. Bligh managed to put off Tynah's requests only by promising that he would ask King George's permission upon his return to England and would then send a bigger ship to fetch the chief and his family.

At the end of March the crew finished getting 774 pots, thirty-nine tubs, and twenty-four boxes of breadfruit plants on board—1,015 plants in all. On April 4, after nearly a year in Tahiti, the *Bounty* sailed. Just before leaving, Bligh wrote the dates of his ship's arrival and departure on the back of Cook's portrait, which Tynah had insisted be kept on board for the duration of the Englishmen's visit.

* * *

Three weeks later, the *Bounty* reached the Tongan Islands, anchoring at Nomuka. One of their first visitors was a man Bligh remembered from his previous voyage, who said Paulaho, Finau, and Topou were all living at Tongatapu and would soon be at Nomuka to greet the Englishmen. Bligh chose not to linger, however, and resumed the voyage after two days.

Five days later, at dawn, Fletcher Christian roused Bligh from his bed at knife point. Unwilling to face the rigors of shipboard life under a difficult, sometimes violent captain after the idyllic months at Tahiti, Christian, the *Bounty*'s master's mate, found support among a disgruntled crew and led twenty-five of his fellows, slightly over half the *Bounty*'s crew, in mutiny. He set Bligh and eighteen men who remained loyal to him, including the botanist Nelson, adrift in one of the ship's small boats. They had with them only about a week's supply of food and water, the logbook and Bligh's journal, a compass, and a few knives; Christian allowed them neither maps nor guns.

Bligh and his crew rigged a makeshift sail and, with that and their oars, managed to get to the coast of New South Wales, through the Great Barrier Reef, and on to Timor in the Dutch East Indies, where, ragged and half starved, they took shelter with Dutch officials until they could get passage back to England. Christian and his supporters sailed the *Bounty* back in the direction of Tahiti.

Christian knew it would be unsafe to stay at Tahiti—he hardly expected Bligh to survive the mutiny, but was certain that the *Bounty*'s failure to return would be investigated—so he headed to Tubuai in the Austral Islands south of Tahiti. Because the island had no animals, the mutineers soon returned to Tahiti. The Tahitians were surprised to see the *Bounty* again so soon, but when Christian told them that Bligh was with Cook on a newly discovered island and had sent the *Bounty* back to Tahiti to get animals, they rushed to fill the ship with gifts for Cook: 460 pigs, fifty goats, chickens, dogs, cats, and the cows Cook had given them. The Englishmen also took eleven women and thirteen men, including Hitihiti. Back at Tubuai, Christian's crew, united only in their dislike of Bligh, got into serious quarrels. A dozen or more staged a mutiny of their own, demanding to be taken back to Tahiti. Christian finally agreed, and in September the *Bounty* made its third appearance in Matavai Bay.

The group narrowly missed just the sort of encounter that Christian had feared: a meeting with another English ship. Barely three weeks before, a commercial ship en route from England to North America on a fur-trading mission had left Tahiti after a brief stop for provisions. The ship's commander was John Henry Cox, a merchant based in Canton, who was among the backers of Meares's and Colnett's voyages and had hosted the Hawaiian chief Kaiana during his stay in China. Cox and his crew learned about Bligh's visit when the Tahitians showed them Cook's portrait with Bligh's inscription on the back, but they were puzzled when several men, including Tynah, told them that the *Bounty,* commanded by an officer named "Titreano" (Christian), had returned recently from an island called "Tootate" (Tubuai).

After a brief stay at Tahiti, Christian and most of his crew sailed the *Bounty* south to remote Pitcairn Island, discovered by Carteret on his voyage in the 1760s. Sixteen men chose to remain at Tahiti, including Hitihiti and Charles Churchill. They were well supplied with weapons, and Tynah, who now called himself Pomare, saw his opportunity to subdue his rivals permanently. The Englishmen were no more able to agree among themselves than were the Tahitian chiefs, however, and Churchill and another man—Matthew Thompson, known as one of the most violent of the *Bounty*'s crew—left Matavai and took up residence with Vehiatua, chief of the Tahiti-iti district and one of Pomare's major rivals. Churchill and Vehiatua exchanged names and became *taios,* and when the chief died shortly afterward without an heir, Churchill assumed Vehiatua's position as *arii rahi* of the district according to Tahitian custom. But Thompson and Churchill quarreled; Thompson killed Churchill; and Churchill's Tahitian followers retaliated, killing Thompson.

Pomare, meanwhile, turned his attention to Moorea and asked the Englishmen remaining at Matavai for help in attacking the smaller island. They refused, but offered to repair Pomare's guns, which proved sufficient to shift the balance of power in the ensuing battle. After defeating the Mooreans, Pomare decided to take on Vehiatua, and this time the Englishmen agreed to help. Again he was victorious, and at the conclusion of the battle he could claim to rule over all Tahiti. With the help of English guns and English manpower, Pomare had become what Cook mistakenly tried to make of him, what had in fact never existed in Tahiti: the single, supreme chief over all the island.

As for Pomare's English allies, Fletcher Christian's fears for them became reality early in 1791 when the next English ship to visit Tahiti, the *Pandora,* dropped anchor in Matavai Bay. Commanded by Edward Edwards and charged with finding the mutineers and returning them to England for trial, the *Pandora*'s crew took just six weeks to capture the fourteen Englishmen remaining on Tahiti. Four of them drowned when the *Pandora* ran aground in Torres Strait. The survivors were court-martialed upon their return to England; four were acquitted on the grounds that they had been forced to join the mutiny against their will, and the rest were found guilty. Of these six, three were eventually pardoned and the others hanged.

By the time Edwards returned to England in mid-1792, Bligh was already on his way back to Tahiti for another try at collecting breadfruit, with a second ship commanded by Nathaniel Portlock. This time Bligh succeeded in delivering his seedlings to the West Indies, but in the long run the experiment was a failure: the West Indian blacks didn't like breadfruit and refused to eat it.

22. *The End of an Era: Vancouver in the Pacific*

By the beginning of the 1790s, European voyages to the Pacific had become commonplace, most of them privately financed and dedicated to commercial purposes. In less than thirty years, a vast, little-known, and seldom traversed ocean was opened for commercial exploitation—largely because of Cook's work in preparing accurate charts, demonstrating the practical use of new techniques for determining longitude, developing a dietary regimen that kept men healthy over long distances at sea, and training a generation of sailors in his methods—in a triumph of the application of scientific discovery to practical uses. Bligh's experiences in Tahiti and Portlock's and others' in Hawaii already pointed to the consequences for newly "discovered" Pacific peoples. The last great eighteenth-century voyage to the Pacific, inspired by a remarkable collage of commercial, political, and scientific motives, demonstrated beyond a doubt that European discovery of the Pacific had irrevocably changed both Europeans' view of the world and the lives of the people of the Pacific.

The voyage originated in a convoluted series of events. In 1788, Richard Cadman Etches, a London merchant who was among the backers of Portlock and Dixon's voyage as well as other fur-trading ventures, proposed that the British government establish a second convict colony on the northwest coast of North America, ostensibly to protect English interests in that area by creating a permanent settlement between Russian and Spanish territory. He also suggested that the colonizing expedition survey the coast between Nootka Sound and Cook Inlet. A settlement on the northwest coast would obviously serve his company's interests; by suggesting that it take the form of a convict colony and adding provision for a survey, Etches hoped to gain government support for the project. The proposal was never taken seriously—Banks was among those who vetoed the idea—but interest in government-sponsored activity on the northwest coast persisted. A few months after Etches made his proposal, Dixon urged government officials to reconsider the idea of a colony on the northwest coast, citing again the need for England to stake out its claim amid the competing interests of Russia, Spain, and the United States. The government was not interested in getting involved in fur traders' concerns, however, or in pursuing the few questions left unanswered by Cook's voyage to the North Pacific; a proposal from Alexander Dalrymple to continue the search for a northwest passage got no farther than Etches's or Dixon's plans.

All this changed early in 1790 when news of the Spanish seizure of James Colnett and his ships at Nootka Sound reached London. Suddenly Nootka and its surrounding region became the focus of great interest, and when England and Spain finally reached a negotiated settlement of the dispute later that year, government officials decided to send out an expedition to receive formal restitution from Spain. Two ships originally commissioned for an exploring mission in the South Atlantic, the *Discovery* and *Chatham*, were refitted for the Pacific; George Vancouver, former midshipman on Cook's second and third voyages, was chosen to command the expedition.

Although diplomacy provided the impetus for Vancouver's voyage, it was clearly wasteful to send two ships halfway around the world merely to settle a point with the Spanish, and ultimately Vancouver's political mission formed only a small part of his instructions. His major task was to survey the entire west coast of America from the Spanish settlement at San Diego to Cook Inlet, searching for a northwest passage; during the winter months he was to explore the Hawaiian Islands more thoroughly. Archibald Menzies, the young Scot who had accompanied Colnett on his first voyage to the northwest coast, would join the expedition as botanist. Banks selected Menzies and, at the government's request, drew up his instructions. By the time Banks was finished, a voyage first suggested by self-interested merchants and mounted in response to a dispute with Spain was transformed into a full-blown scientific expedition designed to complete the work left unfinished on Cook's third voyage.

The *Discovery* and *Chatham* sailed in January 1791 and anchored in Matavai Bay nearly a year later, after stops at the Cape of Good Hope and Dusky Bay, New Zealand. Vancouver, like Bligh, noticed a remarkable change in the Tahitians' behavior; although they still crowded aboard the ships in high spirits, the chiefs were eager to control theft and urged Vancouver, as a precautionary measure, not to allow too many people on board at once.

Pomare, still chief over all of Tahiti and Moorea, was then living on the smaller island, and Vancouver sent two officers to fetch him soon after they had anchored. To Vancouver's dismay, the chief and his substantial entourage settled down on board the *Discovery*, announcing their intention to remain on board for the duration of Vancouver's visit. It was an inconvenient arrangement, but Vancouver could hardly deny hospitality to such a long-standing friend of the English. After some days he managed to get rid of the group by announcing that he would henceforth be required to spend most of his time on shore. At that news, Pomare moved into a house assembled especially for him at Point Venus, where he could remain close to the English camp.

The two men had many questions for each other. Pomare wanted to know the names of the ships, where they had stopped en route to Tahiti, and where they were going from there. He asked if Banks was still alive and if he would visit Tahiti again soon, and expressed regret at hearing the news of Cook's death. Cook's portrait was still much in evidence on the island; it had become the "public register," Vancouver remarked, with inscriptions on the back from each visiting ship captain.

Vancouver, in turn, wanted to know about the political changes that had taken place since Cook's time. Pomare described his conquest of Moorea and the role of the *Bounty* mutineers. More recently, he had extended his power to claim sovereignty over Huahine, Raiatea, and Taaha as well, although his control over the last two islands remained tenuous. Their small stock of guns made Pomare and his family feel "invincible," Vancouver thought;

they even talked of an attempt to conquer Bora Bora, badgering Vancouver to give them more guns and help lead an attack on their archrivals. Vancouver refused the request for guns and manpower, although he did let Pomare have some ammunition. Undaunted, the chief countered by asking Vancouver to have King George send a ship to aid the Tahitians in battle.

Pomare and his subjects were also eager to get English tools and cloth, so much so that competition to buy from the ships' crews became keen. Iron and cloth increased in value the longer the ships stayed, exactly the opposite of Cook's experience. The chiefs took it upon themselves to regulate trade by fixing the price of hogs and warning the Englishmen when they were paying too much, probably to restrict the purchases of the common people and reserve the best items for themselves. Such behavior demonstrated the Tahitians' growing dependence on European goods, as Vancouver himself recognized. The islanders made few tools; what they did make was of inferior workmanship, intended only for trade with visiting Europeans—the original tourist trinkets. They clearly preferred English cloth to their own, and once they got more of it, they would neglect cloth-making too, Vancouver thought. He did not go so far as to suggest that Europeans had corrupted the Tahitians, as some had, but, like James King, he believed that Europeans had a responsibility to visit the islands regularly for trading purposes, "to furnish those wants which they alone have created."

Although theft was not a significant problem for Vancouver's crew, during their last few days on the island several small items were found missing. Vancouver's attitude about native theft being closer to Cook's than to Bligh's, he decided to make an example of the accused thieves and had their heads shaved in the presence of several chiefs. Menzies deplored the decision—it was the first of several occasions when he and Vancouver would disagree about treatment of native peoples—because he, as well as several others, suspected that some of the sailors were the real culprits; clothing was so valuable in trade with the Tahitians that the Englishmen stole from each other to augment their stock of trade goods.

More serious conflict occurred when a bag of linen containing shirts, sheets, and tablecloths disappeared from the *Chatham,* and, at about the same time, a chief who had promised to deliver wood failed to return with either wood or the axes lent to him. Convinced the chiefs were implicated in the theft of the linen,

Vancouver showed his displeasure by canceling a *heiva* planned by the Tahitians and threatening to destroy houses and canoes if the goods weren't returned promptly. In an attempt to head off more theft, he also told his crew to shoot anyone caught stealing, although only in the presence of an officer. The sentries did in fact shoot at two men, but missed both.

Eventually Pomare and his brothers produced a low-ranking Tahitian and accused him of stealing the linen. He in turn accused one of the minor chiefs, and Vancouver, unsure who to believe but reasonably certain that the man before him must have played some part, had him chained and confined on board ship, threatening to hang him if the linen was not returned. He also told Pomare that there would be no more presents until the linen was found. In a scene reminiscent of Cook's voyages, later that day the chiefs disappeared, other men began dismantling houses, and canoes were seen leaving the bay. William Broughton, the *Chatham*'s commander, went alone, against Vancouver's advice, to seek out Pomare in an attempt to make peace. He succeeded in persuading the chief to return to the English camp, but in the meantime, Vancouver attempted to seize the canoes leaving the harbor. The Tahitians defended themselves by throwing stones and some of the Englishmen's muskets misfired, so Vancouver's men managed to capture only one.

The incident ended in stalemate. Another day went by and still the linen was not returned; when the following day brought a favorable wind, Vancouver decided to take advantage of the opportunity to move on, saddened at the turn affairs had taken and unable to explain why, after three weeks of warm, friendly relations with the Tahitians, the visit should have soured as it did.

On March 1, a little over a month after leaving Tahiti, the men of the *Discovery* caught their first sight of Hawaii, and a day later canoes paddled out to the ship with provisions for sale at "exhorbitant prices," in Vancouver's opinion. Kaiana, the chief who had visited China, was one of the first Hawaiians to board the ship; Vancouver remembered reading about him in Meares's account of his voyage. He told Vancouver that Kamehameha now controlled the northern part of Hawaii, while he himself ruled over the southern districts. He also reported that three or four American ships and one from Macao (the last commanded by

Colnett) had visited the island the previous fall, and that the *Discovery* was the first to arrive since then.

Soon after Kaiana left, a man named Tarehooa—or Jack, as he preferred to be called—caught up with the *Discovery*. Even more widely traveled than Kaiana, he had gone to China and then to Boston in an American merchant ship. Tarehooa confirmed Kaiana's stories about recent visitors to the island, but punctured his claim to be chief over half the island; he was in fact a minor chief subject to Kamehameha, as Tarehooa was himself. Having acquired a reasonable command of English on his travels, Tarahooa worked as an interpreter for a chief named Kameeiamoku, managing trade with American ships. Now he wanted to travel with Vancouver, and Kameeiamoku was willing to let him go. Since his potential usefulness seemed obvious, Vancouver took him on.

After three more days of coasting offshore, Vancouver sailed toward Oahu, where he discovered a sheltered bay on the south side of the island. A small and rather subdued group of people, "excessively orderly and docile" in their demeanor, greeted the Englishmen; unlike the inhabitants of every other place Vancouver had visited in the South Pacific, no one here would board the *Discovery* without permission. Such politeness and restraint impressed Vancouver, as did the well-maintained village and carefully cultivated fields he visited; the villagers had even built what Vancouver called "aqueducts" to carry water to their fields. But the supply of water at Oahu was limited, and after a brief stay Vancouver decided to go on to Kauai. A day later he anchored in Waimea Bay, Cook's first landfall in the Hawaiiian Islands.

In contrast to the reception at Oahu, here men and women rushed to board the ships, the men aggressively prostituting their female relatives and the women eagerly cooperating. Vancouver had seen nothing quite like it at Tahiti, nor did he remember such "excessive wantonness" among the people of Kauai on his previous visits. Disgusted at the scene, he blamed the frequent visits of European and American ships for encouraging prostitution, and probably rightly so. As the anthropologist Marshall Sahlins has noted, prostitution of women was the Pacific islanders' chief resource, besides foodstuffs, in trade with Europeans; and since chiefs exerted increasingly tight control over the trade in food, sex was the commoners' principal means of getting European goods. The fact that the women offered themselves mostly to sailors allowed them to bypass the alliance between chiefs and ships' officers altogether, improving their chances of keeping what they got in trade, instead of being forced to surrender it to their chiefs.

The people at Oahu had told Vancouver that he would find Europeans living at Kauai, and shortly after the *Discovery* anchored, a young man named Rowbottom visited the ship. He and two other men living on the island were employed by the American ship captain John Kendrick, one of the most active of the Pacific fur traders. Kendrick wanted to broaden his trade to include sandalwood, which grew abundantly in the Hawaiian Islands, and he left the three men at Niihau to stockpile a cargo of wood while he sailed back to China and then to Boston. They had adjusted well to their isolated existence on Kauai—perhaps too well, Vancouver thought. The second member of the group to visit the *Discovery* had thoroughly adopted the native customs, "particularly in dress, or rather in nakedness. . . . The colour of his skin was little whiter than the fairest of these people." Worst of all, he reveled in what Vancouver disgustedly called "a savage way of life."

Rowbottom warned Vancouver to be on his guard against treachery from the Hawaiian Islanders; they had captured a schooner off Hawaii and murdered some of the crew, and later attempted to seize another ship off Maui. When Vancouver confronted the chiefs of Kauai with this story, they confirmed it but condemned the murders, claiming that Kaiana, whom they counted among their enemies, had instigated the plot. Later Vancouver learned that Rowbottom had his facts a bit confused. What really happened was a good deal more complicated. In 1789, the New York–based ship *Eleanora,* commanded by Simon Metcalf, had anchored off Maui. Angered by the islanders' theft, Metcalf fired into a fleet of three hundred canoes, killing over one hundred people. Coincidentally, about two weeks later Metcalf's son, in command of a small ship called the *Fair American,* stopped off the west coast of Hawaii, where a chief had once been insulted by the elder Metcalf and vowed revenge on the next ship that called. Without knowing that the *Fair American*'s captain was Metcalf's son, he and his followers attacked the ship and killed all the crew except one, a man named Isaac Davis. Soon afterward Davis met John Young, who had been accidentally left on shore when the

Eleanora called at Hawaii. The two remained on the island, eventually becoming advisers to Kamehameha.

Rowbottom's stories were supported by certificates left by ship captains who had stopped at Kauai in recent months. In addition to noting the ships' and captains' names and the dates of their visits, all but one warned subsequent visitors to exercise caution in dealing with the islanders. Vancouver took the warnings to heart. When fires blazed in the hills on their third day at Kauai, he became alarmed, thinking the fires might be a signal to attack. Rowbotom, who understood Hawaiian, picked up nothing threatening in the conversations he overheard, but still Vancouver voiced his suspicions to his officers and threatened the crowd of islanders around them with such obvious signs of anger that they began to retreat, even though they could not understand his words. Several chiefs remained behind and tried to convince Vancouver that the fires had been set merely to burn off old grass and weeds. Menzies believed them—he had been to Kauai with Colnett, and remembered seeing fires set in a similar fashion—but Vancouver decided to get his shore parties back to the ship as quickly as possible. This task proved difficult because of high surf. Several islanders ferried the Englishmen out to their ship by canoe; when the surf upset Vancouver's canoe, he suspected that the islanders had tipped the boat on purpose and refused to climb back in. Instead he swam back to the ship. Some of his men were forced to remain on the beach overnight, and Vancouver ordered the rest of his crew to arm the boats and keep watch in case of attack.

Vancouver's fears proved unfounded, and by the end of a week at Kauai his initial negative reactions and fears of violence abated. The islanders' friendly and considerate behavior contradicted the negative stories of previous visitors, and while Vancouver did not doubt the accuracy of what others had said, he suspected they had probably provoked violence by their own aggressive behavior, compounded perhaps by an imperfect understanding of each other's languages. Vancouver prided himself on his fair treatment of native peoples (at times to the point of excessive self-congratulation), and he believed that he had succeeded where other ship captains had failed in gaining the confidence of the Hawaiian chiefs. He also observed with pride that the people of Kauai had learned to distinguish among the different nationalities of their visitors and favored the English. One chief, in fact, called himself

King George, and did not like to be addressed by his Hawaiian name.*

In mid-March, Vancouver went on to Niihau in search of more provisions, but he had trouble getting what he wanted because the islanders were unwilling to accept anything but guns and ammunition in payment. Every chief at Kauai had begged for guns too, until Vancouver finally silenced them by saying that his ship and everything in it belonged to King George, who had placed a *kapu* on all weapons. Vancouver attributed the chiefs' eagerness for weapons largely to the traders who sold guns to the islanders without regard for the consequences, but he also thought Kaiana deserved a share of the blame. Best known for bringing guns back from China, "his example has produced in every chief of consequence an inordinate thirst for power, and a spirit of enterprize and ambition seems generally diffused among them."

Vancouver was most concerned about the threat posed to Europeans if the islanders acquired a substantial supply of firearms and became more proficient at using them, but he also recognized the potential dangers to the Hawaiians themselves; the very possession of sophisticated weapons encouraged warfare among the islands. Vancouver got a sense of this problem at Hawaii and Oahu, where the chiefs of each island claimed that the others were about to launch an attack; he also observed that the population of Kauai was considerably reduced since the time of Cook's voyage, and blamed it on mortality in war. In fact, war was responsible for only a portion of the mortality, as European-introduced diseases proved much more deadly than European-introduced weapons, a point Vancouver failed to understand.

When the Englishmen arrived in Nootka Sound at the end of August after surveying the entire coastline north of San Francisco, they met the *Daedalus,* recently arrived from England with fresh supplies for the expedition, exactly according to plan. But its crew had tragic news to report: the ship's commander and astronomer had been murdered during a brief stop at Oahu. The deaths only

* Taking the names of famous Englishmen became a fairly common practice in the Hawaiian Islands. By 1793, at least three chiefs had named their sons King George, and later visitors encountered men who called themselves William Pitt and William Cobbett. It was not only English statesmen who were so honored, however; other Hawaiians took the names Washington, Adams, Jefferson, Bonaparte, and even Tom Paine.

confirmed Vancouver's views about selling guns to Pacific islanders, and when he left Nootka after six weeks of negotiating with Spanish officials about the return of disputed territory, it was with the conviction that he must do something to demonstrate to the Hawaiian Islanders that they could not attack Europeans with impunity.

En route to the islands, where Vancouver intended to winter again, he stopped briefly at San Francisco and Monterey to continue discussions with the Spanish and buy cattle and sheep to take to Hawaii; like Cook, he believed in the potential benefits of English agriculture for the people of the Pacific. Aproaching the northern coast of Hawaii in mid-February, he sent the *Chatham* to survey the island's eastern and southern coasts, with a view toward locating a good harbor, while he sailed the *Discovery* south along the now-familiar west coast. To his surprise, only one ventured out to the ship; from its crew Vancouver learned that a general *kapu* kept everyone on shore and, more serious, that Kamehameha had ordered his people not to sell food to European or American visitors for anything except guns and ammunition. At this renewed evidence of the Hawaiian Islanders' passion for European weapons, Vancouver once again denounced the "injudicious conduct of unrestrained commercial adventurers."

When the *Discovery* and the *Chatham* met again, just north of Kealakekua Bay, Vancouver and Peter Puget, who had replaced Broughton as commander of the *Chatham,* debated their next move. Puget had found no harbor better than Kealakekua Bay, but Vancouver was reluctant to anchor at the scene of Cook's death, especially since Puget heard from an American ship captain that Kamehameha had fortified the bay with cannon. Meanwhile the animals deteriorated rapidly, and Vancouver finally sent them ashore in canoes with one of the local chiefs.

Three days later Kamehameha himself visited the ships, and Vancouver began to change his mind about the dangers of stopping at Kealakekua Bay. Recalling the Hawaiian as a tall, overbearing, even "savage" warrior, he was surprised to encounter a man so obviously friendly, sincere, and intelligent. Still he hesitated to take his ships into Kealakekua Bay and sent boats to scout other harbors in the vicinity, despite Kamehameha's assurances that the Englishmen would find no better anchorage. By the time the boats returned to confirm the chief's judgment, Vancouver had learned that the story about cannon had no basis in fact, and

Kamehameha's dignity and goodwill banished what remained of his reluctance to anchor at Kealakekua Bay.

The next day, the Englishmen sailed into the bay to an extravagant welcome. Dozens of canoes greeted them, led by eleven large canoes arranged in a V formation. Kamehameha, resplendent in a printed linen gown Cook had given him and a cloak and helmet of yellow feathers, stood in the prow of a large canoe at the head of the procession. After the boats paraded around the ship in precision formation, the chief went on board to present Vancouver with four helmets, several hogs, and more vegetables than the English could use. Vancouver interpreted the ceremonial greeting and gifts as Kamehameha's desire to establish good relations, and perhaps also as thanks for bringing animals to the island. He was correct on the first point; Kamehameha, unlike most of the other Hawaiian chiefs, had adopted a policy of hospitality and friendly trade toward all visiting European and American ships, and the English had a special significance because of their connection with Cook, who remained a figure of godlike importance to the Hawaiians.

As for the business of introducing animals to Hawaii, Vancouver could take some pleasure in the fact that the Hawaiian chiefs, unlike the Tahitians, seemed genuinely grateful for the cows and sheep; one chief was put out because Vancouver had given the largest cattle to Kamehameha. The chiefs did not value the animals for their use as a potential food supply, however, and Vancouver's belief to the contrary was evidence only that he shared Cook's blind spot about the value of transplanting European agricultural practices to the Pacific. In reality they coveted the animals as status symbols. European and American goods were prized by the Polynesians regardless of their usefulness, and English goods had a special cachet because of the connection with Cook. For similiar reasons, the Hawaiian chiefs vied for Vancouver's attention as well as his gifts, and the captain soon learned that he could not entertain one chief without earning the wrath of others. Puzzled about how to handle the chiefs' jealousy, he finally decided to "pay my principal court" to Kamehameha as "king" of the island. This strategy, of course, played into Kamehameha's hands as he attempted to maintain his only recently achieved hegemony over the island.

Despite the warm reception, it was impossible for Vancouver to escape the fact that his expedition was camped near the scene of

Cook's murder, and that many of these same people who were now so friendly had probably been involved in his death. Well aware of the potential for disaster, both Vancouver and Kamehameha took extraordinary precautions to prevent violent incidents between Englishmen and Hawaiians. Vancouver posted sentries on the ships and at the encampment on shore; he also directed his crew to carry no weapons other than pocket pistols when they left the ship and to keep those hidden from the islanders unless it became absolutely necessary to use them. Otherwise, he feared, given the Hawaiians' great desire for guns, the temptation to steal would be more than they could resist. Kamehameha, for his part, ordered his people not to molest the Englishmen in any way, but many of the islanders were loyal to rival chiefs, and he worried that they might harass the Englishmen as a way of attacking himself. Concerned that Vancouver would retaliate against him regardless of who instigated trouble, Kamehameha asked the captain to permit only the principal chiefs on board ship and to keep the sailors from entering any sacred places or wandering about the countryside unescorted. He would provide guides for anyone wishing to travel away from Kealakekua Bay and would spend much of his time on board the *Discovery* to keep watch over the Hawaiians' behavior.

After about a week at Kealakekua Bay, Vancouver finally visited the spot where Cook had been killed, and soon afterward he met the man who had allegedly committed the murder. Although the Hawaiian expressed sorrow at Cook's death, he displayed no sign of guilt; on the contrary, he was perfectly at ease in the Englishmen's company. His attitude was common among the Hawaiians. Genuinely sorrowful about Cook's death—"they adored Captain Cook," Puget said—they nevertheless felt no remorse at having killed him, since they viewed Cook's death as a fulfillment of the prophecies of their chiefs.

Puget, in conversation with a priest, picked up indications of the ways in which Cook continued to be venerated. Lono had become the principal god—his image accompanied Kamehameha on all official occasions—and the *heiau* dedicated to Lono at Kealakekua Bay, declared *kapu* since Cook's death, was considered so sacred that the priests would not allow the Englishmen to enter it. In addition, the *makahiki* festival had become much longer and more elaborate, an evolution that continued long after Vancouver's expedition departed; nineteenth-century visitors reported that Cook's bones and a feather-covered wooden cross symbolizing the masts and sails of his ship were carried in the annual procession around the island.

During his visit to Kealakekua Bay, Vancouver became increasingly disturbed at the stories he heard about persistent warfare among the Hawaiian Islands. During the first years of Kamehameha's rule, the chief was forced to focus his attention on resisting rival claims to power and consolidating his control over the island. Having largely accomplished this task by the time of Vancouver's visit, he nourished an ambition to extend his power to other islands. His principal opponents were Kahekili and Kaeo, the half brothers Portlock had met several years earlier; they had subsequently joined forces and between them controlled the rest of the islands. In 1790 Kamehameha briefly occupied Maui, but Kahekili and Kaeo soon drove his forces out and retaliated by attacking the coast of Hawaii north of Kealakekua Bay. That battle ended in stalemate, and since then the chiefs had preserved a shaky peace. But now Kamehameha saw his chance to launch a more effective invasion with English help. He was confident of gaining Vancouver's support, since the men from the *Daedalus* had been killed at Oahu, Kaeo's home base, and Kamehameha assumed Vancouver would wish to avenge the deaths.

Much to the chief's surprise, Vancouver refused military support, urging Kamehameha instead to use his influence among the islands to bring about peace. Eventually he managed to convince Kamehameha and most of the other Hawaiian chiefs to take the issue seriously, but further discussions revealed what appeared to be an insurmountable problem: the chiefs' deep-seated mistrust of each other. Any man who went to another island to sue for peace, they were convinced, would be killed immediately. Vancouver countered by offering to take any one of them to Maui and remain there until the peace negotiations were completed. This would not work, the chiefs maintained, because the visitor would be safe only as long as Vancouver was present. They proposed a more elaborate scheme, in which Vancouver would take one of them to Maui, persuade the chiefs there to discuss peace, and then return to Hawaii with one of the Maui chiefs to complete the negotiations. Vancouver thought this a good plan, but felt he didn't have the time to carry it out. As an alternative, the Hawaiian chiefs suggested that he stop first at Maui on his next trip to the islands, pick up one of their chiefs, and bring him to Hawaii.

This scheme seemed reasonable to Vancouver until he asked what the Hawaiians' terms would be for a peace settlement. Kamehameha wanted Maui and Molokai under his jurisdiction; Vancouver was certain the Maui chiefs would never agree. Instead he urged Kamehameha to give up his claim to Maui, pointing out that the chief had enough trouble maintaining control over Hawaii. Eventually the group accepted this plan. Vancouver was to present the idea to the Maui chiefs and then write a letter to John Young advising him of the outcome of the discussions. The Hawaiian chiefs promised to ratify the peace agreement if a chief from Maui delivered the letter.

After two weeks at Kealakekua Bay, the expedition moved on to Maui. Just before the Englishmen sailed, Kamehameha gave Vancouver an intricately crafted cloak of yellow feathers as a gift to King George. It was the most valuable object in the islands, and no one had ever worn it except himself, Kamehameha told Vancouver; he would part with it only for another great king and friend. Vancouver also learned that Kamehameha planned to levy a *kapu* once the ships had left, going into seclusion to purify himself after his association with the English. In addition, everyone who had traded with the English would be required to display his acquisitions, and the chief would exact a certain portion as his due.

Vancouver sailed to Maui with two tasks on his mind: mediating the disputes among the island chiefs and bringing the men responsible for the *Daedalus* murders to justice. The two were not really compatible, since the chiefs he was soon to meet at Maui expected vengeance and were therefore little inclined to trust him.

As the ships approached Maui, a chief boarded the *Discovery* and questioned Vancouver about his reasons for visiting the island. Since the man was obviously worried about the *Daedalus* murders, Vancouver replied that he intended no harm to Kahekili and Kaeo, although he wanted to see the murderers punished. On the chief's advice, Vancouver headed for Lahaina, a village on the west coast, where Kahekili's brother met the ships and guided them to a safe anchorage. He too had the *Daedalus* uppermost in his mind, and hastened to tell Vancouver that Kahekili and Kaeo had not been responsible for the murders; indeed, they had ordered the execution of three of the killers, although three or four others involved had escaped. Vancouver was inclined to believe the story, but he was concerned about the increasing number of attacks on European and American ships and thought stern measures were needed to demonstrate European power and prevent future incidents. He found Kahekili willing to cooperate. When Vancouver proposed that the murderers who remained at large be captured and executed publicly by their own people, the chief agreed, promising to send one of his subordinates to Oahu with the English in search of the criminals.

Kahekili was also receptive to Vancouver's peace proposals. Although the battles with Kamehameha had occurred two years earlier, the people of Maui still had not recovered. The Hawaiian warriors consumed stockpiles of food and ripped up fields, leaving the islanders too demoralized to take up cultivating them again; and Kahekili's need to keep a substantial force in readiness for further battles, along with the demands of visiting ships, contributed to further reductions in food supplies and neglect of the land. Vancouver could see the effects for himself, as the people of Lahaina had little to offer the English in trade. But though he wanted peace, Kahekili insisted that the Hawaiian chiefs could not be trusted, and all Vancouver's efforts to persuade him otherwise failed. Kaeo was more optimistic, although he too suspected Kamehameha's motives. After two days of discussion, the two chiefs finally agreed to an arrangement similar to what the Hawaiians had proposed: Vancouver would return to Maui on his next visit to the islands, taking Kaeo to Hawaii to negotiate personally with Kamehameha.

This task accomplished, Vancouver sailed to Oahu, where he intended to bring the *Daedalus* affair to a conclusion. Kahekili's subordinate acted with dispatch, delivering the alleged murderers within a day after the *Discovery* anchored. Vancouver conducted a sort of trial, in which the local chiefs declared the men's guilt, while the accused persistently maintained their innocence. Menzies believed the victims, because they had not protested when they were arrested—a sign that they were unaware of having committed any crime, in his opinion—while Vancouver was inclined to accept the word of the chiefs, not realizing that they had decided to sacrifice the four men as a way of avoiding more serious retaliation. Still, he was uneasy about the lack of evidence and the unorthodox nature of this "trial"; but his conviction of the need to maintain respect for European power outweighed considerations of justice, and the men were pronounced guilty and executed on the spot.

Vancouver did not linger at Oahu, but went on to Kauai to replenish his provisions before heading north at the end of April. After spending the summer surveying Alaska north of Nootka Sound, he turned south to complete his chart of California, this time covering the coast between San Francisco and San Diego. Shortly after Christmas the *Discovery* was back in Kealakekua Bay, which Vancouver planned to use as a base to finish his work mapping the Hawaiian Islands. Then, after a third summer off the Alaskan coast, his work would be complete.

The Englishmen anchored briefly in a small bay near Kealakekua, where Kamehameha boarded the ship. He urged them to stay a few days, but because the anchorage was poor, Vancouver suggested that Kamehameha accompany them to Kealakekua Bay instead. The chief replied that he could not, because a *kapu* imposed as part of the festival of the new year prevented him from leaving for several days. Vancouver, who like most European visitors to the islands was inclined to view *kapu* as entirely arbitrary, pressed Kamehameha to go with them anyway, challenging the chief not to allow a *kapu* to stand in the way of his friendship for the English. Torn between customary demands and his desire to stay on good terms with Vancouver, Kamehameha agreed to ask the priests' permission to allow him to return to Kealakekua Bay early, but he chided Vancouver for doubting his friendship, and added that the Englishmen should not force him to violate his country's laws. Ironically, the festival was the *makahiki,* in which the memory of Cook played a major ceremonial role.

Once settled at Kealakekua Bay, Vancouver busied himself with distributing livestock—he had brought more cattle from California—and with establishing a base on shore. He wanted to set up the portable observatory near the *heiau* as he had the previous winter, but Kamehameha said he would have to get permission from the priests. When they refused, the chief urged Vancouver to choose another place. Again Vancouver failed to understand the significance of Hawaiian religious customs, insisting on that particular spot; Kamehameha, once again caught between the demands of his country and those of his influential visitors, called together more priests and persuaded them to allow the Englishmen to use the sacred ground.

The group stayed at Kealakekua Bay several weeks, restocking the ship at a leisurely pace and enjoying the islanders' hospitality. Young and Davis were still working for Kamehameha, with a deserter from an American ship who had recently joined them. The three were trying to build a European-style boat for Kamehameha, with little success, since none was a carpenter, and Vancouver redeemed himself with the chief by directing his crew to finish the job.

At least three or four other Europeans and one Chinese man had also taken up residence on the island in the past year; Vancouver estimated that eleven foreigners were living on the island. Most were employed by chiefs, who found them valuable for their ability to use firearms, but to Vancouver they were nothing more than a "banditti of renegadoes" who would contribute to strife in the islands by feeding the chiefs' expansionist ambitions. A letter from another English captain, William Brown, waiting for Vancouver at Kealakekua Bay, said several men of similarly dubious character were living on Oahu and Kauai; at Kauai, they had supported a chief in an attack on Kaeo and Kahekili and conspired with the islanders (unsuccessfully) to capture an American ship. The situation had become so serious, according to Brown, that it was no longer safe for small ships to visit Kauai and Oahu. Vancouver tried to convince Kamehameha that the beachcombers would cause trouble sooner or later, but none of the chiefs would get rid of them. Vancouver could only hope that Young and Davis, whose abilities he respected, could keep the others in line.

Nor was Vancouver having any better luck in his efforts to bring peace to the islands. Mistrust among the chiefs remained strong, leading him to conclude that a truce could be effected only by traveling back and forth among the islands himself, carrying one counterproposal after another until an agreement was hammered out; and although he had most of the winter to spend in the islands, Vancouver continued to believe he did not have time to play the role of envoy. As an alternative, he and Kamehameha began to discuss a different strategy for bringing peace to the islands: ceding Hawaii to England, placing it permanently under the protection of European power.

It is not clear who first suggested the idea. According to Vancouver, Kamehameha had raised the possibility during the *Discovery*'s previous visit, but most of the other chiefs had been opposed. According to Menzies, the idea was Vancouver's, and Kamehameha refused to go along unless Vancouver left one of his ships behind to protect the island. Whoever first suggested the plan, by this time both Kamehameha and Vancouver were prepared to consider it seriously.

Vancouver's views were colored by his experiences negotiating England's and Spain's rival claims to Nootka Sound. Traditionally, European nations claimed sovereignty over previously unknown territory based on the principle of first discovery, but Spain had used a local chief's agreement to cede his lands as justification for claiming Nootka Sound despite the fact that Cook had discovered it. Hoping to prevent a similar conflict over the Hawaiian Islands, Vancouver encouraged Kamehameha's interest in granting Hawaii to England as a way of cementing his nation's claim to the island and its neighbors. His belief, developed over the course of three visits to the islands, that the chiefs needed English help to resolve their internecine warfare provided further justification for the plan. On this point Vancouver was a bit ahead of his time. The notion that indigenous peoples, whether they be in Asia, Africa, or the Pacific, could not govern themselves without European intervention would become a common justification for imperialism in subsequent decades; but at the end of the eighteenth century, the English government did not want to get involved in the internal affairs of distant countries, a position the Earl of Shelburne summed up when he said, "We prefer trade to dominion." (Shelburne was First Lord of the Treasury at the time.) Vancouver understood this point of view, recognizing that England was unlikely to attempt to colonize or govern Hawaii. Even so, he thought it important to establish a strong claim as insurance against interference from other nations, which might endanger England's trading activity in the islands.

Kamehameha, for his part, increasingly felt the need for British protection, but not against his enemies within the islands; his main concern was the growing number of unscrupulous foreigners visiting Hawaii each year. Many refused to pay for goods purchased or sold the islanders faulty goods, notably adulterated gunpowder and guns that broke after a few firings. In addition, Kamehameha and his fellow chiefs were beginning to comprehend the nature of the rivalry among European nations—they had been visited by the ships of four countries and understood that many others existed as well—and believed it was only a matter of time before their islands fell under the control of one of them. Under the circumstances, Kamehameha preferred England to the rest. He also expected that a formal alliance with England would give him what Vancouver had denied: the support of English ships and guns in conquering the other islands. Vancouver took credit for England's favored position among the Hawaiians, believing that they distinguished between the motives of his expedition and those of the traders and appreciated the useful objects he gave them. In fact the Hawaiians' attachment to England was bound up with the legends surrounding Cook; and, on a practical level, if Kamehameha was to turn over his island to anyone, he wanted to choose his time and his terms.

On February 25, in a formal ceremony, the chiefs ceded their island to Britain; a week later, Vancouver sailed out of Kealakekua Bay for the last time.

Just a few weeks later, the fragile truce among the chiefs of the Hawaiian Islands shattered, triggered by Kahekili's death. The conflicts that followed went on for more than a year and featured all the elements Vancouver had worried about—rivalries among chiefs, interference from foreign traders, attacks on foreign ships—ending only when Kamehameha finally achieved his ambition of extending his control over Maui, Oahu, and Molokai in addition to Hawaii. His success encouraged him to attempt taking over Kauai and Niihau as well, and with the help of Europeans living on Oahu, who built him a forty-ton ship, he launched an attack in the spring of 1796. The attempt failed, and local conflicts on Hawaii kept Kamehameha at home mending his own fences for several years afterward. He did not give up, however, but took advantage of the continued influx of foreigners in the islands to build up a fleet of about thirty western-style ships, including a 175-ton vessel obtained in an exchange with an American trader. In 1810 he finally gained control over Kauai when an American ship captain helped negotiate an arrangement under which the chief of Kauai acknowledged Kamehameha's supremacy, while Kamehameha allowed him to continue to rule his island. With this agreement, the Hawaiian Islands were united under a single chief.

Vancouver and his crew, meanwhile, went on to finish their survey of the Alaskan coast. In September 1794 they headed for

Monterey, where Vancouver finally received the documents officially ceding Nootka Sound to England, and then on to England, arriving in late August 1795, after four and a half years at sea.

Vancouver's expedition effectively completed Cook's work, producing a chart of the west coast of North America from San Diego to the Alaskan peninsula that rivaled Cook's finest efforts in accuracy and detail, and establishing English influence in the Hawaiian Islands much as Cook had in Tahiti. As a result of his work and that of his predecessors, the outlines of the Pacific and its principal island groups were known. The final answers to the two major questions that had inspired Cook's voyages—the existence of a southern continent and a northwest passage—remained unanswered, to be sure; but Cook had demonstrated that no usable land existed in the extreme southern Pacific, and Vancouver, following on the work of Cook's third voyage, laid to rest any notions of reaching Asia from Europe by way of a passage across North America.

Not that exploration was finished. Much remained to be learned about the extremes of the North and South Pacific and the land masses bordering the ocean, and the western powers would continue to sponsor scientific expeditions to the Pacific at intervals throughout the nineteenth century, from Matthew Flinders's circumnavigation of New Holland in 1801–03, to Darwin's voyage in the *Beagle* and the United States' expedition to Antarctica in the 1830s and 1840s, to the voyage of the *Challenger* to study the ocean's depths in 1870s. But these were isolated, specialized undertakings, hardly comparable to the sustained effort to explore and map the Pacific that began with Byron, reached its apogee in Cook's three voyages, and ended with Vancouver—an effort concentrated in just three decades, at a time when science, national pride, and the demands of economic growth briefly coincided to bring a little-known part of the globe within reach of European influence and understanding.

The effects were far-reaching, both for Europe and for the Pacific. Both Bligh and Vancouver were struck by how much Tahiti and Hawaii changed in the few years between Cook's voyages and their own: population declined, the islanders became increasingly dependent on European products, interisland warfare became more common. Bligh, at least, was astute enough to understand the extent to which Europeans were responsible for these changes. In both island groups, European attention to certain chiefs—Tu/Tynah/Pomare in Tahiti and Kamehameha in Hawaii—gave them a degree of influence and prestige that played on their ambitions and encouraged them to extend their influence far beyond what traditional Polynesian political structure permitted. In later years, European and American visitors sold the chiefs guns and built them ships that allowed them to wage war on a scale, and with a level of destruction, unknown to Polynesia. The favored chiefs, especially Kamehameha, gained an enormous advantage over their rivals through these contacts with foreigners, and while the islands' social hierarchy lent itself to the eventual centralization of political power, it is unlikely that either Pomare or Kamehameha would have been able to achieve control over an entire island group without foreign help.

Continued foreign influence, expanded to include missionaries as well as traders, would encourage the chiefs to embrace western religious beliefs and eventually western forms of law in preference to native traditions—not because they believed these customs superior, but because traditional ways had been undermined after years of contact with outsiders. In the second decade of the nineteenth century, the sons of Pomare and Kamehameha both abolished the *tabu/kapu* system—Pomare's son as a means of gaining the support of the missionaries (and through them foreign ship captains) in his bid to regain power over Tahiti and Moorea, and Kamehameha's son in response to pressures from his own people, who had watched Europeans ignore *kapu* for years without the ill effects that Hawaiian custom decreed would befall anyone who defied the traditional laws.

But although foreign contact brought enormous change to the Pacific islands, the Hawaiians' and Tahitians' ability to adapt to those changes, along with the highly organized nature of their social and political systems, kept them from being overrun by western powers. (The same was true of the Tongan Islands, which along with the Society and Hawaiian Islands had the most complex social organization in Polynesia.) Europeans and Americans visiting these islands met stable societies with strong native leaders who were resistant to domination by outsiders and clever at turning the foreign presence to their own uses; the visitors also came to know and appreciate the Polynesians as individuals with interests and aspirations much like their own, which itself worked against any tendency toward conquest. When England, France, and the United States eventually did impose their rule over the islands later in the nineteenth century, they were able to do so only with the involvement and support of native leaders. The Polynesians' experience contrasted sharply with that of less highly structured and highly developed groups, notably the Australian aboriginals, who were perceived by the English settlers of New South Wales as so primitive that they existed essentially without government or society, so savage that they seemed almost subhuman. Hence, as Banks himself had pointed out, they posed no obstacle to settlement, in English eyes, and were ruthlessly suppressed when they created trouble for the early colonists.

The opening of the Pacific brought change—more subtle, but significant nevertheless—to Europeans as well as to the people of the Pacific. At the most fundamental level, voyages of exploration changed Europeans' knowledge of their world, replacing geographical myths and theories with fact and introducing them to cultures much different from their own. Interest in the distant corners of the world did not abate with the end of the era of great voyages; indeed, Europeans' appetite for the exotic only seemed to increase, as the publication of geographies and travel books flourished. Books and plays about Cook's voyages and those of his contemporaries remained enormously popular, while publishers further capitalized upon Europeans' interest in distant cultures by introducing a new kind of travel book combining a descriptive narrative with drawings of representative men and women from various parts of the world, based largely on the illustrations in Cook's voyages. Thousands of English men and women viewed the substantial collection of artifacts from the third voyage purchased by Sir Ashton Lever and displayed in his museum of natural history in London. In a manner reminiscent of earlier travel literature, interest in Cook voyage artifacts (which commanded high prices whenever they came on the market) was so great that genuine items could not fill the demand, and enterprising manufacturers found a ready market for fake "Tahitian" trinkets. Other entrepreneurs turned out mass-produced copies of engravings illustrating the various accounts of Cook's voyages, as well as wallpaper depicting scenes from the South Pacific. Some well-to-do Englishmen even went so far as to add Tahitian-style verandas to their country houses and artificial lagoons to their gardens.

More fundamentally, knowledge of the Pacific challenged Europeans' views about the nature of humanity by introducing them to people vastly different from themselves, whose differences made them both attractive and repellent, at once the embodiment of a classical ideal of nature—innocents pure in their simplicity and free from the corruptions of civilization—and savages addicted to thievery, practitioners of infanticide and cannibalism, the killers of the great Cook. In their very "differentness," Pacific peoples demonstrated the full range of the human condition, exposing both the ideals to which humanity might aspire and the evil to which it might succumb, a vision profoundly disturbing to some Europeans and potentially liberating to others.

Depending on their point of view, Europeans (and Americans as well) reacted to knowledge of Pacific cultures by attempting to differentiate themselves from these more "primitive" people, impose western customs and values on them, or seek common truths about all of humanity from their way of life. Explorers and scientists, for example, continued the work of studying nonwestern peoples, but with an increasing emphasis on differentiation and classification that, while rooted in scientific method, had the effect of defining those cultures as inferior. Johann Forster had established the precedent in his attempts to define a hierarchy of cultures and trace the evolution of civilization through a progression of cultural traits. His work was enormously influential, but in contradictory ways: in his methods of gathering information and his theoretical interest in comparing cultures, Forster helped launch ethnographic studies as a field of serious scientific inquiry, but in his manner of classifying the people he studied, his work had the effect of supporting racial theories of human evolution.

Rejecting any notion of cultural relativism, Protestant evangelicals, whose numbers grew rapidly in the late eighteenth and early nineteenth centuries, viewed newly discovered Pacific peoples as heathen savages—but not so deeply rooted in evil that they could not be saved, if converted to Christian faith and values. The London Missionary Society led the way in 1797 when it sent thirty missionaries, some with wives and children, to Tahiti, the Marquesas, and the Tongan Islands; although they accomplished little at first (apart from contributing to local political rivalries) and about half went home within the first year, the society continued to send new recruits. American missionary groups were the first to set up shop in the Hawaiian Islands, in 1820, and eventually missionaries extended their reach to all the major Pacific island groups, as the early frustrations gave way to success in converting the islanders and persuading them to enact laws prohibiting behavior the missionaries considered offensive, such as extramarital sex, dancing, and Sunday game-playing.

If the missionaries saw only negatives in the Pacific islanders' culture and felt it necessary to impose European values on them, many artists and intellectuals took quite the opposite view, finding inspiration in the islanders' lives and in the very notion of the discovery of others. The English romantic poets, born in the late eighteenth century, read about Cook's exploits as children; they used the motif of the voyage and images drawn from travel literature to symbolize self-discovery. In sharp contrast to those who would remake Pacific peoples in the image of Europeans, the poets wrote of men seeking universal truths by shedding the limitations of European culture.

A few men among later generations of artists and writers became voyagers themselves, finding inspiration even in islands already changed by decades of western contact and missionary influence. On a whaling expedition in the early 1840s, Herman Melville jumped ship in the Marquesas, where he lived several weeks among the local residents before signing on with another ship; his experiences provided the material for his first novel, *Typee,* which explores the meaning of savagery and civilization. Melville's protagonist, surrounded by a tribe reputed to be vicious cannibals, discovers they are in fact gentle, welcoming people who take him in as one of their own; and yet, despite their idyllic life, he seeks only to escape and eventually returns to the hardships of a sailor's life. His experiences in the Pacific provided the setting for several of Melville's later novels as well, notably, of course, *Moby Dick;* but even more significantly, the psychological dimension of his work—his exploration of the darker side of human nature—owed much to his early experiences with "civilized" but cruel ship captains and supposedly "savage" but kind Pacific islanders. Later in the nineteenth century, the painter Paul Gauguin would seek escape from the responsibilities of modern life and inspiration for his art in Tahiti, while Robert Louis Stevenson, in an escape of a different sort, took family and household trappings with him as he sought respite from chronic sickliness in Hawaii, Tahiti, the Marquesas, and finally the remote Samoan Islands, still relatively unscathed by western influence.

Ultimately European fascination with the Pacific had conflicting results: the tendency to differentiate and classify people, combined with the urge to "civilize" and Christianize them, defined Pacific natives as inferior to Europeans and justified their exploitation in the name of God and progress, while the artistic imagination used their culture to explore the common humanity shared by all men and women and to fabricate a modern version of the myth of the noble savage, whose life might prove instructive to those willing to listen.

This paradisiacal image of the Pacific has remained remarkably pervasive despite decades of irreversible exploitation that destroyed much of what attracted Europeans to the Pacific in the first place. As late as the first decade of the twentieth century, Jack London followed in Melville's and Stevenson's footsteps on a voyage to some of the islands they had described in their writings; on a pilgrimage of a different sort, Thor Heyerdahl retraced a hypothetical route of ancient Polynesian voyagers on a raft equipped with sails in 1947; on a more mundane level, millions of tourists visit Pacific islands every year seeking a brief escape from the pressures of everyday life. Cook, who thought he had finished with the South Pacific and questioned whether the nations of Europe would ever find much use for its then-remote islands and coasts, would no doubt have deplored both the exploitation and the sentimentalization of his discoveries. Driven by his belief in the importance of scientific discovery, and imbued with the eighteenth century's conviction that scientific knowledge should be put to practical use to benefit humanity, he could not have been expected to foresee the often unfortunate use made of his achievements.

References

ABBREVIATIONS USED IN REFERENCES

Books

Banks	*The* Endeavour *Journal of Joseph Banks 1768–1771*, ed. J. C. Beaglehole, 2 vols. (Sydney, 1962).
Bligh Log	*The Log of the Bounty*, ed. Owen Rutter, 2 vols. (London, 1937).
Bligh Voyage	William Bligh, *A Voyage to the South Sea* (Dublin, 1792).
Cook	*The Journals of Captain James Cook*, ed. J. C. Beaglehole. Volume 1: The Voyage of the *Endeavour*, 1768–1771 (Cambridge, 1955). Volume 2: The Voyage of the *Resolution* and *Adventure*, 1772–1775 (Cambridge, 1961). Volume 3: The Voyages of the *Resolution* and *Discovery*, 1776–1780 (Cambridge, 1967).
J Forster	*The* Resolution *Journal of Johann Reinhold Forster 1772–1775*, ed. Michael Hoare (London, 1982).
G Forster	George Forster, *A Voyage Round the World in His Britannic Majesty's Sloop*, Resolution ... *During the Years 1772, 3, 4, and 5*, 2 vols. (London, 1786).
Parkinson	Sydney Parkinson, *A Journal of a Voyage to the South Seas, in His Majesty's Ship the* Endeavour (London, 1784).
Robertson	*The Discovery of Tahiti: A Journal of the Second Voyage of H.M.S.* Dolphin *Round the World by George Robertson*, ed. Hugh Carrington (London, 1948).

Sparrman — *A Voyage Round the World with Captain James Cook in H.M.S.* Resolution (London, 1944).

Vancouver — *A Voyage of Discovery to the North Pacific Ocean, and Round the World; . . . Performed in the Years 1790, 1791, 1792, 1793, 1794, and 1795 . . . Under the Command of Captain George Vancouver* (London, 1798).

Institutions

AT — Alexander Turnbull Library, Wellington.
BL — British Library, London.
DL — Dixson Library, Library of New South Wales, Sydney.
ML — Mitchell Library, Library of New South Wales, Sydney.
NAL — National Library of Australia, Canberra.
NMM — National Maritime Museum, London.
PRO — Public Record Office, London.
RS Archives — Archives of the Royal Society, London.

Preface

9 O.H.K. Spate: *The Spanish Lake* (Minneapolis, 1979), 2.
10 "the drama of exploration": *ibid.*
11 Effects of discovery: See especially Wilcomb E. Washburn, "The Intellectual Assumptions and Consequences of the Geographical Exploration of the Pacific," in Herman R. Friis, ed., *The Pacific Basin: A History of Its Geographical Exploration* (New York, 1967), 321–34, and Bernard Smith, *European Vision and the South Pacific: A History of Art and Ideas* (London, 1960). "The fatal impact": Alan Moorhead, *The Fatal Impact: An Account of the Invasion of the South Pacific 1767–1840* (New York, 1966).

Prologue

16 Royal Society petition to Crown: Royal Society Council Minutes, Vol. 5, 292–95, RS Archives.
17–18 Dalrymple: Alexander Dalrymple to the Council of the Royal Society, read to Council Dec. 18, 1767, quoted in Cook 1, 512.
18 Wales and Green: Royal Society Council Minutes, Dec. 18, 1767, *ibid.*

Chapter 1

23–25 Early knowledge of Pacific: O.H.K. Spate, "The Pacific as Artefact," in Neil Gunson, ed., *The Changing Pacific: Essays in Honour of H. E. Maude* (Melbourne, 1978), 36–38.

24–25 Southern continent idea: O.H.K. Spate, *The Spanish Lake* (Minneapolis, 1979), 5; James A. Williamson, *The Age of Drake* (London, 1938), 147–48. The Biblical story of Ophir is in II Chronicles 8:18. Mandeville: Spate, *Spanish Lake,* 56; J. H. Parry, *The Age of Reconnaissance: Discovery, Exploration, and Settlement, 1450–1650* (London, 1963), 8; John Noble Wilford, *The Mapmakers* (New York, 1981), 39–40.

25 Maps of southern continent: R. A. Skelton, "Map Compilation, Production, and Research in Relation to Geographical Exploration before 1600," in Herman R. Friis, ed., *The Pacific Basin: A History of Its Geographical Exploration* (New York, 1967), 47; Lawrence C. Wroth, "The Early Cartography of the Pacific," *Papers of the Bibliographical Society of America* 38:2 (1944), 169–70.

25–26 Robert Thorne: J. C. Beaglehole, *The Exploration of the Pacific,* 3rd ed. (London, 1966), 59. Search for northwest passage: Wilford, *Mapmakers,* 140–41.

26 "Spanish Lake": the phrase is taken from the title of Spate's book, cited above.

26–27 English-Spanish rivalry: Williamson, *Drake,* 150–52; Kenneth R. Andrews, *Trade, Plunder and Settlement: Maritime Enterprise and the Genesis of the British Empire, 1480–1630* (Cambridge, 1984), 139–41.

27–29 Drake's voyage: Williamson, *Drake,* 150–52, 166–73, 178–86; Beaglehole, *Exploration,* 60; Parry, *Reconnaissance,* 196; Andrews, *Trade,* 142–43, 154–59; Robert C. Ritchie, *Captain Kidd and the War Against the Pirates* (Cambridge, Mass., 1986), 12; Richard I. Ruggles, "Geographical Exploration by the English," in Friis, *Pacific Basin,* 228–29.

29–30 Political changes: Ritchie, *Captain Kidd,* 13, 29.

30–34 Dampier: James A. Williamson, Introduction to William Dampier, *A Voyage to New Holland,* esp. xxxi–xxxv, xliii, liii–iv; Willard H. Bonner, *Captain William Dampier: Buccaneer-Author* (Palo Alto, 1934), esp. 22–26; Clennell Wilkinson, *Dampier: Explorer and Buccaneer* (New York, 1929); *Dampier's Voyages,* ed. John Masefield, 2 vols. (New York, 1906); Glyndwr Williams, " 'The Inexhaustible Fountain of Gold': English Projects and Ventures in the South Seas, 1670–1750," in John E. Flint and Glyndwr Williams, eds., *Perspectives of Empire* (London, 1973), 32–33; Thrower, "Art and Science of Navigation," 34.

34 Dampier and Rogers's voyage: Woodes Rogers, *A Cruising Voyage Round the World* (London, 1928; orig. pub. 1712).

35 Travel books: Bonner, *Dampier,* 31–32; O.H.K. Spate, *Monopolists and Freebooters* (Minneapolis, 1983), 157; J. H. Plumb, *England in the Eighteenth Century* (New York, 1950). John Harris: P. J. Marshall and Glyndwr Williams, *The Great Map of Mankind: Perceptions of New Worlds in the Age of Enlightenment* (Cambridge, Mass., 1982), 48–49; Bonner, *Dampier,* 59.

36 Circulation of books: J. H. Plumb, "Reason and Unreason in the Eighteenth Century: The English Experience," in Plumb and Vinton A. Dearing, *Some Aspects of Eighteenth-Century England* (Los Angeles, 1971), 15; Richard D. Altick, *The English Common Reader: A Social History of the Mass Reading Public, 1800–1900* (Chicago, 1957), 47, 60–62; Marshall and Williams, *Great Map,* 49, 52–53; Ray McKeen Wiles, "The Relish

for Reading in Provincial England Two Centuries Ago," in Paul J. Korshin, ed., *The Widening Circle: Essays on the Circulation of Literature in Eighteenth-Century Europe* (Philadelphia, 1976), 98–99; Bonner, *Dampier,* 66.

37 Defoe's description of South Pacific island: *A New Voyage Around the World by a Course Never Sailed Before* (New York, 1974), 109–14. Defoe's fake travel books: Percy Adams, *Travelers and Travel Liars 1660–1800* (Berkeley and Los Angeles, 1962), 110–14. Imitations of *Robinson Crusoe:* Bonner, *Dampier,* 182–83; Spate, *Monopolists,* 157.

37–38 Religious books: Bonner, *Dampier,* 62–64.

38 Travel writers' promotion of exploration: Williams, "Fountain of Gold," 33–34.

39–40 Anson's voyage: *ibid.,* 46–49; Marshall and Williams, *Great Map,* 54–55.

Chapter 2

41–42 John Campbell: Howard T. Fry, *Alexander Dalrymple and the Expansion of British Trade* (Toronto, 1970), 95–96; Marshall and Williams, *Great Map,* 49.

42 "make us masters": Richard Walter, *Anson's Voyage Round the World* (London and Boston, 1928), 87.

42–43 Northwest passage myths: Adams, *Travel Liars,* 64–72, 135; Wilford, *Mapmakers,* 142; Henry W. Wagner, "Apocryphal Voyages to the Northwest Coast of America," *Proceedings of the American Antiquarian Society* 41 (1931), 179–234.

43–44 De Brosses and Callender: Marshall and Williams, *Great Map,* 260.

44–45 Dalrymple's arguments for a southern continent: Fry, *Dalrymple,* 125.

46 Planning Byron's voyage: *Byron's Journal of His Circumnavigation,* ed. Robert E. Gallagher (Cambridge, 1963), "Introduction," xx–xxi.

47 "the key to the whole": Earl of Egmont to Duke of Grafton, July 20, 1765, in *Byron's Journal,* 160–61.

47–48 Byron's experiences with Anson: Walter, *Anson's Voyage,* 137–45; *Byron's Narrative of the Loss of the* Wager (London, 1832), 8–26.

49 Description of Patagonians: [Charles Clerke], *A Voyage Round the World, in His Majesty's Ship the* Dolphin, *Commanded by the Honourable Commodore Byron* (London, 1767), 43–44. "The most amicable": *ibid.* "the nearest," "a mere shrimp": *Byron's Journal,* 46–48.

49–50 Magellan in Patagonia: Account by Antonio Pigafetta, in *Magellan's Voyage Around the World: Three Contemporary Accounts,* ed. Charles E. Nowell (Evanston, Ill., 1962), 100–07.

50 Drake's account of Patagonians: Adams, *Travel Liars,* 19, 22–23, 27. "all in the Navy": *Byron's Journal,* 60. In Falkland Islands: Clerke, *Voyage,* 73.

52 In Strait of Magellan, "all our Beef": *Byron's Journal,* 133.

53 Controversy over Falkland Islands: *ibid.,* Introduction, xxxviii–xlii; Andrew Sharp, *The Discovery of the Pacific Islands* (Oxford, 1960), 62–64.

54 Hawkesworth on Patagonians: John Hawkesworth, *An Account of the Voyages Undertaken by the Order of His Present Majesty for Making Discoveries in the Southern Hemisphere* (London, 1773), 1, 27–29. Reactions to Patagonians: Helen Wallis essay in *Byron's Journal,* 186; "legally stamped," "give a little respite," "intends a visit": quoted in *Byron's Journal,* 204, 207–08.

55 Clerke's letter to Royal Society: in *ibid.,* 211–12. Publicity about Byron's voyage: Adams, *Travel Liars,* 19–33.

Chapter 3

56 Byron's observations: *Byron's Journal,* 104–05.

57 French and Spanish reactions to proposed voyage: *Carteret's Voyage Round the World 1766–1769,* ed. Helen Wallis (Cambridge, 1965), 2, 289, 296–99, 303.

58 "in order to fire": Robertson, 23.

59 "carefully wrapt upp": Robertson, 26. Wallis and Patagonian myth: Helen Wallis essay in *Byron's Journal,* 196; *Carteret's Journal,* 2, 315–20.

60 "very pale": Samuel Wallis, Log Book of the *Dolphin,* May 31, 1767, ADM 55/35, PRO. First signs of land: Robertson, 116–21, 135.

61 "within pistol shot": Robertson, 136. First meeting with Tahitians: Robertson, 137.

62 "the most populoss": Robertson, 140. Problems getting on island: Robertson, 143–44.

63 "insolent": *ibid.* "they Laughd at us": Robertson, 146.

64 "bon-bon": *ibid.* "every lewd action": Wallis Log, June 22, 1767.

65 "As our men": Francis Wilkinson Log, June 24, 1767, PRO.

65–66 Tahitian attack: Robertson, 154–56.

66 "nails and Toys": Robertson, 156–57. "If any of us": Robertson, 157.

67 "Honesty is the Best": Wilkinson Log, June 25, 1767. Furneaux's landing party: Robertson, 160; Wallis, Logbook, June 25, 1767. "as if a Great Gun": Robertson, 161.

67–68 Tahitians' reaction to English flag: W. H. Pearson, "European Intimidation and the Myth of Tahiti," *Journal of Pacific History* 4 (1969), 209.

68 English fear of another attack: Wallis, Logbook, June 26, 1767; Robertson, 162–65. "necessary to conquer": Wallis, *loc. cit.* Exchange of cloth: Wallis, Logbook, June 27, 1767; Robertson, 165–66.

69 "the old men": Robertson, 166. "all the sailors": Robertson, 167.

69–70 First visits to island: Robertson, 172.

70 "treacherous people": Robertson, 175. "a dear Irish boy," "for not beginning," "Honour of having": Robertson, 180. "not so fair": Wilkinson Log, June 24, 1767.

71 "A Breed of English Men": *ibid.,* early July 1767. Prostitution of Tahitian women: Robertson, 185.

71–72 Theft of nails: Robertson, 207–09, quote from 109; Wallis, Logbook, July 21, 1767.

72 "made Signs": Robertson, 188. Entertaining "Jonathan": Robertson, 193.

73 Entertaining Purea: Robertson, 203–05; "Seemd to Surprize": *ibid.* Wallis visit to Purea, "with as much ease": Wallis, Logbook, July 22, 1767.

74 "Sleep with her": Robertson, 213. "Smart Sensable people": Robertson, 206.

Chapter 4

75 Wallis's failure to explore: Robertson, 231, 233–34.

76–77 Cook's instructions: in Cook 1, cclxxxii; Morton's "hints": in Cook 1, 514–19.

78–79 Cook's decision to enter Navy: J. H. Plumb, *England in the Eighteenth Century* (New York, 1950).

82 Capacity of ships: James A. Williamson, *The Age of Drake* (London, 1938), 354–55. "heavy sailer": Banks 1, 153.

83–84 Composition of crew: Cook 1, cxxvi. Names, ages, birthplaces of crew: Cook 1, 588–600; average age based on fifty men whose ages are known, out of eighty-two total.

84 "Every blockhead": Banks 1, 23.

85–86 Provisions for ship: Victualling Board minutes, June 15, 1768, in Cook 1, 613.

86–87 Scurvy preventives: James Watt, "Medical Aspects and Consequences of Cook's Voyages," in Robin Fisher and Hugh Johnston, *Captain James Cook and His Times* (Seattle, 1979), 144; Cook correspondence with Sick and Hurt Board, in Cook 1, 615, 618.

87 Further provisions, including scientific instruments: Cook correspondence with Navy Board, in Cook 1, 615–17.

88–89 Methods of determining longitude: E.G.R. Taylor, "Navigation in the Days of Captain Cook," National Maritime Museum Monographs and Reports No. 18 (London, 1974); Eric G. Forbes, "The Birth of Navigational Science," National Maritime Museum Monographs and Reports No. 10 (London, 1980); Derek Howse,"Navigation and Astronomy in Eighteenth-Century Voyages of Exploration," paper presented at the William Andrews Clark Memorial Library Seminar, 1984; D. W. Waters, "Navigation at the Time of Cook's Voyages," paper presented at Cook Conference, Simon Fraser University, 1978. Lunar-distances method: Taylor, "Navigation," 3; Neville Maskelyne, *The British Mariner's Guide* (London, 1763), Chapter 2.

89 Longitude of Tahiti: Waters, "Navigation," 16. "Dr. Masculines Method": William Hutchinson, *A Treatise on Practical Seamanship* (Liverpool, 1777), 85–92.

Chapter 5

94 Banks's experiments: Banks 1, 168, 175; "a Method": Cook 1, 74. "perhaps as miserable": Cook 1, 44–45.

96 "till we know": Banks 1, 240. "as uneven as a piece": Parkinson, 13. "of the inferior sort": John Gore Journal, April 13, 1769, ADM 51/4548, PRO. "was the truest vision": Banks 1, 252.

97 Cook's regulations: Cook 1, 75. "they clime like Munkeys": Cook 1, 77. "I never beheld": Parkinson, 14, 18.

98 "but as there were," "they were much less": Banks 1, 254. First meetings with Tahitians, encounters with women, including quotes: Banks 1, 255. Theft of spyglass, including quotes: Banks 1, 256.

99 English names for Tahitians: Robert Molyneux Journal, in Cook 1, 551; Banks 1, 260, 264; quotes from Banks. Tahitians' names for English, including quotes: Banks 1, 256.

100 "obeyed with the greatest," "If we quarreled": Parkinson, 15. Banks's views on incident: Banks 1, 265–67. "Their dispositions": Parkinson, 16. "My airy dreams": Banks 1, 257–58.

101 Setting up camp: Molyneux Journal, in Cook 1, 552–53.

102 Tahitians' stories about other Europeans: Banks 1, 286–87.

102–03 Origins of venereal disease: Howard M. Smith, "The Introduction of Venereal Disease into Tahiti: A Reexamination," *Journal of Pacific History* 10 (1975), 38–45. "little satisfaction": Cook 1, 99.

103 Tahitian imitation of English: Parkinson, *Journal,* 18. "More handily": Banks 1, 260. Tobacco incident: Banks 1, 268.

103–04 English-Tahitian trade: Cook 1, 82.

104 Dog meat: Cook 1, 103, 122, quote from 122; Parkinson, 20. Cook's taste: James King, "An Account of the Late Captain Cook, and Some Memoirs of His Life," *Universal Magazine,* July 1784, 33–40, Ms. Q144, DL; James Trevenen, "Notes on Passages from December 1777 to October 1779 Published in 1784 Quarto Edition of Captain Cook's Third Voyage," AT; quote from Trevenen. "Great and small," Banks 1, 264.

105 "with a countenance": *ibid.* Henry Jeffs incident: Banks 1, 264–65, Molyneux Journal, in Cook 1, 554.

105–06 Meeting with Purea, including quotes: Banks 1, 266; Cook 1, 85; Parkinson, 21.

106 "Obereas right hand man": Banks 1, 270. Gift of doll: Cook 1, 525–26; Banks 1, 266. Incidents since *Dolphin*'s visit: Cook 1, 84–85; Banks 1, 266; Molyneux Journal, in Cook 1, 554.

106–07 Description of fort: Parkinson, facing 16.

107 "I now thought," "It was a matter": Cook 1, 86–87. Theft of quadrant: Cook 1, 88–92; Banks 1, 270–73.

108 "for it is": Cook, 1, 88. "We are for the first time": Banks 1, 271. "We now begin": Banks 1, 274.

109 Transit of Venus: Richard Woolley, "The Significance of the Transit of Venus," in G. M. Badger, ed., *Captain Cook: Navigator and Scientist* (Canberra, 1970), 118–35; Cook's notes, in Cook Ms. 200, AT.

Chapter 6

110 Cook's views on Tahiti as base: Cook 1, 134.

111 "I fear I shall," "ladies in waiting," Banks 1, 255, 276.

112 "Her majesties person": Banks 1, 279. Comparison of Tahitian women to Greek statues: Banks to Prince of Orange, 1773, Ms. 9, NAL. Funeral procession, including quotes: Banks 1, 288–89.
113 Meeting with Amo: Cook 1, 103–04, quote from 104.
113–14 Trip around Tahiti: Cook 1, 105–13; Banks 1, 294–305.
114 "I stuck close," "jilted": Banks 1, 300; Banks's views on Tahitian social structure: Banks 1, 384.
115–16 Tahitian social structure: Robert W. Williamson, *The Social and Political Systems of Central Polynesia* (Cambridge, 1924), I, 189–96; Douglas Oliver, *Ancient Tahitian Society* (Honolulu, 1974) 1, 749–77; Edwin N. Ferdon, *Early Tahiti: As the Explorers Saw It, 1767–1797* (Tucson, 1981), 27–37; Irving Goldman, *Ancient Polynesian Society* (Chicago, 1970), 9, 14–15; and, in general, Marshall Sahlins, *Social Stratification in Polynesia* (Seattle, 1958), 37–45.
116 Purea's *marae*, including quote: Banks 1, 303–04.
117 "They have often": Banks 1, 307. "especially as they are": Cook 1, 123.
117–18 Tahitians' disgust at English customs: Banks 1, 348. Writing as tattooing: [James Magra], *A Journal of a Voyage Round the World, in His Majesty's Ship* Endeavour (London, 1771), xiii–xiv.
118 "Neither was at all": Banks 1, 289.
119 Theft of rake: Banks 1, 291–94; Cook 1, 102–03. "so as in some": Banks 1, 292.
120 Tehau i Ahurai and pistol, including quote: Banks 1, 276–77. Tehau i Ahurai and nails: Banks 1, 278–79, quote from 278.
121 Attack on English sailors: Banks 1, 289–90, 292.
122 "it appear'd to be": Cook 1, 93–94. "as if a change": Parkinson, 25.
122–23 Banks on Tahitian labor: Banks 1, 341; Banks to Prince of Orange. Cook on Tahitian culture: Cook 1, 129, 131.
123 "The great facility": Banks 1, 352. "They seem . . . contented": Parkinson, 22–24.
124 Deserters: Cook 1, 114–16, quote from 116; Molyneux, in Cook 1, 562–63.
124–25 "I do not know why": Banks 1, 312. "clamourous weeping": *ibid.*, 313.

Chapter 7

126 Tupaia's map: Cook 1, 291–94. At Huahine: Cook 1, 143–44, quote from 144.
127 Huahine relations with Bora Bora: Richard Pickersgill Journal, July 17, 1769, Adm 51/4547, PRPO; Parkinson, 73. Cook's views on southern continent: Cook 1, 137.
127–28 Tasman's voyage: J. C. Beaglehole, *The Exploration of the Pacific* (London, 1966), 143–57; "remaining unknown part": quoted in *ibid.*, 143.
128 "journalizing," "we livd like": Banks 1, 396, 388.
128–29 Comet, discovery of land: Robert Molyneux Log, Aug. 31, Sept. 7, Sept. 22, 1769, Adm 55/39, PRO.

129 "Continent we are": Banks 1, 399. First landing at New Zealand: Cook 1, 168–69; Banks 1, 400. New Zealanders' first impressions of English: Maori description, quoted in Harrison M. Wright, *New Zealand, 1769–1840* (Cambridge, Mass., 1959), 7.

129–30 War dance: John Gore Journal, Oct. 9, 1769, Adm 51/4548.

130 Meeting with New Zealanders, including quotes: William Monkhouse Journal, in Cook 1, 566–67. Theft of weapons: Cook 1, 169–70; Banks 1, 400–02; Monkhouse, in Cook 1, 566; first quote from Monkhouse, second from Banks, p. 401.

131 Capturing canoe: Cook 1, 170–71; Banks 1, 402–04.

131–32 "humane men," "but as they did": Cook 1, 171. "Black be the mark": Banks 1, 403.

132 Conversation with boys from canoe: Cook 1, 171–72, quote from 172; Banks 1, 403–05; Monkhouse, in Cook 1, 570. Leaving Poverty Bay: Cook 1, 173–74; Banks 1, 406.

132–33 Description of New Zealanders: Banks 1, 408; Parkinson, 90; Monkhouse Journal, in Cook 1, 572–74.

133 Migration to New Zealand: Peter Bellwood, *Man's Conquest of the Pacific* (New York, 1979), 326, 381–400; Janet Davidson, *The Prehistory of New Zealand* (Auckland, 1984), 26; Joan Metge, *The Maoris of New Zealand* (London, 1967), 4; Michael King, *Maori: A Photographic and Social History* (Auckland, 1983), 40–41; J. C. Beaglehole, *The Discovery of New Zealand* (London, 1961), 4–7; Wright, *New Zealand,* 10; Peter H. Buck, *The Coming of the Maori* (New Plymouth, New Zealand, 1929), 8–23; Eric Schwimmer, *The World of the Maori* (Wellington, 1974), 12; Raymond Firth, *Economics of the New Zealand Maori* (Wellington, 1959), 84. The question of whether the migration was planned and whether any return voyages were made to central Polynesia remains controversial; the current prevailing view is that the voyages were planned, but one-way only, although there is some archaeological evidence of at least one return voyage.

134 "this . . . courage": Pickersgill Journal, Oct. 10, 1769.

135 Cape Kidnappers: Cook 1, 177–78; Banks 1, 412; Monkhouse Journal, in Cook 1, 589; Parkinson, 94.

135–36 Anaura Bay: Banks 1, 417–18; Monkhouse Journal, in Cook 1, 583–85; Magra, *Voyage,* 78; quotes from Banks.

136 "Such fair appearances": Banks 1, 416.

137 "as if done," "unless the head": Parkinson, 98. New Zealand carving: Schwimmer, *World of Maori,* 93–95.

137–38 Bay of Plenty: Banks 1, 424; Cook 1, 191, quote from Banks.

138 Banks' views on New Zealand government: Banks 1, 424. New Zealanders' view of the English: Keith Sinclair, *A History of New Zealand* (New York, 1969), 32–33.

138–39 Mercury Bay: Banks 1, 425–27; Cook 1, 193–94; Pickersgill Journal, Nov. 4, 1769; John Bootie Journal, Nov. 4, 1769, Adm 51/4546, PRO; Parkinson, 103.

139 Description of *pa:* Cook, 198–200, quote from 200. "loth . . . to beleive": Banks 1, 443.

139–40 Violence at Mercury Bay: Cook 1, 196; Banks 1, 429; *Further Papers Relative to the Affairs of New Zealand* (London, 1854), 181.

140 "We knew that he was": quoted in John White, *Ancient History of the Maori* (Wellington, 1887–91), 121–30. "a little too severe": Cook 1, 196.

141 "all hands were": Banks 1, 449.

141–42 Landing on South Island: Cook 1, 234–35; Banks 1, 452–54. Description of people, including quotes:. Cook 1, 246–7; Parkinson, 119; Pickersgill Journal, Feb. 4, 1770.

142 Population of South Island: Metge, *Maori*, 4; Bellwood, *Man's Conquest*, 381; Firth, *Economics*, 114; Elsdon Best, *The Maori as He Was* (Wellington, 1924), 28. Cannibalism: Cook 1, 236; Banks 1, 455, quote from Cook.

143 "The natives seemed": Parkinson, 116. "Barbourous": Peter Briscoe Logbook, Jan. 20, 1770, DL. "Hatefull": Pickersgill Journal, same date. "Well pleasd": Banks 1, 455. Trade in human heads: Banks 1, 457–58; Cook 1, 237. Banks and people on South Island: Banks 1, 458–59, 463; quote from 459.

144 Naming Queen Charlotte Sound: Cook 1, 242; Bootie Journal, Jan. 31, 1770. English consuming food: Cook 1, 244. At Cape Turnagain: Cook 1, 249–50; Banks 1, 465.

144–45 "We once more," "was supposd," "carried us round": Banks 1, 469, 471, 472.

145 "mountains piled on mountains": Banks 1, 472.

146 "of a prodigious height," "as is very common": Cook 1, 270. "Continent mongers": Banks 1, 472. "I cannot do better": Julien Crozet, quoted in John Noble Wilford, *The Mapmakers* (New York, 1981).

Chapter 8

147 Cook's decision to sail to New Holland: Cook 1, 272–73; Banks 2, 38–39.

148 Plan for second voyage: Banks 2, 40–41, quote from 41.

148–49 Early exploration of Australia: William Dampier, *A Voyage to New Holland*, ed. James A. Williamson (London, 1939), xiii–xvi; J. C. Beaglehole, *The Exploration of the Pacific* (London, 1966), 102–03, 114–21, 157–59.

149–50 Maps: Lawrence C. Wroth, "The Early Cartography of the Pacific," *Papers of the Bibliographical Society of America* 38:2 (1944), 198–99; Emmanuel Bowen, *A Complete System of Geography* (London, 1747), 1, facing pp. viii and 1.

150 First sight of New Holland: Cook 1, 298–99.

150–51 "enormously black," "that we fancied": Banks 2, 50. "the Inhabitants of this Country": *Dampier's Voyages*, ed. John Masefield (New York, 1906), I, 453.

151 First landing: Banks 2, 53–54.

151–52 Attempts to make contact with native people: Cook 1, 304–08; Banks 2, 54–55.

152 "All they seem'd": Cook 1, 306. "the most wretched sett": Richard Pickersgill Journal, May 5, 1770, Adm 51/4547.

153 Bank's and Parkinson's work: Banks 2, 58, 62. Name Botany Bay: Cook 1, 310. Cook's and Banks's impressions of the land: Cook 1, 308–09. Sailing through shoals: Cook 1, 342–44, quote from 343.
154 Accident on reef: Banks 2, 77–80, quote from 77; Cook 1, 346.
155 Sailing damaged ship to shore: Cook 1, 346–49.
155–56 Activities at Endeavour River: Banks 2, 83, 85, 89; Cook 1, 351–53.
156 Kangaroo: Pickersgill Journal, 73a. Cook's attempts to find a safe passage: Cook 1, 354.
157 First sight of native men: Banks 2, 88–91; Cook 1, 357–59.
157–58 Description of native men: Parkinson, 146; Banks, 92–93.
158 Men's interest in English clothes: Parkinson, 153; Banks 2, 98; quotes from Banks. Incident with turtles: Cook 1, 361–62; Banks 2, 95–97, quote from Cook.
159 Sailing north: Cook 1, 368–70. On Lizard Island: Cook 1, 372–73; Banks 2, 102–03.
160 Outside the reef: Banks 2, 104; Cook 1, 375. "a wall of coral": Banks 2, 105. "All the dangers," "happy once more": Cook 1, 378, 380. "so much do great dangers": Banks 2, 109. Naming of New South Wales: Cook 1, 385–87. "having been already": Cook 1, 391.
161 "The world will hardly admit," "was it not for the pleasure": Cook 1, 380.
162 "indefatigable," "would Sea officers": Cook 1, 392. Description of New Holland: Cook 1, 392, 397, quote from 397; Banks 2, 112–13.
162–64 Description of people: Cook 1, 392–99, quote from 396; Banks 2, 99, 112–37, quote from 99.
164 "in reality they are": Cook 1, 399. "but one degree," "Providence seems to act": Banks 2, 130.
165 "we can hardly tell": Cook 1, 413. "they have gone through": Cook 1, 501.
166 Unhealthiness of Batavia: Banks 2, 194n.
167 Cook's conversations at Table Bay: Cook 1, 460–61. Bougainville: Beaglehole, *Exploration,* 222–23. Concern about publishing results of voyage: Banks 2, 249. Bougainville and Royal Society: Howard T. Fry, "Alexander Dalrymple and Captain Cook: The Creative Interplay of Two Careers," in Robin Fisher and Hugh Johnston, *Captain James Cook and His Times* (Seattle, 1979), 47. De Surville: John Dunmore, *French Explorers in the Pacific* (Oxford, 1965), I, 148–49.
168 Meeting with whaling ships: Cook 1, 474.

Chapter 9

170 "many hundred Leagues": *Public Advertiser,* Aug. 6 and 19, 1771. "the savages," "undertook to climb": *General Evening Post,* July 29, 1771. Accounts of Tahitian women: *General Evening Post, Middlesex Chronicle,* July 29, 1771; *London Evening Post,* Aug. 28, 1771; quote from *Post.*
171 "despotic without controul," description of *marae: General Evening Post, Middlesex Chronicle, loc. cit.* "of enormous size": *London Evening Post,* July

30, 1771. "hospitable, ingenious": *Public Advertiser, General Evening Post,* July 27, 1771. "some of the richest": *Public Advertiser, General Evening Post, Bingley's Journal,* all July 27, 1771. " 'tis expected": *Public Advertiser,* Aug. 21, 1771.

171–72 Franklin's comments: Franklin to Bishop Shipley, Aug. 19, 1771, AT.

172 Johnson's comments: quoted in Helen Wallis, "Postscript to the Voyages: Some New Sources and Assessments," paper presented at the Cook Conference, Simon Fraser University, 1978, 5. Banks cartoons: reprinted in Charles Lyte, *Sir Joseph Banks: Eighteenth Century Explorer, Botanist, and Entrepreneur* (London and North Pomfret, Vt., 1980), facing 127, 142.

172–73 Attention focused on Banks: *Gazeteer and New Daily Advertiser,* Aug. 5 and 24, 1771; *Public Advertiser,* Aug. 24, 1771; *London Evening Post,* Aug. 6 and 24, 1771.

173 Cook's reports to the Royal Society: Council Minutes, 6, 101–02, 107–10, RS Archives. Reports to Navy: Cook to Admiralty secretary and to Victualling Board, both July 12, 1771, in Cook 1, 631–33.

173–74 Cook's collection of artifacts: Cook to Admiralty secretary, Aug. 13, 1771, in Cook 1, 638.

174 Cook's reflections on voyage, including quote: Cook to John Walker, Aug. 17, 1771, Cook Ms. 200, AT.

174–75 Cook's summary of voyage, including quote: Cook to John Walker, Sept. 13, 1771, reprinted by Council of the Library of New South Wales (Sydney, 1970), 18–19.

175–76 Hawkesworth's engravings: Bernard Smith and Rüdiger Joppien, *The Art of Captain Cook's Voyages* (New Haven, 1985), I,. 16–19.

176 Cook's reaction to Hawkesworth, including quote: Cook 1, ccxlvi. "the entertaining matter": quoted in Wallis, "Postscript," 5. "An old black": *Letters of Horace Walpole,* ed. Mrs. Paget Toynbee (Oxford, 1904), 8, 292–93.

185 Pamphlets satirizing Tahiti: *An Epistle from Mr. Banks, Voyager, Monster-hunter, and Amoroso, to Oberea, Queen of Otaheite* (London, 1773); *An Epistle (Moral and Philosophical) from an Officer at Otaheite. To Lady GR*S**N*R* (London, 1774); *Mimosa: or, the Sensitive plant; A Poem, Dedicated to Mr. Banks* (London, 1779); *An Epistle from Oberea, Queen of Otaheite, to Joseph Banks, Esq.* (London, 1774); *A Second Letter from Oberea Queen of Otaheite, to Joseph Banks, Esq.* (London, 1774). Wesley's reactions: quoted in Cook 1, ccli. Dalrymple's attacks on Hawkesworth: quoted in Cook 1, cclii.

186–87 "Noble savage" idea: John Dunmore, *French Explorers in the Pacific* (Oxford, 1965), 1, 109–11; L. David Hammond, ed., *News from New Cythera: A Report of Bougainville's Voyage 1766–69* (Minneapolis, 1970), 53–54; Peter Gay, *The Enlightenment: An Interpretation* 2, 196–99; J. H. Parry, *Trade and Dominion: The European Overseas Empires in the Eighteenth Century* (London, 1971), 309–10; Hoxie Neale Fairchild, *The Noble Savage: A Study in Romantic Naturalism* (New York, 1928), 2–12; Henri Baudet, *Paradise on Earth: Some Thoughts on European Images of Non-European Man,* trans. Elizabeth Wentholt (Westport, Conn., 1976), 10–36, 55–57;

Antonello Gerbi, *The Dispute of the New World: The History of a Polemic, 1750–1900,* trans. Jeremy Moyle (Pittsburgh, 1973), 82, 170–71; Geoffrey Symcox, "The Wild Man's Return: The Enclosed Vision of Rousseau's *Discourses,*" in Edward Dudley and Maximillian E. Novak, *The Wild Man Within: An Image in Western Thought from the Renaissance to Romanticism* (Pittsburgh, 1972), 229; Hayden White, "The Noble Savage Theme as Fetish," in Fredi Chiapelli, ed., *First Images of America: The Impact of the New World on the Old* (Berkeley and Los Angeles, 1976), 130–33; Bernard Smith, *European Vision and the South Pacific, 1768–1850: A Study in the History of Art and Ideas* (London, 1960), 1–7.

187 "Mr. Banks is to have": *Gazetteer and New Daily Advertiser, Public Advertiser,* Aug. 29, 1771. "the celebrated Mr. Banks": *Westminster Journal,* Aug. 31–Sept. 7, 1771; *Middlesex Journal,* Sept. 3, 1771.

188 Requests for favors from Banks: Banks 1, 64–67.

189 "It was thought," "clever & Excentric," "quiet and inoffensive": John Elliott, "Memoir," n.p., NMM.

190 Cook's supervision of preparations: George Forster, *Cook the Discoverer,* ed. Gerhard Steiner, trans. P. E. Klarwill, 40. Typescript in AT.

190–91 Provisioning ships: Ms. 6, NAL.

191 Plan to take livestock to Pacific: Fry, "Alexander Dalrymple," 48.

192 Problems with modification of ship: Cook 2, 5–6. Party on board: Elliott, "Memoir." "By God I'll go": Clerke to Banks, May 15, 1772, Banks Papers A78–1, ML.

193 Banks's complaints to Sandwich, including quote: Banks to Sandwich, n.d. (but sometime in late May 1772), Banks Papers, ML. "He *swore & stomp'd":* Elliott, "Memoir."

194 "Mr. Banks seems": Memorandum from Navy Board, June 3, 1772, Banks 2, 344–45. "for I now had": quoted in Lyte, *Joseph Banks,* 154. "Some cross circumstances": Nov. 18, 1772, *Historical Records of New South Wales,* 372–73.

194–95 Biographical information on Forster: J Forster 1, 1–51.

195 "A clever, but litigious": Elliott, "Memoir."

Chapter 10

200 Catching birds: *ibid.* "Gave it for a meal," "cruel & illnatured": J Forster 1, 159.

201 Forsters at Cape: J. Forster to Thomas Pennant, Nov. 19, 1772, Banks Papers, ML. Cook learns about French explorers: Cook 2, 50. "half-cooked salt meat": Sparrman, 22–23.

202 "Extreemly fatigued": Furneaux Journal, in Cook 2, 729. "The sea beat": Richard Pickersgill Journal, Dec. 11, 1772, NMM. Forster's description of ice: J Forster 2, 193. "Plumpers": Furneaux, in Cook 2, 729. "but here if": Pickersgill Journal, Dec. 1772, NMM.

203 Trying to get around ice: Cook 2, 59–60. "to satisfy my self": Cook 2, 67.

203–04 Forster theories about ice: J Forster 2, 199–202.

204 Melting ice for water: James Burney Log, 9–10, 14, Adm 51/4523, PRO; Cook 2, 74, quote from Cook. Cook's experiment with melting ice: G Forster 1, 108. Forster puzzling about ice and fresh water: J Forster 2, 194. Forster on crew: J Forster 2, 196.

205 "seeing the people": Cook 2, 66. "in the English fashion," "throw light": Sparrman, 30–31, 34.

205–06 Continued study of ice: Cook 2, 77; J Forster 2, 215; Johann Forster, *Observations Made During a Voyage Round the World* (London, 1778), 76–97.

206 "Now have no excuse," "a thing that was": Cook 2, 78, 77–78. Surrounded by ice: Cook 2, 80–81; Furneaux, in Cook 2, 730; quote from Furneaux. "If my friend Monsieur": Charles Clerke Journal, Feb. 6, 1773, Adm 55/103, PRO.

207 "in some measure": Cook 2, 98–99. "After cruising four months": Cook 2, 106. "Even the Catts": Richard Pickersgill Journal, March 25, 1773, NMM.

208 "everybody that was able": Pickersgill Journal, *ibid.* "the real good taste," "and perhaps looked at us": G Forster 1, 124, 128. "more trees": Williams Wales Journal, in Cook 2, 777–78.

209 "this spot": G Forster 1, 177–79. First meeting with people: J Forster 2, 242–44; Cook 2, 113; Robert Cooper Journal, March 29, 1773, Adm 55/104, PRO: quote from Cooper.

210 Forsters' work: J Forster 2, 146–47, 156–57. Forster's complaints about his cabin: J Forster 2, 251. "Hope, Fear, Dispair": Pickersgill Journal, 1773, NMM. "openness and honesty": G Forster 1, 171–72.

211 Cook's meeting with New Zealanders: G Forster 1, 137–38; J Forster 2, 148–49; Cook 2, 116; Elliott, "Memoir"; first quote from J Forster, 249; second quote from Elliott.

211–12 Later meetings: G Forster 1, 140–42, 149–50; Wales Journal, in Cook 2, 780–81; Bowles Mitchell Journal, April 11, 1773, Adm 51/4555, PRO; quotes from Wales.

212–13 New Zealand population: see above, sources for page 133.

213 "richness of the furr": G Forster 1, 162. "mightily pleased": J Forster 2, 259. "though perhaps held": G Forster 1, 163.

213–14 New Zealanders' visit to the ship: Cook 2, 122; J Forster 2, 258–59; G Forster 1, 160–63; Mitchell Journal, April 19, 1772; Wales Journal, in Cook 2, 781.

214 "state of barbarism": G Forster 1, 173. "why do they not": Cook 2, 134. Planting seeds, Cook 2, 126–27. "we can by no means": Cook 2, 131.

215 At Queen Charlotte Sound: Cook 2, 161–69; quote from 167.

215–16 Visits with people of Queen Charlotte Sound: Cook 2, 171–72, quote from 171.

216 Differences from previous visit, including quotes: Cook 2, 174–75.

217 "moral characters," "If these evils": G Forster 2, 212–13. "*desperate, fearless*": Elliott, "Memoir." "Being going to leave": Wales, in Cook 2, 790.

218 "found delight": Sparrman, 49–50. "The more we consider": G Forster 1, 230. Cook's plans to sail to Tahiti: Cook 2, 172–73.

Chapter 11

219 Sailors' jokes: G Forster 1, 234. Eating dog: G Forster 234–35; Sparrman, 61.
220 "experimental": J Forster 2, 309–10. "is too important": Cook 2, 189. Cook on Bougainville: Cook 2, 193–95, quote from 195. "little paltry islands": Charles Clerke Journal, Aug. 11, 1773, Adm 55/103, PRO.
221 Geology of islands: J Forster 2, 324. Near-wreck at Vaitepiha: Sparrman, 67; J Forster 2, 326.
222 Story about another ship: *The Private Journal of James Burney,* ed. Beverley Hooper (Canberra, 1975), 64–65; J Forster 2, 333; G Forster 1, 302.
222–23 Regulations on trade: Burney, *Journal,* 62–63.
223 "how little the ideas": G Forster 1, 255–56. "took the opportunity": G Forster 1, 263. "the People were": Cook 2, 201. "a little more shy": G Forster 1, 274.
224 Forsters at Tahiti-iti, including quotes: G Forster 1, 257.
224–25 Visit to a Tahitian family: G Forster 1, 285–92.
225 G. Forster's reflections on Tahitian politics: G Forster 1, 296–97, quote from 296.
226 "The riches of his subjects," "before the corruption," "fat chief": G Forster 1, 302–03. Vehiatua's visit to the ship: J Forster 2, 333–34; G Forster 1, 306–09; Cook 2, 204. Arrival at Matavai: Cook 2, 205; G Forster 1, 319–20.
227 G. Forster's assessment of Tahitians, including quotes: G Forster 1, 321–22. "much hurt": G Forster 1, 337–38.
228 Meeting with Tu: Cook 2, 205; J Forster 2, 338–39. "Timerous Prince": Cook, *ibid.* "encreasing his bulk": G Forster 1, 332. Tu's interest in English objects: G Forster 1, 333–34; Cook 2, 207. Attempts to get hogs: Pickersgill Journal, in Cook 2, 769–71; Cook 2, 210, 212–13; G Forster 1, 357–60.
229 Cook comparing observations with Bougainville's: Cook 2, 234. Forsters' analysis of Tahiti, including quotes: G Forster 1, 365–66. Burney's comments on government: *Journal,* 69.
230 G. Forster on evolution of Tahitian government, including quotes: G Forster, 366–67. Burney on appearance; *Journal,* 67. "If the knowledge": G Forster 1, 368. *"Paradice":* Elliott, "Memoir."
231 Wales on Tahitian women, including quotes: Journal, in Cook 2, 796–97. "Incontency in unmarried people": Cook 2, 236. Marra's views on Tahitians: [John Marra], *Journal of the Resolution's Voyage, in 1772, 1773, 1774, and 1775* (London, 1775), 45.
232 "The more one is acquainted": Cook 2, 236. Taking Tahitians to England: Cook 2, 211; Clerke Journal, Sept. 1, 1773; Elliott, "Memoir." Mai's motives: G Forster 1, 391–92.
233 "a spirit of ambition": G Forster 1, 392. Porio: G Forster 1, 374–75. At Huahine: Cook 2, 217. Attack on Sparrman: Sparrman, 91–93.
234 Attack on G. Forster: J Forster 2, 363. Dispute between Cook and J. Forster: J Forster 2, 369–70, quote from 369.

Chapter 12

236 Arriving at Eua: G Forster 1, 425–26; J Forster 3, 377; Cook 2, 245.

236–37 Description of Eua: William Wales Journal, in Cook 2, 810; Charles Clerke Journal, in Cook 2, 757; G Forster 1, 431; first quote from Wales, second from Clerke, third from Forster.

237 Forsters at Eua: G Forster 1, 425–29; J Forster 3, 377–79; Cook 2, 248. Arriving at Tongatapu: Cook 2, 248–49; J Forster 3, 382.

238 "one of the most": Cook 2, 252. Exploring Tongatapu: Cook 2, 250–51; J Forster 3, 384; G Forster 1, 446–47. Wales's shoes: Wales Journal, in Cook 2, 812. Theft of books: J Forster 3, 387; William Bayley, private journal of the second voyage, Oct. 1773, AT; G Forster 1, 464–65.

239 Treatment of thieves: G Forster 1, 465. Comparison of Tongans and Tahitians: Cook 2, 267; J Forster 3, 390.

240 G. Forster's analysis of Tongans: G Forster 1, 469–77; quotes from 469.

240–41 Cook and J. Forster's views of Tongans: Cook 260–74, quote from 271; J Forster 3, 393–97, quote from 395.

242 Off Cape Kidnappers: Sparrman, 110; Cook 2, 279; quote from Sparrman. Arrival at Queen Charlotte Sound: Cook 2, 287; J Forster 3, 418; quote from Cook. J. Forster's work at Queen Charlotte Sound: J Forster 3, 424.

243 Wales's work: Records of the Board of Longitude, Vol. 42, NMM. Theft: Cook 2, 292, 288; quote from 292.

243–44 Reactions to cannibalism: Wales Journal, in Cook 2, 818–19; Cook 2, 292–94; J Forster 3, 426–27; Richard Pickersgill Journal, Nov. 24, 1773; Adm 51/4553, PRO; Charles Clerke Journal, same date, Adm 55/103, PRO; G Forster 1, 511–12; Sparrman, 115–17. Quotes: "whose hardened Soul" from J Forster 3, 426; "he became perfectly" from Wales, 819; "savage" from Cook 2, 294; "increasing civilization" from G Forster 1, 511–12. Palliser Bay: J Forster 3, 429; G Forster 1, 523; quote from G Forster.

245 Sailing south from New Zealand: Cook 2, 302; J Forster 3, 408. J. Forster's views on significance of voyage: J Forster 3, 409, 439.

246 Conditions on board ship: J Forster 3, 438; Cook 2, 304, 308–09; Clerke Journal, Dec. 22, 1773; first quote from Forster, second quote from Clerke, third quote from Cook, 304. Hitihiti: G Forster 1, 529–30, 536–37, quote from 536–37.

246–47 Christmas: G Forster 1, 535–36; Elliott, "Memoir"; Cook 2, 310; Clerke Journal, Dec. 26, 1773; Pickersgill Journal, Dec. 30, 1773; quote from Pickersgill.

247 Cook's decision to turn south: Cook 2, 312–13. Morale on board: Elliott, "Memoir"; J Forster 3, 434–44, quote from 444.

247–48 Cook's refusal to consult: Elliott, "Memoir"; G Forster 1, 540–42; quote from G Forster, 542.

248 "God knows how far": J Forster 3, 450. Cook's view on extent of ice: Cook 2, 321–22. "It is so far": J Forster 3, 451. "I will not say": Cook 2, 322.

249 Cook's plans for continuing voyage: Cook 2, 325–27; first quote from 327, second from 325. Cook's illness: Cook 2, 333.

Chapter 13

251 Arrival at Easter Island: Cook 2, 339. Exploring the island: Clerke Journal, in Cook 2, 760; Cook 2, 349–51; quote from Clerke. "It is extraordinary": Cook 2, 354.

252 Description of Easter Island: J Forster 3, 475.

252–53 Stone statues: Pickersgill Journal, March 16, 1774; Clerke Journal, in Cook 2, 760; Alfred Metraux, *Ethnology of Easter Island* (1940); Thomas S. Barthel, *The Eighth Land: The Polynesian Discovery and Settlement of Easter Island* (Honolulu, 1974).

253 "No nation will ever contend": Cook 2, 354. "the people were good": G Forster 1, 601.

253–54 Sailing to Marquesas: Cook 2, 363; J Forster 3, 480, quote from Forster.

254 "How much I wished": Wales Journal, in Cook 2, 828–29. violent encounter with Marquesans: Cook 2, 365–67; Wales Journal, in Cook 2, 829; G Forster 2, 11, 13; quote from Cook, 365. "The fine prospect": Cook 2, 369.

255 "as fine a race": Cook 2, 372–73. G. Forster's views on Marquesans: G Forster 2, 31–35. Marquesan social structure: Marshall Sahlins, *Social Stratification in Polynesia* (Seattle, 1958), 72–74.

255–56 Arrival at Tahiti: Cook 2, 381–83; J Forster 3, 496; G Forster 2, 58.

256 Hitihiti's welcome: G Forster 2, 49, 53, 75–78; quote from 53. "The existence of": G Forster 2, 57–58. "The excesses of the night": G Forster 2, 54.

265 Women's manipulation of men: [John Marra], *Journal of the Resolution's Voyage, in 1772, 1773, 1774, and 1775* (London, 1775), 176. War canoes: Cook 2, 390; G Forster 2, 105–06.

265–66 Estimates of Tahitian population: Cook 2, 384–87; J Forster 3, 498–99; G Forster 2, 60–66; Douglas Oliver, *Ancient Tahitian Society* (Honolulu, 1974), I, 33.

266 Towha's visit to ship: G Forster 2, 68–69. Towha's speech on theft: Cook 2, 388–89; G Forster 2, 78–79.

267 Dinner with Towha: G Forster 2, 79–80. J. Forster's excursions: J Forster 3, 500–506. Incident involving sentry's gun: Cook 2, 393–98; J Forster 3, 507; Wales Journal, in Cook 2, 837; quotes from Cook, 398.

268 Tu's request to Forster and Hodges: G Forster 2, 103. Tahitians wanting to go to England: Cook 2, 400–02.

269 Desertion: Cook 2, 403–04; G Forster 2, 112; first quote from Cook, second from Forster. Attacks at Huahine: Cook 2, 414–15, 418; J Forster 3, 515–18; G Forster 2, 117; first quote from Cook, second from J Forster.

270 Ori's request to Cook: Cook 2, 415–17; "a sort of Banditti," *ibid.*, 415. Performance of *arioi:* Wales Journal, in Cook 2, 840–43. "Eatooa": J Forster 3, 524.

271 G. Forster's explanation of *arioi,* including quotes: G Forster 2, 131–35. Report about Furneaux and Banks: Cook 2, 427. "Tattaow some Parou": Cook 2, 426.

272 Arrival at Nomuka: Cook 2, 441–43. "Custom house officer": Cook 2, 445. "and thought by that means": Cook 2, 444.

273 Quarrel between Wales and J. Forster: J. Forster 3, 544–55. "Friendly Archipelago": Cook 2, 449.

Chapter 14

275 Arrival at Malekula: J Forster 4, 559–60; G Forster 2, 201–02, 205–06; Cook 2, 460.
276 Description of Malekulans: G Forster 2, 206–07; J Forster 4, 565; Cook 2, 464–65; first two quotes from Cook, 464, third quote from G. Forster. Malekulans' first visit to ship: G Forster 2, 208.
276–77 Learning language: G Forster 2, 213–15, quote from 214.
277 G. Forster's views of Malekulans, including quotes: G Forster 2, 219, 227–35. Trade with Malekulans, including quotes: William Wales Journal, in Cook 2, 851.
278 "We are to float": J Forster 4, 578. Landing at Eromanga: Cook 2, 477–80; J Forster 4, 581-82.
279 People of Eromanga: Cook 2, 480; Charles Clerke Journal, Aug. 4, 1774, Adm 55/103, PRO. Landing at Tanna: Cook 2, 480–83; Wales Journal, in Cook 2, 851–52; quote from Cook, 482. Second day at Tanna: Cook 2, 484; Clerke Journal, Aug. 1774; quote from Cook.
279–80 Meeting on beach: Cook 2, 485–86; Wales Journal, in Cook 2, 853; G Forster 2, 271–73; J Forster 4, 585–86.
280 Islanders' views of English: G Forster 2, 267; Cook 2, 484–85n. Islanders leave: J Forster 4, 596. Name of island: J Forster 4, 586. "which I thought": Wales Journal, in Cook 2, 854.
280–81 Description of Tannese: J Forster 4, 586; G Forster 2, 274–77; quote from J Forster.
281 Dinner with young man: J Forster 4, 593–94; Cook 2, 488–89; G Forster 2, 288–89; quote from G. Forster.
281–82 Language: Cook 2, 504; J Forster 4, 596–97 and note; G Forster 2, 310–13; quote from Cook.
282 Islanders' attempts to control English movements: J Forster 4, 591–94; G Forster 2, 299–301, 316–17; Cook 2, 489, 493, 501; first quote from Cook, 501, second from 493.
283 Visit to volcano: Cook 2, 491–93; Wales Journal, in Cook 2, 858–59; J Forster 4, 603–04.
283–84 Views of Tannese people: G Forster 2, 360–64; Cook 2, 506. First quote from Forster, 360; second from Cook; third from Forster, 361.
285 "I firmly believe": Clerke Journal, "Account of Great Cyclades," ff. Aug. 31, 1774. "The study of nature": G Forster 2, 373. "We were in want": Cook 2, 516.
286 Arrival at New Caledonia: G Forster 2, 379; Cook 2, 531. Description of inhabitants: G Forster 2, 378, 393; Clerke Journal, in Cook 2, 763; first quote from Clerke, second from Forster, 393.
286–87 J. Forster's conflict with Cooper: J Forster 4, 646–47.

287 "this isle therefore": J Forster 2, 419–20. "Nature has been": Cook 2, 543. Description of land: *ibid.*, 538.

288 Views of people: G Forster 2, 386–87, 393; Wales Journal, in Cook 2, 867. Pickersgill's excursion, including quote: G Forster 2, 418–19.

289 Hodges's Melanesian portraits: Bernard Smith and Rüdiger Joppien, *The Art of Captain Cook's Voyages* (New Haven, 1985), 2, 99.

291 "the different characters": G Forster 2, 429. J. Forster's theories on Pacific migration: *Observations Made During a Voyage Round the World* (London, 1778), 228–44, 358–59, 575.

291–92 Modern theories about Pacific migration: Peter Bellwood, *Man's Conquest of the Pacific* (New York, 1979), 262–64, 297–318; Janet Davidson, *The Prehistory of New Zealand* (Auckland, 1984); Michael King, *Maori: A Photographic and Social History* (Auckland, 1983).

292 Forster's ranking of civilizations: *Observations,* 228–44, 285–86, 303–04.

293 Forster's context in European thought: Antonello Gerbi, *The Dispute of the New World: The History of a Polemic, 1750–1900,* trans. Jeremy Moyle (Pittsburgh, 1973), 35–40; Geoffrey Symcox, "The Wild Man's Return: The Enclosed Vision of Rousseau's *Discourses,* in Edward Dudley and Maximillian E. Novak, *The Wild Man Within: An Image in Western Thought from the Renaissance to Romanticism* (Pittsburgh, 1972), 230–31. J. Forster's opposition to noble savage idea: *Observations,* 302–03, first quote from 302, second from 303.

293–94 G. Forster's rejection of noble savage: George Forster, *Cook the Discoverer,* ed. Gerhard Steiner, trans. P. E. Klarwill, typescript, AT, 60–62, quote from 62; Erwin H. Ackerknecht, "George Forster, Alexander von Humboldt, and Ethnology," *Isis* 46 (1955), 86–87.

294 "communicate intellectual": *Observations,* 306. G. Forster's doubts, including quote: G Forster 2, 410, Ackerknecht, "George Forster," 87.

294–95 Pillars: Cook 2, 550–54; J Forster 4, 658–61; quote from Cook, 554.

295–96 Incident with ships at Queen Charlotte Sound: Cook 2, 572–73, 576; G Forster 2, 456.

296 Establishing longitude: Cook 2, 578–80; J Forster 4, 682; quote from Forster. "I never was makeing," "I have now done": Cook 2, 587.

297 Ice, including quotes: Cook 2, 622. Cook's theory about continent: Cook 2, 637.

298 "My people were yet": Cook 2, 647. "Savage," "wild," "doomed": Cook 2, 622, 638. "the intention of the voyage": Cook 2, 643.

299 "The risk one runs": Cook 2, 637–38. News about *Adventure:* Cook 2, 653.

299–300 Story about *Adventure:* James Burney Journal, in Cook 2, 749–52; Burney account of massacre in "Holograph Letters and Documents of and Relative to Captain James Cook," AT; Bayley Journal, Dec. 17, 1773, AT; quote from Burney Journal, 751.

300 "Brave, Noble, Open": Cook 2, 653. At St. Helena: Cook 2, 661–62. Four deaths: Cook to Phillip Stephens, about Aug. 1, 1775, in Cook 2, 682, 695. "Their never was": Elliott, "Memoir."

Chapter 15

301 "What, won't he *eat me*": *London Chronicle,* Aug. 4–6, 1774.

302 "How do ye do?": *London Daily Advertiser,* July 23, 1774. "Sir, You are": quoted in E. H. McCormick, *Omai, Pacific Envoy* (Auckland, 1977), 96. English interest in the exotic: McCormick, *Omai,* 76–77; T. B. Clark, *Omai: First Polynesian Ambassador to England* (San Francisco, 1940), 29.

302–03 Later reactions to Mai: *London Magazine,* August 1774, 363; *Gentlemen's Magazine,* Aug. 25, 1774, 388; *London Evening Post,* Aug. 6–9, 1774; quotes from *London Magazine.*

303 Fanny Burney's reactions, including quotes: *The Early Diary of Frances Burney, 1768–1778,* ed. Annie Raine Ellis (London, 1913), I, 334. Mai's appearance: *London Magazine,* Aug. 1774, 363; Burney, *Diary,* I, 334; Daniel Solander to an unknown correspondent, Aug. 19, 1774, "Holograph Letters and Documents of and Relative to Captain James Cook," AT; quotes from *London Magazine.*

304 "an understanding far superior": Burney, *Diary,* 1, 336–37. "It might be well," "barbarous," "We practise those virtues": *London Chronicle,* July 28–30, 1774.

304–05 Using Mai to criticize English society: *An Historic Epistle, from Omiah, to the Queen of Otaheite; Being His Remarks on the English Nation* (London, 1775), quote from 12–13.

306 Mai's reactions to England: McCormick, *Omai,* 118. Motives for going to England: Daniel Solander letter, Aug. 19, 1774, Holograph Letters, AT.

306–07 Mai's travel in England: McCormick, *Omai,* 101, 112, 116.

307 *Resolution*'s return: *London Public Advertiser, London Evening Post,* Aug. 1, 1775; *Lady's Magazine,* Aug. 1775; *London Magazine,* 1775, 434. "our continent hunting": Charles Clerke to Joseph Banks, July 30, 1775, in Cook 2, 953.

308 Solander's meeting with Cook: Solander to Joseph Banks, Aug. 1 and 22, 1775, in Cook 2, 957, 960. Cook's election to Royal Society: Council Minutes, 6, 309, RS Archives.

308–09 Cook's reports on voyage: to Admiralty secretary, Aug. 1, 1775, in Cook 2, 954–55; to John Walker, Sept. 14, 1775, in Cook 2, 696. Rumors of another voyage: Burney, *Diary,* 2, 38; Daniel Solander to Joseph Banks, Aug. 14, 1775, in Cook 2, 958.

309 Cook's appointment to Greenwich Hospital: Admiralty Minutes, Aug. 19, 1775, in Cook 2, 958. "Months ago," "a fine retreat," "ease and retirement": Cook to John Walker, Aug. 19, 1775, in Cook 2, 960.

310 Unofficial publications: Cook to Admiralty Secretary, Sept. 18, 1775, in Cook 2, 961. "inaccuracies," "vulgar expressions": quoted in J. C. Beaglehole, *The Life of Captain James Cook* (London, 1974), 462.

311 Douglas's alterations of Cook: Cook 2, 466; *A Voyage Towards the South Pole, and Round the World. Performed in His Majesty's Ships the* Resolution *and* Adventure, *in the Years 1772, 1773, 1774, and 1775* (London 1777), 2, 34; Bernard Smith and Rüdiger Joppien, *The Art of Captain Cook's Voyages* (New Haven, 1985), 2, 92. "unexceptionable to the nicest": Cook to John Douglas, Jan. 10, 1776, Egerton Ms. 2180, BL. Cook-Forster

conflict over publication, including quotes: Cook to Douglas, June 11, 1776, Egerton Ms. 2180, BL.

311–12 Sales of Cook's account: Alan Frost, "New Geographical Perspectives and the Emergence of the Romantic Imagination," in Robin Fisher and Hugh Johnston, eds., *Captain James Cook and His Times* (Seattle, 1979), 6–7.

312 Forster publications: J. Forster 1, 77, 82–83, 86, 90–91, 113. "with so much": William Wales, *Remarks on Mr. Forster's Account of Captain Cook's Last Voyage Round the World* (London, 1778), 2.

313 G. Forster's defense: *Reply to Mr. Wales's Remarks* (London, 1778).

313–14 Later lives of Forsters: Michael Hoare, *The Tactless Philosopher: Johann Reinhold Forster (1729–98)* (Melbourne, 1976), 190–203, 216, 240–42, 265–68, 277, 285–86, 299–302; Erwin H. Ackerknecht, "George Forster, Alexander von Humboldt, and Ethnology," *Isis* 46 (1955), 84–86.

314–15 Plans for another voyage: Daniel Solander to Joseph Banks, Aug. 14, 1775, in Cook 2, 958; Charles Burney to John Lind, Aug. 12, 1775, Banks Papers Q160, DL; Admiralty to Navy Board, Aug. 17, Sept. 13, and Dec. 13, 1775, Adm A/2694, A/2695, and A/2698, PRO; Navy Board to Admiralty, Jan. 23, 1776, Adm B/191, PRO.

315 Sandwich dinner with Cook: Andrew Kippis, *Captain Cook's Voyages: with an Account of His Life During the Previous and Intervening Periods* (New York, 1925, originally pub. 1788), 324–25. "It is certain": Cook to unknown correspondent, Feb. 14, 1776, Cook Ms. Q143, DL.

317 Trevenen: *A Memoir of James Trevenen*, ed. Christopher Lloyd and R. C. Anderson (London, 1959), 10–11, 14. Banks's recommendation of Nelson: Charles Lyte, *Sir Joseph Banks: Eighteenth Century Explorer, Botanist, and Entrepreneur* (London and North Pomfret, Vt.: 1980), 173.

317–18 Boswell and Cook: *Boswell: The Ominous Years, 1774–76*, ed. Charles Ryskamp and Frederick A. Pottle (New York, 1963), 308–10, 341; first quote from 308, second from 341.

318 "apparatus for recovering": Cook to Admiralty Secretary, April 9, 1776, Adm 1/1611; Admiralty Secretary to Cook, same date, Adm 2/734, PRO. Chronometer: Board of Longitude to Cook, May 25, 1776 and to William Bayley, same date; Admiralty to Cook, June 5, 1776, all Adm 2/191, PRO. Supplies for Mai: *Boswell: The Ominous Years*, 310; list of ship's provisions, Ms. 9, NAL; David Samwell Journal, in Cook 3, 989.

318–19 Concern about practicality of Mai's goods: Boswell, *loc. cit.;* G Forster 1, xvii. Attempt to teach him religion: McCormick, *Omai*, 165–67.

319 Departure: Admiralty Secretary to Cook, July 6, 1776, Adm 2/1333, PRO.

Chapter 16

323 Cook's French charts: Cook 3, 25n.

324 First meeting with people at Queen Charlotte Sound: Cook 3, 59, 62, 68–69, quote from 69.

324–25 Trading with New Zealanders: James Burney Journal, Feb. 24, 1777, Adm 51/4528, PRO; David Samwell Journal, in Cook 3, 995; quote from Burney. Prostitution: John Williamson Journal, ff. Feb. 12, 1777, Adm 55/117, PRO; Samwell Journal, in Cook 3, 995; Cook 3, 61; quotes from Samwell and Williamson.

325 Views of New Zealanders: *Captain Cook's Final Voyage: The Journal of Midshipman George Gilbert,* ed. Christine Holmes (Honolulu, 1982), 26; Thomas Edgar Log, n.d., Adm 55/21, PRO; William Anderson Journal, in Cook 3, 809–13; James King Log, Feb. 1777, Adm 55/116, PRO; first quote from Gilbert, second from Edgar.

326 Stories about *Adventure:* Cook 3, 63–64; Anderson Journal, in Cook 3, 798–99; Samwell Journal, in Cook 3, 998–99; *John Ledyard's Journal of Captain Cook's Last Voyage,* ed. James Kenneth Munford (Corvallis, Oregon: 1963), 18–21.

326–27 Visit to Mangaia: Cook 3, 78–80. Mangaians' play: *World of Polynesians,* 357–61, quote from 358. The play was seen by the missionary William Gill about 1860.

327 Decision to go to Tongan Islands: Cook 3, 91. Red feathers: Cook 3, 97; Samwell Journal, in Cook 3, 1014.

328 Theft: Cook 3, 101. First meeting with Finau: Samwell Journal, in Cook, 1015; James Burney Journal, May 4, 1777, Adm 51/4528, PRO; first quote from Samwell, second from Burney.

328–29 Gifts of food and entertainment: Cook 3, 106–08, Burney Journal, May 17, 1777; quote from Cook, 108.

329 Marines exercises and dancing: Cook 3, 109; King Journal, in Cook 3, 1361; quote from Cook. Fireworks: King Journal, in Cook 3, 1361–62, quote from 1361.

329–30 Plot to kill Cook: John Martin, *An Account of the Natives of the Tonga Islands . . . Compiled . . . from . . . Mr. William Mariner* (London, 1818), 71–72.

330 Finau's story: Cook 3, 111–12. Meeting with Paulaho: Cook 3, 115–16; Anderson Journal, in Cook 3, 880–81.

330–31 Paulaho's discussion with English: King Journal, May 28, 1777; Anderson Journal, in Cook 3, 882; Cook 3, 880–81; quotes from King.

331 Meeting with Finau at Nomuka: Cook 3, 120; Anderson Journal, in Cook 3, 886–87; Clerke Journal, in Cook 3, 1303–04; quote from Clerke. Meeting with Maealiuaki: Clerke Journal, in Cook 3, 1304–05; Cook 3, 127–28; first quote from Clerke, 1305, second from Cook, 128.

332 Tongan dancing: Anderson Journal, in Cook 3, 894–98; Cook 3, 131; first quote from Cook; second and third from Anderson, 895, 898.

333 "Had our seamen": Cook 3, 135–36.

333–34 Problems with theft: Burney Journal, June 18, 1777; Samwell Journal, in Cook 3, 1028–29; William Griffen Journal, late June 1777, DL.

334 Cook's severe punishment of thief: John Williamson Journal, ff. June 15, 1777; Samwell Journal, in Cook 3, 1029; King Journal, in Cook 3, 1362–63. Guards assigned to sailors, "insolent": Thomas Edgar Log, June 20, 1777, Adm 55/21, PRO. Attack on Samwell: Samwell Journal, in Cook 3, 1030.

334–35 Further attacks, conflict over Cook's policies: Cook 3, 136–37; Williamson Journal, in Cook 3, 1342–43; King Journal, in Cook 3, 1362.

335 "we could not go": Ledyard, *Journal,* 38.

335–37 English learn about Tongan politics: Bayley Journal, June 1777; Burney Journal, June 1777; King Journal, June 16, 1777; Anderson Journal, in Cook 3, 950–53; Clerke Journal, in Cook 3, 1309–10.

337 Modern views of Tongan political system: H. G. Cummins, "Tongan Society at the Time of European Conquest," in Neil Rutherford, ed., *Friendly Islands: A History of Tonga* (Melbourne, 1977), 64–66; Elizabeth Bott, *Tongan Society at the Time of Captain Cook's Visits* (Wellington, 1982), 20–29; Marshall Sahlins, *Social Stratification in Polynesia* (Seattle, 1958), 22–25.

337–38 Tongans' view of eclipses: Williamson Journal, ff. July 16, 1777.

338 Tongan's view of astronomical observations: King Journal, July 5, 1777. Fiji: Bayley Journal, July 9, 1777; King Journal, late June or July 1777; Gilbert, *Journal,* 35. Cook's failure to explore further: Gilbert, *ibid.*

338–39 Nature of *inasi:* Bott, *Tongan Society,* 39.

339 Description of *inasi:* Cook 3, 144–53. Williamson's complaint: Journal, ff. July 16, 1777.

339–40 Leaving animals: Cook 3, 133–34; Anderson Journal, in Cook 3, 901; Clerke Journal, in Cook 3, 1306; King Journal, July 20, 1777; quote from Clerke.

340–41 English opinions of Tongans: Clerke Journal, in Cook 3, 1308; Gilbert, *Journal,* 37; King Journal, in Cook 3, 1365–68; Bayley Journal, July 12, 1777; Anderson Journal, in Cook 3, 925–59; Cook 3, 160–80; first quote from Gilbert; second quote from Anderson, 959.

Chapter 17

342 First impressions of Tahiti: David Samwell Journal, in Cook 3, 1052. Reactions to Mai: *ibid.,* 1053; Cook 3, 186; James King Journal, in Cook 3, 1370; quote from Cook.

342–43 Cook and Mai: King Journal, in Cook 3, 1368–69.

343 Reports of Spanish ships: Charles Clerke Journal, in Cook 3, 1313–15; Samwell Journal, in Cook 3, 1055; William Bayley Log, Aug. 12, 1777, Adm 55/20, PRO; Cook 3, 224; quotes from Clerke, 1314, 1315.

343–44 Effect of Cook's return on stories about Spanish: Samwell Journal, in Cook 3, 1055; William Anderson Journal, in Cook 3, 973–74; Cook 3, 223; quote from Anderson, 974.

344 "pressing invitation": King Journal, in Cook 3, 1372. "proportioned their Caresses," "disgusting": King, 1373–74.

345 Mai and Tu: Cook 3, 192–93; Samwell Journal, in Cook 3, 1059; King Journal, in Cook 3, 1375; quote from Cook, 193. Story about Spanish ship: Clerke Journal, Aug. 27, 1777, Adm 55/22, PRO; Thomas Edgar Log, Aug. 28, 1777, Adm 55/21, PRO. Continuing discussion of Spanish: Bayley, Log and Journal, n.p., section on Tahiti, AT; Samwell Journal, in Cook 3, 1054–55; Anderson Journal, in Cook 3, 972–73; quote from Samwell.

346 Cook's views on colonization: Cook 3, 222–24, quote from 224.
346–47 Spanish visits to Tahiti: David Howarth, *Tahiti: A Paradise Lost* (New York, 1983), 109–11, 120–30.
347 Tahitians' dispute with Moorea: Cook 3, 197–99; Anderson Journal, in Cook, 977–80.
347–48 Visit to Atehuru: Cook 3, 199–206; Anderson Journal, in Cook 3, 981–85; first and second quotes from Anderson, 982, 985; third from Cook, 206.
348 Riding horses: Cook 3, 209; King Journal, in Cook 3, 1376. Tahitian football: Samwell Journal, in Cook 3, 1063. Distributing livestock: Cook 3, 194–95, quote from 194.
349 Cook's massage: Cook 3, 106–07, 212–13, quote from 212–13. *Taios:* Samwell Journal, in Cook, 1061. Sex with *arii* women: Williamson Journal, in Cook 3, 1344. Reduction in theft: Cook 3, 221.
349–50 Spyglass trick: Cook 3, 219.
350 Preparations for attack on Moorea: Alexander Home, fragments of journal, typescript, NAL. Towha's plans for revenge: Cook 3, 213–18.
351 Meeting with Maheine: Cook 3, 326–27.
351–53 Theft of goat: Cook 3, 228–31; *Zimmermann's Account of the Third Voyage of Captain Cook 1776–1780,* trans. U. Tewsley (Wellington, 1926), 23; Bayley Journal, Oct. 7, 1777, Adm 55/20, PRO; Edgar Log, Oct. 10, 1777; Clerke Log, Oct. 11, 1777.
352 "to the people": Cook 3, 229.
353 "Thus this troublesome": Cook 3, 231. "We were all again": Cook 3, 232.
353–54 Criticism of Cook's handling of this incident: Edgar Log, Oct. 10, 1777; King Journal, in Cook 3, 1383; Gilbert, *Journal,* 46–47; Clerke Log, Oct. 11, 1777; Samwell Journal, in Cook 3, 1068–69.
354 Punishment at Huahine, including quote: King Journal, in Cook 3, 1383. Arranging a home for Mai: Cook 3, 234–35; King Journal, in Cook 3, 1384.
354–55 Stolen sextant incident: Cook 3, 236–38, quote from 236; King Journal, in Cook 3, 1384–85; Edgar Log, Oct. 23, 1777; Samwell Journal, in Cook 3, 1070–73; Henry Roberts Journal, Oct. 30, 1777, DL; Clerke Log, Oct. 22–29, 1777; Gilbert, *Journal,* 48.
355 "that a baked hog": Cook 3, 237. Guns for Mai: Cook 3, 238–39.
356 Beads for Mai: Cook 3, 244. Mai's farewell: Cook 3, 240–41.
357 Tahitians' use of English flag: Bayley Log, Oct. 1777. King's criticism of Society Islanders: Journal, in Cook 3, 1390–91.

Chapter 18

359 Catching turtles: *A Memoir of James Trevenen,* ed. Christopher Lloyd and R. C. Anderson (London, 1959), 18. New islands: Cook 3, 263–64.
360 "I never saw": Cook 3, 265. Theories about discovery of Hawaiian Islands: Abraham Fornander, *An Account of the Polynesian Race,* 2 (Lon-

don, 1880), 158; O. H. K. Spate, *The Spanish Lake* (Minneapolis, 1979), 109; James King Log, Jan. 20, 1778, Adm 55/116, PRO.

360–61 Tour of *Resolution:* King, *ibid.*

361 Williamson incident: John Williamson Journal, in Cook 3, 1348–49; first quote from 1348, second and third from 1349.

362 Cook's reaction to the incident: William Griffen Journal, Jan. 1778, DL. Other officers' reactions: King Log, Jan. 20, 1778; James Burney Journal, same date, Adm 51/4528, PRO. Cook's first landing: Burney Journal, Jan. 21, 1788; Cook 3, 269. Trade with islanders: Charles Clerke Log, Jan. 22, 1778, Adm 55/22, PRO; Cook 3, 272. Concern about venereal disease: King Log, Jan. 20, 1778; David Samwell Journal, in Cook 3, 1083–84; Thomas Edgar Log, n.d., description of Kauai, Adm 55/21, PRO.

363 Chiefs' visit to ship, including quotes: Clerke Log, Jan. 24, 1778. Comparison of Hawaiians and Tahitians: King Log, Jan. 1778; Cook 3, 270, 279–84; Burney Journal, Feb. 2, 1778; Clerke Journal, in Cook 3, 1320–21.

364 Cook on coast of North America: Cook 3, 289–94. "a very desirable place": Anonymous account of Cook's third voyage, 6, DL. First meeting with Nootka: Cook 3, 295–96; Clerke Log, March 30, 1778; quote from Clerke.

365 Trade with Nootka: Cook 3, 295–97; King Journal, in Cook 3, 1393–96; Clerke Log, March 20, 1788. "we soon became": Clerke, *loc. cit.* "They were as": Cook 3, 297–98.

366–67 First impressions of Nootka: King Journal, in Cook 3, 1396–97, 1405–06; Clerke Journal, in Cook 3, 1323; Burney Journal, April 1, 1778; Samwell Journal, in Cook 3, 1098; Trevenen, *Memoir,* 21; quote from Samwell.

367 Nootka as traders: Cook 3, 298–99, 306; Samwell Journal, in Cook 3, 1091; Clerke Log, April 18, 1778; quote from Cook, 306.

367–68 Nootka monopolizing trade for themselves: Cook 3, 299; King Journal, in Cook 3, 1398; Samwell Journal, in Cook 3, 1091; quote from Cook.

368–69 Prostitution: Clerke Journal, in Cook 3, 1326; Samwell Journal, in Cook 3, 1095; King Journal, in Cook 3, 1413; William Bayley Journal, late April 1778, DL; Robin Fisher, "Cook and the Nootka," in Robin Fisher and Hugh Johnston, *Captain James Cook and His Times* (Seattle, 1979), 94–96; quote from Samwell.

369 Visit to villages: Cook 3, 303–05; King Journal, in Cook 3, 1404–14; quotes from King, 1406.

369–70 King on difficulty of understanding other cultures, including quotes: King Journal, in Cook 3, 1406–07.

370 "we sailed out": Edward Riou Log, ff. April 26, 1778, Adm 51/4529, PRO. Maps of Arctic: Cook 3, 338–43; King Log, May 11, 1778; Gerhardt Müller, *Voyages from Asia to America, for Completing the Discoveries of the North West Coast of America,* trans. Thomas Jefferys (London, 1761); Jacob von Stählin, *An Account of the New Northern Archipelago, Lately Discovered by the Russians in the Sea of Kamtschatka and Anadir,* trans. Matthew Maty (London, 1774); Glyndwr Williams, "Myth and Reality:

James Cook and the Theoretical Geography of Northwest America," in Fisher and Johnston, *Captain James Cook,* 68–75.

370–71 Sandwich Sound: Cook 3, 351–52.

371 Along coast of Alaska: King Log, May 24, 1778. "had every other": Cook 3, 361.

371–72 Exploring inlet: Cook 3, 361–68. Continuing along coast: Cook 3, 373–88.

372 "had acquired": Cook 3, 390. Unalaska: Cook 3, 390–91. "They are to the last": Clerke Log, late June 1778. "who seem'd": King Journal, in Cook 3, 1426.

373 Samwell's visit to village: Journal, in Cook 3, 1122–25, quotes from 1123. Anderson's death: Cook 3, 406; King Journal, in Cook 3, 1429–30. Into Arctic Sea: Cook 3, 414; King Log, Aug. 13, 1778.

374 First sight of ice: King Log, Aug. 17 and 18, 1778; Cook 3, 418; quote from King, Aug. 17. Walruses: Cook 3, 419–20; Samwell Journal, in Cook 3, 1134–35; Clerke Log, Aug. 20, 1778; King Log, same date; Gilbert, *Journal,* 90–92; Riou Log, Aug. 20, 1778; first quote from King, second from Gilbert. "so as to make": Cook 3, 427.

375 Cook's uncertainty about his maps: Cook 3, 435–41. Meeting with Ismailov: Cook 3, 450–53; King Journal, in Cook 3, 1446–47, 1450. "What could induce him": Cook 3, 456.

Chapter 19

376 Approaching Hawaii: *Captain Cook's Final Voyage: The Journal of Midshipman George Gilbert,* ed. Christine Holmes (Honolulu, 1982), 99. Trading offshore: Cook 3, 474–76. Venereal disease: Cook 3, 474; James King Journal, in Cook 3, 498; Edward Riou Log, Nov. 29, 1778, Adm 51/4529, PRO.

377 Trading with islanders: Cook 3, 483.

377–78 Complaints from crew: *John Ledyard's Journal of Captain Cook's Last Voyage,* ed. James Kenneth Munford (Corvallis, Oregon, 1963), 102; David Samwell Journal, in Cook 3, 1153. Cook's response: Cook 3, 479; John Watts Proceedings, Dec. 10–12, 1778, Adm 51/4559, PRO; King Journal, in Cook 3, 503–04; quotes from Watts.

378 "every innovation": Cook 3, 479–80.

378–79 Cook's illness: James Watt, "Medical Aspects and Consequences of Cook's Voyages," in Robin Fisher and Hugh Johnston, eds., *Captain James Cook and His Times* (Seattle, 1979), 155.

379 Eagerness to trade: Cook 3, 486; Samwell Journal, in Cook 3, 1154.

380 "jaded & very": King Journal, in Cook 3, 503. Hawaiians on board ship: King Journal, in Cook 3, 503–04, quote from 503. Welcoming ceremony for Cook: Samwell Journal, in Cook 3, 1161–62; King Journal, in Cook 3, 506–07n. "a long, & rather": King, 506.

381 "Is looked upon": Samwell, 1162. For a contrasting point of view, see Peter H. Buck, "Cook's Discovery of the Hawaiian Islands," *Bishop Museum Bulletin* 186 (1945), 26–44. Respect shown Cook: Ledyard, *Journal*, 105; Charles Clerke Journal, in Cook 3, 596; quote from Clerke. Parea on ship: Thomas Edgar Log, Jan. 20, 1779, Adm 55/21, PRO.

381–82 Sailors stealing nails: Samwell Journal, in Cook 3, 1164.

382 "the bishop": Clerke Journal, in Cook 3, 597. Attempts to bribe priest: Samwell Journal, in Cook 3, 1161. First meeting with Kalaniopuu: Samwell Journal, in Cook 3, 1168–69; King Journal, in Cook 3, 512; quote from King. Chiefs' ambivalence about English: Marshall Sahlins, *Historical Metaphors and Mythical Realities: Structure in the Early History of the Sandwich Islands Kingdom* (Ann Arbor, 1981), 36.

383 Hawaiian social structure: Marshall Sahlins, *Social Stratification in Polynesia* (Seattle, 1958), 13–17, 47. Chiefs' role in keeping subjects under control: King Journal, in Cook 3, 515, 524–25.

384 English view of Hawaiians: King Journal, in Cook 3, 612; Clerke Journal, in Cook 3, 597–98. Hawaiian's attempt to forge iron: King Journal, in Cook 3, 625–27. Hawaiians' invitation to King, including quote: King Journal, in Cook 3, 519.

384–85 Hawaiians' reaction to Englishmen's return: King Journal, in Cook 3, 528; Gilbert, *Journal*, 104; Ledyard, *Journal*, 141; Samwell Journal, in Cook 3, 1190; *Zimmermann's Account of the Third Voyage of Captain Cook 1776–1780*, trans. U. Tewsley (Wellington, 1926), 37; quote from King.

385 "every day produc'd": Clerke Journal, in Cook 3, 531–32. Theft of tongs: King Journal, in Cook 3, 528–30; Samwell Journal, in Cook 3, 1191–93; Edgar Log, in Cook 3, 1359–60; Edgar Log, Feb. 13, 1779; Clerke Journal, in Cook 3, 532–33.

385–86 Theft of *Discovery*'s boat: Samwell Journal, in Cook 3, 1194; King Journal, in Cook 3, 550.

387 Cook's death: Samwell Journal, in Cook 3, 1196–1200; Edgar Log, Feb. 14, 1779; William Bayley Log, same date, Adm 55/29, PRO; Clerke Journal, in Cook 3, 536–37. "ever too tender": Samwell Journal, in Cook 3, 1196–97. Phillips urging Cook to retreat: Gilbert, *Journal*, 106.

388 "a general silence": Gilbert, *Journal*, 107–08.

388–89 Crew's view of Cook: *A Memoir of James Trevenen*, ed. Christopher Lloyd and R. C. Anderson (London, 1959), 21; Zimmermann, *Account*, 41–42. Cook's self-denial: James King, "An Account of the Late Captain Cook, and Some Memoirs of His Life," *Universal Magazine*, July 1784, 38.

388 "won for him": George Forster, *Cook the Discoverer*, ed. Gerhard Steiner, trans. P. E. Klarwill, typescript, AT, 67.

389 "converse familiarly": Trevenen, *Memoir*, 20. "We all felt": Zimmermann, *Account*, 43. Similar sentiment expressed by Isaac Smith, in letter to Hawke Locker, Oct. 3, 1800, Cook Documents Safe 1/83, ML. "I, as well as": Trevenen, *Memoir*, 23. "instinctive knowledge": Zimmermann, *Account*, 42. "his mind": George Forster, *Cook the Discoverer*, 60.

390 "done with the Southern": Cook 2, 587. "His temper might": King, "Account of Cook," 38. Evacuation of men on shore: Clerke Journal, in Cook 3, 534; King Journal, in Cook 3, 550–52.

390–91 Opinions about Cook's death: Clerke Journal, in Cook 3, 538–39; William Harvey Log, Feb. 14, 1779, Adm 55/121, PRO.

391 Story heard by later ship captain: Peter Puget, Log of the *Chatham,* Feb. 26, 1793, Adm 55/17, PRO.

391–93 *Makahiki* festival and its connection with Cook's death: Gavan Daws, "Kealakekua Bay Revisited: A Note on the Death of Captain Cook," *Journal of Pacific History* 3 (1968), 21; Marshall Sahlins, *Islands of History* (Chicago, 1985), 104–35; Sahlins, *Historical Metaphors,* 7–25.

393–94 Recovering Cook's body: King Journal, in Cook 3, 552–55; Edgar Log, Feb. 15, 1779; Zimmermann, *Account,* 39; Samwell Journal, in Cook 3, 1206; first quote from King, 555; second from Edgar; third from Zimmermann; fourth from King, 555.

394 Arguments about retaliation: King Journal, in Cook 3, 558. King's anger at Koa: King Journal, in Cook 3, 559–61.

395 Priests' visit to ship: Bayley Log, Feb. 15, 1779; Samwell Journal, in Cook 3, 1209; Ledyard, *Journal,* 151. Hawaiians' treatment of Cook's remains: Sahlins, *Historical Metaphors,* 25. Common people's views: Daws, "Kealakakua Bay," 23.

395–96 Watering party: Bayley Log, Feb. 15, 1779; Samwell Journal, in Cook 3, 1211; King Journal, in Cook 3, 562, 565; Edgar Log, Feb. 18, 1779; Watts Proceedings, same date.

396 Truce: Edgar Log, Feb. 18, 1779; Bayley Log, Feb. 17, 1779. Burial: Bayley Log, Feb. 20 and 21, 1779; Samwell Journal, in Cook 3, 1216–17; Clerke Journal, in Cook 3, 548–49.

397 At Kauai: King Journal, in Cook 3, 588; Trevenen, *Memoir,* 25; Samwell Journal, in Cook 3, 1223; Clerke Journal, in Cook 3, 575; quote from King. "That idea of turning Indian": Clerke Journal, in Cook 3, 578.

398 At Kamchatka: Clerke Journal, in Cook 3, 646–47, 658; King Journal, in Cook 3, 653–54, 671–72; Samwell Journal, in Cook 3, 1242.

399 Decision to turn back: King Journal, in Cook 3, 676–91. Clerke's death: James Burney Journal, in Cook 3, 698–701; King Log, Aug. 22, 1779; Bayley Log, Aug. 29, 1779; Edgar Log, in Cook 3, 708. Strengthening the ship: Ledyard, *Journal,* 199; Gilbert Journal, in Cook 3, 712–13. Orders not to molest Cook: J. H. du Magellan to Royal Geographic Society, June 23, 1779, Cook Manuscripts, photostat in ML; Benjamin Franklin to "all captains and Commanders of armed ships acting by Commission from the Congress of the UC," March 10, 1779, Cook Ms. Q140, DL; Joseph Banks to Andrew Kippis, Aug. 15, 1795, photostat in ML; Trevenen, *Memoir,* 31. Sale of sea otter skins: Gilbert, *Journal,* 154; Ledyard, *Journal,* 70; James Cook and James King, *A Voyage to the Pacific Ocean in the Years 1776, 1777, 1778, 1779, and 1780* (London, 1784), 4.

400 Delays in Scotland: Gilbert Journal, in Cook 3, 716–17; Trevenen, *Memoir,* 32–33.

Chapter 20

401 News of Cook's death: Earl of Sandwich to Joseph Banks, Jan. 10, 1780, Cook Ms. Q158, DL; *London Gazette, Daily Advertiser,* Jan. 11, 1780. "It is almost": *London Evening Post, Public Advertiser,* Jan. 12, 1780. "would deserve the gratitude": John Meares, *Voyages Made in the Years 1788 and 1789, from China to the North West Coast of America* (London, 1790), 277·538.

402 "always studied": *Gentlemen's Magazine,* Jan. 1780, 45. "savage," "civilized": Anna Seward, *Elegy on Captain Cook,* excerpted in *Annual Register,* 1780, 195. Younger officers' aspirations: *A Memoir of James Trevenen,* ed. Christopher Lloyd and R. C. Anderson (London, 1959), 31–33.

403 Douglas on Cook's journal: John Douglas autobiographical notes, Egerton Ms. 2181, BL. Problems of publication: Correspondence between Earl of Sandwich and Joseph Banks, Sept.–Oct. 1782, Banks Papers Q158, DL.

403–04 Division of profits: Banks to Sandwich, Sept. 28, 1782; James King to Sandwich, Sept. 11, 1783, both in John Montagu, Earl of Sandwich, Papers relating to Cook, 1771–1790, AT; Draft ms. by Banks, July 28, 1785, Banks Papers, DL.

404 "the only Person": Douglas notes, Egerton Ms. 2181, BL. Summary of Marra's account: *Gentlemen's Magazine,* 1781, 231–33, 278–79. Forster's translation: Michael Hoare, *The Tactless Philosopher: Johann Reinhold Forster (1729–98)* (Melbourne, 1976), 236–37.

404–05 Sales of Cook's account: P. J. Marshall and Glyndwr Williams, *The Great Map of Mankind: Perceptions of New Worlds in the Age of Enlightenment* (Cambridge, Mass., 1982), 56; Alan Frost, "The Pacific Ocean: The Eighteenth Century's 'New World,' " *Studies on Voltaire and the Eighteenth Century,* 152 (1976), 783–85; J. C. Beaglehole, *The Life of Captain James Cook* (London, 1974), 691–92.

405 Later sales: George Nicol to Elizabeth Cook, Jan. 7, 1795; Nicol to an unknown correspondent, Jan. 14, 1801, photostats in ML. Continued popularity of books: Frost, *ibid.*

405–06 Royal Society Cook medal, including quotes: Council Minutes, 6, 393–94; Cook Medal Papers; RS Archives.

406 Praise of Cook for his humanitarianism: Seward, *Elegy;* W. Fitzgerald, *An Ode to the Memory of the Late Captain James Cook* (London, 1790). *Gentlemen's Magazine* article: April 1791, quotes from 319.

406–07 Kippis biography: *Captain Cook's Voyages: With an Account of His Life During the Previous and Intervening Periods* (New York, 1925, orig. pub. 1788), 370–72, quote from 372.

407 Praise for Cook elsewhere in Europe: Benjamin Thompson to Joseph Banks, July 24, 1984, Kenneth Webster Collection, University Research Library, University of California at Los Angeles; Bernard Smith, *European Vision and the South Pacific, 1768–1850: A Study in the History of Art and Ideas* (London, 1960), 100. Continental eulogies: Bernard Smith, "Cook's Posthumous Reputation," in Fisher and Johnston, *Captain James*

Cook, 162–68. Plays: *The Death of Captain Cook; A Grand Serious-Pantomimic Ballet, in Three Parts* (London, 1789); *La Mort du Capitaine Cook* (Paris, 1788); *La Cokiada: Tragedia Nueva en Tres Actos* (Málaga, 1796); *Omai, or, a Trip Round the World* (London, 1785).

408 Staging the play *Omai:* Rüdiger Joppien, "Philippe Jacques de Loutherbourg's Pantomime *Omai, or a Trip Round the World* and the Artists of Captain Cook's Voyages," in *Captain Cook and the South Pacific* (London, 1979), 81–86.

408–09 Paintings of Cook's death: Smith, "Cook's Posthumous Reputation," 169–79; Adrienne Kaeppler, *"Artificial Curiosities": An Exposition of Native Manufactures Collected on the Three Pacific Voyages of Captain James Cook* (Honolulu, 1978), 22–24; Edgar Wind, "The Revolution of History Painting," *Journal of the Warburg Institute* 2:2 (1938), 116–27; Joseph Burke, *English Art, 1714–1800* (Oxford, 1976), 245–53, 304–06.

410 Apotheosis paintings: Smith, "Cook's Posthumous Reputation," 174–77. General memorialization of Cook: *ibid.,* 172–73; Thomas Vaughan and A.A.St.C.M. Murray-Oliver, *Captain Cook, R.N.: The Resolute Mariner* (Portland, Oregon, 1974), 87.

411 *'The Injured Islanders': The Injured Islanders; or, the Influence of Art upon the Happiness of Nature* (London, 1779).

Chapter 21

415 "The object of all": Bligh Voyage, 7.

415–16 British activities in Pacific: David Mackay, *In the Wake of Cook: Exploration, Science, and Empire, 1780–1801* (London, 1985), 116, and "A Presiding Genius of Exploration: Banks, Cook, and Empire, 1767–1805," in Robin Fisher and Hugh Johnston, eds., *Captain James Cook and His Times* (Seattle, 1978), 21–39.

416 Banks and science: Mackay, "A Presiding Genius"; "the general director": Mackay, *Cook,* 20.

416–17 Transportation of convicts: David Mackay, *A Place of Exile: The European Settlement of New South Wales* (Melbourne, 1985), 9–18.

417–18 Banks's testimony: reprinted in Owen Rutter, ed., *The First Fleet: the Record of the Foundation of Australia from Its Conception to the Settlement at Sydney Cove* (London, 1937), 11–12.

418 Government action: Mackay, *Exile,* 40–42; Alan Frost, *Convicts and Empire: A Naval Question, 1776–1811* (Melbourne, 1980), 8–9. Matra's testimony: James Matra to Joseph Banks, July 28, 783, reprinted in Rutter, *First Fleet,* 9–10.

418–19 Matra's proposal: Aug. 23, 1783, reprinted in Rutter, *First Fleet,* 23–27, quote from 24.

419 Matra's revised proposal: "Plan for sending convicts to New South Wales," reprinted in Rutter, *First Fleet,* 28–29, quote from 29; Mackay, *Exile,* 6, 19. Search for alternative sites: Frost, *Convicts,* 16–19; Mackay, *Exile,* 29.

419–20 Young's proposal: "Proposal for a settlement at Botany Bay," Jan. 1785, reprinted in Rutter, *First Fleet,* 32–35, quote from 34–35.

420 Banks and Matra's testimony: in Rutter, *First Fleet,* 37–41. Das Voltas Bay: Frost, *Convicts,* 35–44; Mackay, *Exile,* 48–56. Decision for New South Wales: Mackay, *Exile,* 6–7, 25. Historians have disagreed on the relative importance of commercial and strategic motives in the decision to establish a colony in New South Wales; for an interpretation emphasizing the importance of commercial motives, see Frost, *Convicts.* "a reckless act": Mackay, *Exile,* 57. "The Finest Harbour": Arthur Phillip to Lord Sydney, May 15, 1788, reprinted in Rutter, *First Fleet,* 135.

420–21 Establishing colony: "Heads of a plan for effectually disposing of Convicts . . . by the Establishment of a Colony in New South Wales," in Rutter, *First Fleet,* 45–47.

421 Linking of commercial and political motives for voyages: Mackay, *Cook,* 25.

421–22 Cook's crew and fur trade: James Cook and James King, *A Voyage to the Pacific Ocean in the years 1776, 1777, 1778, 1779, and 1780* (London, 1784), 246–50.

422 Ventures sponsored by King George's Sound Company: Richard Cadman Etches to Joseph Banks, May 21, 1792 and July 29, 1788, Sutro Library, San Francisco; Mackay, *Cook,* 62–63.

422–23 Portlock's voyage: Ernest S. Dodge, *Beyond the Capes: Pacific Exploration from Captain Cook to the* Challenger *1776–1877* (Boston, 1971), 44; Nathaniel Portlock, *A Voyage Round the World; but More Particularly to the North-West Coast of America* (London, 1789), 4–5.

422 "furnished philosophy," "for the cultivation": Portlock, *Voyage,* 1–2.

423 Portlock on Hawaii: *ibid.,* 61–65. On Oahu: *ibid.,* 77. Political changes in Hawaiian Islands: Ralph S. Kuykendall, *The Hawaiian Kingdom, 1778–1854: Foundation and Transformation* (Honolulu, 1938), 30–38.

424 Portlock and Dixon on coast of North America: Portlock, *Voyage,* 302–08, 359–61; John Meares, *Voyages Made in the Years 1788 and 1789, from China to the North West Coast of America* (London, 1790), 9; Dodge, *Beyond the Capes,* 52–56; Ernest S. Dodge, *Islands and Empires: Western Impact on the Pacific and East Asia* (Minneapolis, 1976), 50, 76.

425 Colnett's voyage: Mackay, *Cook,* 68–69; Dodge, *Beyond the Capes,* 70–72. French and Spanish activities: Dodge, *Beyond the Capes,* 30–42, 114–20; Dodge, *Islands,* 56, 76; J. C. Beaglehole, *The Exploration of the Pacific* (London, 1966), 318–19; W. P. Morrell, *Britain in the Pacific Islands* (Oxford, 1960), 26.

426 Practice of stopping in Hawaiian Islands: Portlock, *Voyage,* 314. Chiefs' control of trade: Marshall Sahlins, *Islands of History* (Chicago, 1985), 142. Chiefs' desire for weapons: James Colnett, Log of the *Prince of Wales,* March 3, 1788, Adm 55/146, PRO; Sahlins, *loc. cit.*

427 Europeans teaching chiefs to use guns: Colnett, Log, Jan. 21, 1788, March 1788, quote from Jan. 21. Europeans living in islands: H. E. Maude, *Of Islands and Men: Studies in Pacific History* (Melbourne, 1968), 139–40; Kuykendall, *Kawaiian Kingdom,* 22–23.

428 Interest in practical applications of science: Mackay, *Cook,* 125–26.

429 Breadfruit proposal: Mackay, *Cook,* 127–32; Richard Hough, *The Bounty* (New York, 1984), 43–45; Dodge, *Beyond the Capes,* 79.

429–30 *Bounty*'s arrival at Tahiti: Bligh Voyage, 85–93, first quote from 85, second from 93. *Lady Penrhyn:* James Watt's narrative, in *The Voyage of Governor Phillip to Botany Bay* (Dublin, 1790), 272–85.

430 Routine at Tahiti: Bligh Voyage, 93–96. Tu's name change: Bligh Log 1, 384. "The Otaheiteans": Bligh Log 1, 371. "flew into a rage": Bligh Log 1, 372. Stolen buoy: Bligh Voyage, 102.

431 Punishment of crew: Bligh Log 2, 7–8; Bligh Voyage, 119–20. Attitude about crew: Bligh Log, 381. Changes at Matavai Bay: Bligh Log 1, 379.

431–32 Tynah's desire for gifts: Bligh Voyage, 109, 119; Bligh Log 1, 381, quote from 119.

432 Storing gifts: Bligh Log 1, 374, 401; Bligh Voyage, 138. "Whatever good": Bligh Voyage, 109. Exchange of gifts: Bligh Log 1, 377, 389; Bligh Voyage, 101–02.

433 Strategy to get breadfruit: Bligh Voyage, 110–11. Collecting animals: Bligh Voyage, 124–25, 135–36, 142–43, 167–70.

434 Move to Pare: Bligh Log 1, 414–15; Voyage, 156–59. Desertion: Bligh Voyage, 163–64, 170–71; Bligh Log 2, 23; quote from 164. Activities at Tahiti: Bligh Log 2, 23, 26, first quote from 23, second and third from 26.

435 Cutting cable: Bligh Voyage, 180–83.

435–36 English obligations to Tahitians: Bligh Voyage, 177, 194; Bligh Log 2, 28, quotes from 28.

436 Tahitian dependence on European goods: Bligh Log 2, 63–64, quote from 64; James King, quoted in Morrell, *Britain,* 26. Tynah's desire to go to England: Bligh Voyage, 176–77; Bligh Log 2, 28, 61; quote from 28. Departure from Tahiti: Bligh Voyage, 202–04.

437 At Nomuka: Bligh Voyage, 216–21. Mutineers' return to Tahiti: David Howarth, *Tahiti: A Paradise Lost* (New York, 1983), 153.

438 Cox's visit to Tahiti: George Mortimer, *Observations and Remarks Made During a Voyage to the Islands of . . . in the Brig* Mercury, *Commanded by John Henry Cox, Esq.* (London, 1791), 33. Mutineers and Tahitian chiefs: Howarth, *Tahiti,* 155–56.

Chapter 22

440–41 Origins of Vancouver's voyage: Richard Cadman Etches to Joseph Banks, July 17 and 18, 1788, Sutro Library, San Francisco; Barry M. Gough, *Distant Dominion: Britain and the Northwest Coast of North America, 1579–1809* (Vancouver, 1980), 81–84; Ernest S. Dodge, *Beyond the Capes: Pacific Exploration from Captain Cook to the Challenger 1776–1877* (Boston, 1971), 135.

441 Commissioning Vancouver's voyage: Dodge, *Beyond the Capes,* 136; W. P. Morrell, *Britain in the Pacific Islands* (Oxford, 1960), 27–28; David Mackay, *In the Wake of Cook: Exploration, Science and Empire, 1780–1801* (London, 1985), 41–45.

441–42 Plan for Vancouver's voyage: Mackay, *Cook,* 100–102. Stop at Matavai Bay: Vancouver 1, 252–53.

442 Pomare's welcome: Vancouver 1, 265–66. Pomare's questions: Archibald Menzies Journal of Vancouver's Voyage 1790–1794, 115, Add. Ms. 32641, BL. "public register": Vancouver 1, 262. Pomare's political conquests: Vancouver 1, 324–28; quote from 326.

443 Pomare's request for a ship: Vancouver 1, 328–29. Tahitians' desire for English goods: Vancouver 1, 332–43, quote from 333–34. Vancouver's attitude toward theft: Menzies Journal, 152.

443–44 Theft of linen: Vancouver 1, 300–314; Menzies Journal, 152–55.

444 Arrival at Hawaii: Vancouver 1, 348–51; quote from 349.

445 Tarahooa: Vancouver 1, 354–56. Impressions at Oahu: Vancouver 1, 365–67, quote from 367.

445–46 Reception at Kauai: Vancouver 1, 377–78, quote from 378. Significance of prostitution: Marshall Sahlins, *Historical Metaphors and Mythical Realities: Structure in the Early History of the Sandwich Islands Kingdom* (Ann Arbor, 1981), 50.

446 Rowbottom: Vancouver 1, 378–79. "particularly in dress": Vancouver 1, 384. Rowbottom's warnings: Vancouver 1, 380–81. *Fair American* incident: Ralph S. Kuykendall, *The Hawaiian Kingdom, 1778–1854: Foundation and Transformation* (Honolulu, 1938), 24–25.

447 Ship captains' certificates: Vancouver 1, 383. Vancouver's concern at Kauai: Menzies Journal, 188–90.

448 Vancouver's views of Kauai: Vancouver 1, 398–400. Hawaiians taking British names: Sahlins, *Historical Metaphors,* 29. Chiefs' desire for guns: Vancouver 1, 391–92. Kaiana: Vancouver 1, 403–04. Vancouver's concern about firearms: Vancouver 1, 358–61, 404–06.

449 Arrival at Hawaii: Vancouver 3, 165, 169, 181. "injudicious conduct": Vancouver 3, 184. Reluctance to anchor at Kealakekua Bay: Vancouver 3, 198–201. Kamehameha's visit: Vancouver 3, 203–04; Menzies Journal, 252r.

450 Welcome at Kealakekua Bay: Vancouver 3, 205–214. Kamehameha's attitude toward foreigners: Sahlins, *Historical Metaphors,* 26–27. Animals: Vancouver 3, 214–17, quote from 217.

451 Precautions to prevent violence: Vancouver 3, 220–25. Meeting with Cook's murderer: Vancouver 3, 250; Menzies Journal, 267r; Peter Puget, Journals of Proceedings of the *Chatham,* Jan. 1794–Sept. 1795, Jan. 1794, Add. Ms. 17548, BL; quote from Puget.
Puget's conversation: Log of the *Chatham,* Feb. 26, 1793, Adm 55/17, PRO.

451–52 *Makahiki* festival: Gilbert F. Mathison, *Narrative of a Visit to Brazil, Chile, Peru, and the Sandwich Islands, During the Years 1821 and 1822* (London, 1825), 431–32; John Martin, *An Account of the Natives of the Tonga Islands, in the South Pacific Ocean . . . Compiled . . . from . . . Mr. William Mariner* (London, 1818), 2, 62–64; William Ellis, *A Narrative of a Tour Through Hawaii, or Owhyhee; with Remarks on the History, Traditions, Manners, Customs and Language of the Inhabitants of the Sandwich Islands* (Honolulu, 1917), 101–02.

452 Political rivalries in Hawaiian Islands: Kuykendall, *Hawaiian Kingdom,* 30–38. Vancouver's attempts to negotiate peace: Vancouver 3, 261–66.

453 Kamehameha's gift to Vancouver: Vancouver 3, 268. *Kapu:* Sahlins, *Historical Metaphors,* 54.

453–54 Arrival at Maui: Vancouver 3, 291–93. Vancouver's concern about attacks on Europeans: Vancouver 3, 296–98, 300. Kahekili's willingness to cooperate: Vancouver 3, 308–09, 341.

454 Effects of war on Maui: Vancouver 3, 300–301. Chiefs' lack of trust: Vancouver 3, 306–07, 317–21. "Trial" at Oahu: Vancouver 3, 346–53; Menzies Journal, 284–284r.

455 Return to Hawaii: Vancouver 5, 6–9. At Kealakekua Bay: Vancouver 5, 10–11, 17–18, 20–21, 29–30.

456 Europeans in the islands: Vancouver 5, 112–15, quote from 112. Attempts to negotiate peace: Vancouver 5, 82–84. Cession of Hawaii to England: Menzies Journal, 269r.

457 English government's reluctance to get involved in islands, including quote: J. H. Parry, *Trade and Dominion: The European Oversea Empires in the Eighteenth Century* (London, 1971), 273. Kamehameha's views: Vancouver 5, 47–53, 91–96.

458 Kamehameha's efforts to gain control over islands: Kuykendall, *Hawaiian Kingdom,* 23, 29, 44–50.

460 European influence on Polynesian politics: Marshall Sahlins, *Social Stratification in Polynesia* (Seattle, 1958), 28. Abolition of *tapu/kapu:* David Howarth, *Tahiti: A Paradise Lost* (New York, 1983), 193–98; Kuykendall, *Hawaiian Kingdom,* 65–69; Gavan Daws, *Shoal of Time: A History of the Hawaiian Islands* (New York, 1968), 59.

460–61 Polynesian social structure and comparison with other Pacific groups: Marshall Sahlins, "Poor Man, Rich Man, Big Man, Chief: Political Types in Melanesia and Polynesia," *Comparative Studies in Society and History* 5 (1963), 288. Europeans' feeling for Polynesians as individuals: cf. Tzvetan Todorov, *The Conquest of America: The Question of the Other,* trans. Richard Howard (New York, 1984), 127–30, on Spanish in Mexico.

461–62 Interest in travel books, artifacts: Adrienne Kaeppler, *"Artificial Curiosities": An Exposition of Native Manufactures Collected on the Three Pacific Voyages of Captain James Cook, R.N.* (Honolulu, 1978), 47; Rüdiger Joppien, "Philippe Jacques de Loutherbourg's Pantomime 'Omai, or a Trip Round the World' and the Artists of Captain Cook's Voyages," *Captain Cook and the South Pacific* (London, 1979), 94–95, 111; Leverian Museum, *A Companion to the Museum* (London, 1790); Rüdiger Joppien, "The Artistic Bequest of Captain Cook's Voyages," in Robin Fisher and Hugh Johnston, eds., *Captain James Cook and His Times* (Seattle, 1979), 187–210; Gavan Daws, *A Dream of Islands: Voyages of Self-Discovery in the South Seas* (New York, 1980), 11.

462–63 Missionaries: Morrell, *Britain,* 31; Howarth, *Tahiti,* 160, 175–76, 183–84; Kuykendall, *Hawaiian Kingdom,* 65–69.

463 Romantic poets: Alan Frost, "The Pacific Ocean: The Eighteenth Century's 'New World,'" *Studies on Voltaire and the Eighteenth Century,* 152–3 (1976), 803–09.

463–64 Voyages by writers and artists: Todorov, *Conquest,* 248–49; Daws, *Dream,* 71–128, 163–270; Bernard Smith, *European Vision and the South Pacific, 1768–1850: A Study in the History of Art and Ideas* (London, 1960), 1–10; Alan Frost, "New Geographical Perspectives and the Emergence of the Romantic Imagination," in Fisher and Johnston, *Captain James Cook,* 5–20.

464 Jack London: *The Cruise of the Snark* (New York, 1911); Thor Heyerdahl, *Kon-Tiki: Across the Pacific by Raft,* trans. F. H. Lyon (Chicago, 1950.)

Index